高等学校Java课程系列教材

JSP程序设计（第2版）

耿祥义　张跃平　编著

清华大学出版社
北京

内容简介

JSP(Java Server Pages)是一种动态网页技术标准，它可以无缝地运行在UNIX、Linux和Windows操作平台上。利用这一技术可以建立安全、跨平台的先进动态网站。

本书详细讲解了JSP语法和基本的程序设计方法。全书共分10章，内容包括JSP简介，JSP页面与JSP标记，Tag文件与Tag标记，JSP内置对象，JSP文件操作，在JSP中如何使用MySQL、SQL Server、Oracle等常用数据库，JSP与JavaBean，Java Servlet基础，以及MVC模式等重要内容。本书所有知识都结合具体实例进行介绍，力求详略得当，突出JSP在开发Web动态网站方面的强大功能及在开发商务网站方面的应用，使读者快速掌握和运用JSP的编程技巧。

本书不仅可以作为高等院校计算机及相关专业的选修课教材，也可作为自学者及网站开发人员的参考书。

图书在版编目(CIP)数据

JSP程序设计/耿祥义，张跃平编著.--2版.--北京：清华大学出版社，2015(2020.8重印)
(高等学校Java课程系列教材)
ISBN 978-7-302-37236-3

Ⅰ.①J… Ⅱ.①耿… ②张… Ⅲ.①JAVA语言－网页制作工具 Ⅳ.①TP312 ②TP393.092

中国版本图书馆CIP数据核字(2014)第153406号

责任编辑：魏江江 王冰飞
封面设计：杨 兮
责任校对：时翠兰
责任印制：杨 艳

出版发行：清华大学出版社
网 址：http://www.tup.com.cn，http://www.wqbook.com
地 址：北京清华大学学研大厦A座 **邮 编**：100084
社 总 机：010-62770175 **邮 购**：010-62786544
投稿与读者服务：010-62776969，c-service@tup.tsinghua.edu.cn
质量反馈：010-62772015，zhiliang@tup.tsinghua.edu.cn
课件下载：http://www.tup.com.cn，010-83470236

印 装 者：北京富博印刷有限公司
经 销：全国新华书店
开 本：185mm×260mm **印 张**：22 **字 数**：536千字
版 次：2009年8月第1版 2015年1月第2版 **印 次**：2020年8月第15次印刷
印 数：73001～73800
定 价：39.50元

产品编号：060267-01

前　言

JSP是由Sun公司倡导、许多公司参与，于1999年推出的一种动态网页技术标准。JSP是基于Java Servlet以及整个Java体系的Web开发技术，利用这一技术可以建立安全、跨平台的先进动态网站，这项技术还在不断地更新和优化中。JSP以Java技术为基础，又在许多方面做了改进，具有动态页面与静态页面分离，能够脱离硬件平台的束缚，以及编译后多平台运行等优点，JSP已经成为Internet上的主流开发工具。

本书第2版对一些例子和部分内容做了适度的调整、更新和修改，考虑到数据库在Web设计中的重要作用，本书第2版加强了数据库的知识容量，特别采用了目前在Web设计中占主导地位的MySQL数据库作为主要的数据库来讲解有关知识点。本书分为10章，第1章主要介绍Tomcat的安装与配置，通过一个简单的JSP页面初识JSP概貌。第2章详细讲解JSP的基本语法，包括程序片，页面指令等重要内容。第3章主要讲解Tag文件与标记，特别重点强调了怎样使用Tag文件实现代码复用。第4章主要讲解JSP的内置对象，特别重点讲解了session会话对象。第5章讲解输入/输出流技术，重点介绍了文件的上传与下载以及怎样使用Tag标记实现文件的读/写操作。第6章涉及的内容是数据库，也是Web应用开发的非常重要的一部分内容，采用MySQL数据库讲解主要知识点，也特别介绍了各种数据库的连接方式。第7章讲解JavaBean的使用，是JSP技术中很重要的内容，即怎样使用JavaBean分离数据的显示和处理，给出了许多有一定应用价值的例子。第8章讲解Servlet，对servlet对象的运行原理给予了详细的讲解。在第9章对Servlet在MVC开发模式中的地位给予了重点介绍，并按照MVC模式给出了易于理解MVC设计模式的例子，本章中的许多例子都是大多数Web开发中经常使用的模块。第10章是一个完整的网站，完全按照MVC模式开发设计，其目的是掌握一般Web应用中常用基本模块的开发方法。

希望本教材能对读者学习JSP有所帮助，并请读者批评指正。

编者

2014年10月

目　录

第1章 JSP 概 述

本章导读

主要内容

- 什么是 JSP
- JSP 引擎与 Tomcat 服务器
- JSP 页面与 Web 服务目录
- JSP 运行原理

难点

- JSP 的运行原理
- 设置 Web 服务目录

关键实践

- 上机编写、保存、运行一个简单的 JSP 页面

1.1 什么是 JSP

网络应用中最常见的模式是 B/S 模式，即需要获取信息的用户使用浏览器向服务器发出请求，服务器对此做出响应，将有关信息发送给用户的浏览器。在 B/S 模式中，服务器上必须有所谓的 Web 应用程序，服务器通过运行这些 Web 应用程序来响应用户的请求。因此，基于 B/S 模式的网络程序的核心就是设计服务器端的 Web 应用程序。

随着网络的迅速发展，服务器不仅要和用户动态、安全地交互更多的信息，而且对 Web 应用程序的规模、难度和维护都提出了更高的要求。JSP(Java Server Pages)正是在这一背景下诞生的优秀的 Web 开发技术，利用这一技术可以建立安全、跨平台、易维护的 Web 应用程序。JSP 的跨平台、易维护和安全性得益于 Java 语言，这是因为 Java 语言具有不依赖于平台、面向对象、安全等优良特性，已经成为网络程序设计的佼佼者，许多和 Java 相关的技术得到了广泛的认可和应用，JSP 就是其中之一。读者可能对 Microsoft 的 ASP(Active Server Pages)比较熟悉，ASP 也是一项 Web 开发技术，可以开发出动态的、高性能的 Web 应用程序。JSP 和 ASP 技术非常相似，ASP 使用的是 VBScript 脚本语言，而 JSP 使用的是 Java 编程语言。与 ASP 相比，JSP 以 Java 技术为基础，又在许多方面做了改进，具有动态页面与静态页面分离，能够脱离硬件平台的束缚，以及编译后运行等优点，完全克服了 ASP 的脚本级执行的缺点。JSP 已经成为开发动态网站的主流技术。

需要强调的一点是：要想真正地掌握 JSP 技术，必须有较好的 Java 语言基础，以及较

好的 HTML 语言方面的知识。

注意:在 Web 设计中,用户一词通常代表客户端计算机上驻留的浏览器,即使浏览器在客户端计算机的多个窗口中被打开,服务器也认为这是一个用户。

1.2 JSP 引擎与 Tomcat 服务器

一个服务器上可以有很多基于 JSP 的 Web 应用程序,以满足各种用户的需求。这些 Web 应用程序必须有一个软件来统一管理和运行,将这样的软件称作 JSP 引擎或 JSP 容器,将安装 JSP 引擎的计算机称作一个支持 JSP 的 Web 服务器。

JSP 的核心内容之一就是编写 JSP 页面(有关 JSP 页面的内容见 1.3 和 2.1 节),JSP 页面是 Web 应用程序的重要组成部分之一。一个简单 Web 应用程序可能只有一个 JSP 页面,而一个复杂的 Web 应用程序可能由许多 JSP 页面、JavaBean 和 servlet 组成(见第 7 章、第 8 章、第 9 章)。当用户请求 Web 服务器上的 JSP 页面时,JSP 引擎负责运行 JSP,并将运行结果返回给用户,有关 JSP 的运行原理将在 1.4 节讲解。

自从 JSP 发布以后,出现了各式各样的 JSP 引擎。目前,比较常用的 JSP 引擎包括 Tomcat、JRun 和 Resin,其中以 Tomcat 的使用最为广泛。Tomcat 由 Apache 和 Sun 公司共同开发而成,是一个免费的开源 JSP 引擎。可以登录 http://tomcat.apache.org/免费下载 Tomcat。目前,Tomcat 的最新版本是 Tomcat 8.0,登录之后,可以在 Download 里选择 Tomcat 8.0,然后在 Binary Distributions 的 Code 中选择 Zip 或 32-bit/64-bit Windows Service Installer 即可。如果选择 Zip 将下载:apache-tomcat-8.0.3.zip;如果选择 32-bit/64-bit Windows Service Installer 将下载:apache-tomcat-8.0.3.exe。

本章重点讲述在 Windows 7/Windows XP 操作系统下 Tomcat 的安装与配置,将安装了 Tomcat 的计算机称作一个 Tomcat 服务器。

1.2.1 安装 JDK

安装 Tomcat 之前,首先安装 JDK,这里安装 JDK1.7。假设 JDK 的安装目录是:

```
D:\jdk1.7
```

安装 JDK 之后需要进行几个环境变量的设置。对于 Windows 7/Windows XP,用鼠标右键单击“计算机”/“我的电脑”,在弹出的快捷菜单中选择“属性”命令,弹出“系统特性”对话框,再单击该对话框中的“高级系统设置”/“高级选项”,然后单击“环境变量”按钮,分别添加如下的系统环境变量:

```
变量名:Java_home,变量值:D:\jdk1.7
变量名:Path,变量值:D:\jdk1.7\bin
```

如果曾经设置过环境变量 Java_home 和 Path,可单击该变量进行编辑操作,将环境变量需要的值加入即可。需要注意是,在编辑环境变量的值时,新加入的值,如果不是环境变量取值范围中的第一个值或最后一个值,那么新加入的值要和已有的其他值用分号分隔,如果是最后一个值需要和前面的值用分号分隔,如果是第一个值需要和后面的值用分号分隔。如图 1-1 和图 1-2 所示。

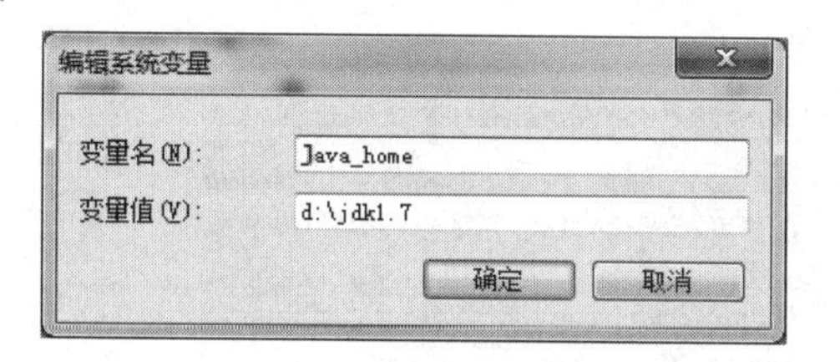

图 1-1　设置 Java_home

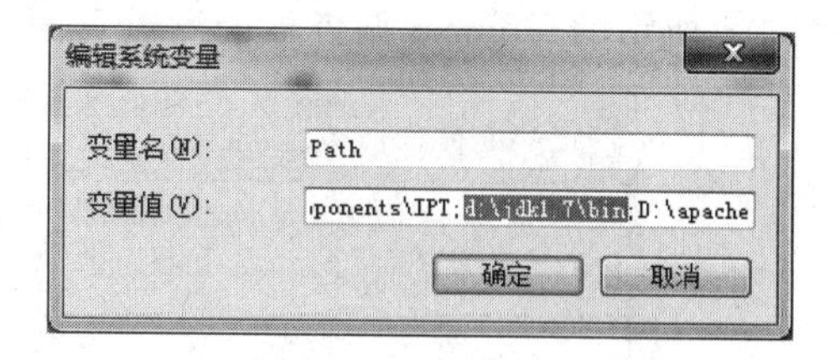

图 1-2　编辑 Path

1.2.2　安装与启动 Tomcat 服务器

1. apache-tomcat-8.0.3.zip 的安装

将下载的 apache-tomcat-8.0.3.zip 解压到磁盘某个分区，比如解压到 D:\，解压缩后将出现如图 1-3 所示的目录结构。

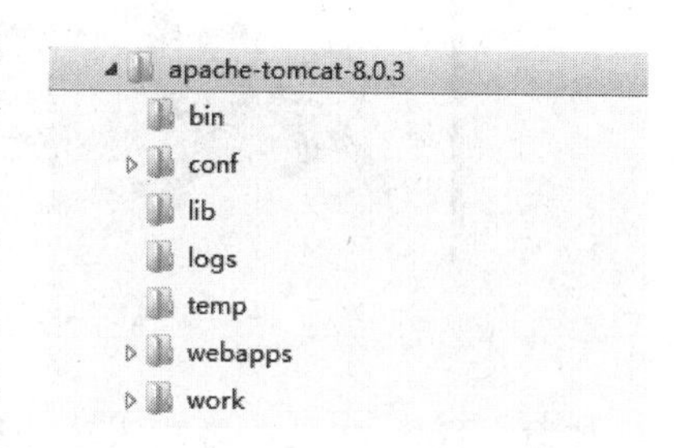

图 1-3　Tomcat 服务器的目录结构

执行 Tomcat 安装根目录中 bin 文件夹中的 startup.bat 启动 Tomcat 服务器。执行 startup.bat 启动 Tomcat 服务器会占用一个 MS-DOS 窗口(如图 1-4 所示的界面)，如果关闭当前 MS-DOS 窗口，将关闭 Tomcat 服务器。使用 startup.bat 启动 Tomcat 服务器，Tomcat 服务器将使用 Java_home 环境变量设置的 JDK。

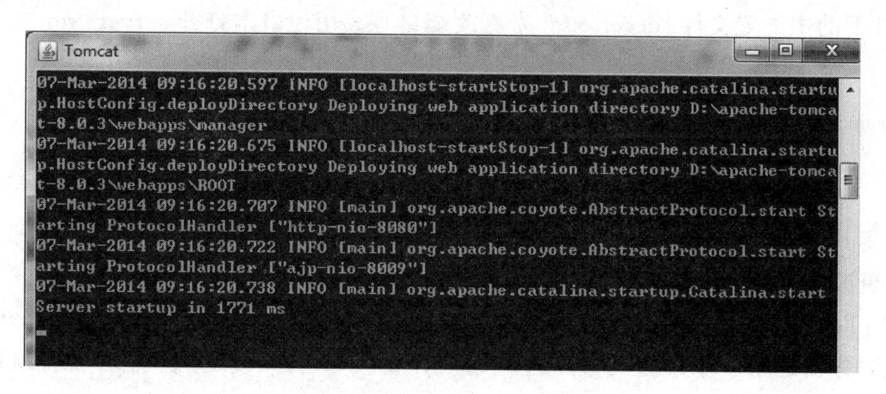

图 1-4　启动 Tomcat 服务器

2. apache-tomcat-8.0.3.exe 的安装

apache-tomcat-8.0.3.exe 文件是针对 Windows 的安装版，安装后形成的目录结构和 apache-tomcat-8.0.3.zip 安装目录相同。

双击下载的 apache-tomcat-8.0.3.exe，将出现"安装向导"界面，单击其中的 Next 按钮，接受授权协议后，将出现选择"安装方式"的界面。在"安装方式"界面中选择 Norma、Minimun、Custom 和 Full 之一，然后按着安装向导的提示进行安装即可，不再赘述。

3. 测试 Tomcat 服务器

在浏览器的地址栏中输入：http://localhost:8080 或 http://127.0.0.1:8080，会出现

如图 1-5 所示的 Tomcat 服务器的测试页面。

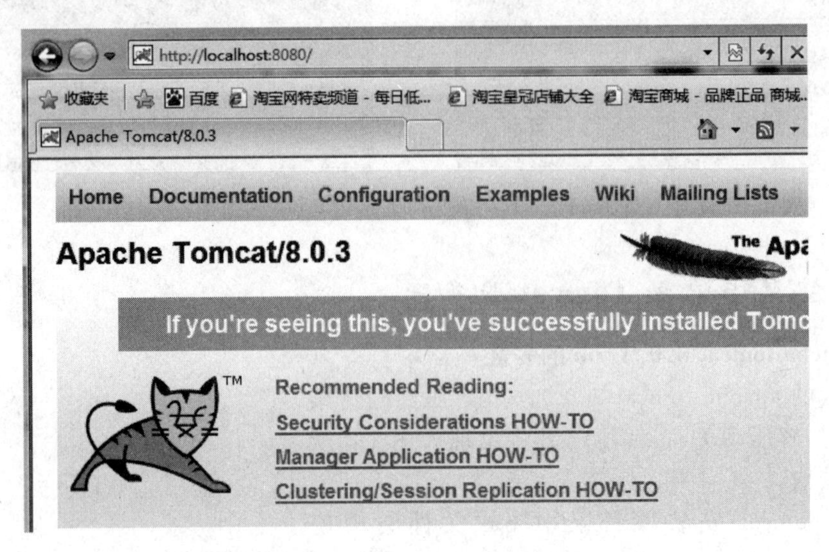

图 1-5　测试 Tomcat 服务器

注意：Tomcat 服务器默认地占用 8080 端口，如果 Tomcat 所使用的端口已经被占用，Tomcat 服务器将无法启动，有关端口号的配置稍后讲解。

4. 配置端口

8080 是 Tomcat 服务器默认占用的端口。可以通过修改 Tomcat 服务器安装目录中 conf 文件下的主配置文件 server. xml 来更改端口号。用记事本打开 server. xml 文件，找到出现

```
<Connector port = "8080" protocol = "HTTP/1.1"
           connectionTimeout = "20000"
           redirectPort = "8443" />
```

的部分，将其中的 port＝“8080”更改为新的端口号，并重新启动 Tomcat 服务器即可，例如将 8080 更改为 9090 等。

如果 Tomcat 服务器所在的计算机没有启动占用 80 端口号的其他网络程序，也可以将 Tomcat 服务器的端口号设置为 80，这样用户在访问 Tomcat 服务器时可以省略端口号。例如：

```
http://127.0.0.1/
```

1.3　JSP 页面与 Web 服务目录

1.3.1　JSP 页面

在后面的章节里将详细地学习编写 JSP 页面的语法。简单地说，一个 JSP 页面除了普通的 HTML 标记符外，还可以使用标记符号“＜％”，“％＞”加入 Java 程序片。一个 JSP 页面按文本文件保存，扩展名是. jsp。例如，使用文本编辑器“记事本”编辑 JSP 页面，在保存

JSP页面时,必须将"保存类型"选择为"所有文件",将"编码"选择为"ANSI"。如果保存JSP页面时,计算机系统总是给文件名额外加上".txt",那么在保存JSP页面时可以将文件名用双引号括起,如图1-6所示。

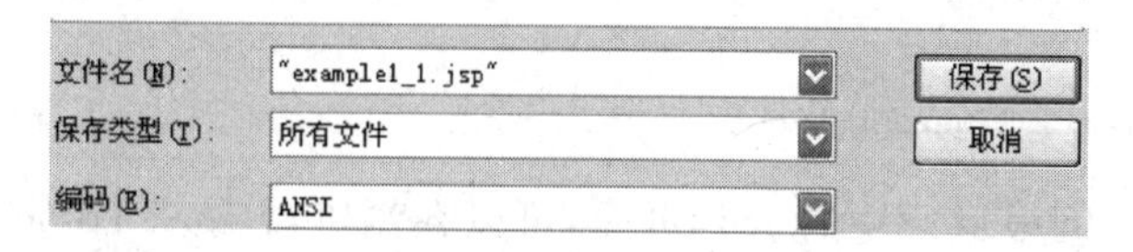

图1-6　JSP页面的保存

在保存JSP页面时,文件的名字必须符合标识符规定(名字可以由字母、下划线、美元符号和数字组成,并且第一个字符不能是数字)。需要注意的是,JSP技术基于Java语言,名字区分大小写,Boy.jsp和boy.jsp是名字不相同的JSP文件。

为了明显地区分普通的HTML标记和Java程序片段以及JSP标签,用大写字母书写普通的HTML标记符号(这不是必须的,HTML标记不区分大小写)。可以用记事本或更好的文本编辑器来编辑JSP页面的源文件。

下面的例1-1是一个简单的JSP页面,页面的运行效果如图1-7所示。

图1-7　简单的JSP页面

例1-1

example1_1.jsp

```
<%@ page contentType = "text/html;charset = GB2312" %>
<HTML>
<BODY BGCOLOR = yellow>
<FONT Size = 3>
<P>这是一个简单的JSP页面
<%        //这是Java程序片
   int i,sum = 0;
   for(i = 1;i <= 100;i++){
      sum = sum + i;
   }
%>
<P> 1到100的连续和是：<% = sum %>
</FONT>
</BODY>
</HTML>
```

1.3.2　Web服务目录

必须将编写好的JSP页面文件保存到Tomcat服务器的某个Web服务目录中,只有这样,远程的用户才可以访问该Tomcat服务器上的JSP页面。人们常说的一个网站,实际上就是一个 Web 服务目录。

1. 根目录

如果 Tomcat 服务器的安装目录是 D:\ apache-tomcat-8. 0. 3,那么 Tomcat 的 Web 服务目录的根目录是:

```
D:\apache-tomcat-8.0.3\webapps\Root
```

用户如果准备访问根目录中的 JSP 页面,可以在浏览器输入 Tomcat 服务器的 IP 地址(或域名)、端口号和 JSP 页面的名字即可(必须省略 Web 根目录的名字),例如,Tomcat 服务器的 IP 地址是 192. 168. 1. 100,根目录中存放的 JSP 页面的名字是 example1_1. jsp,那么用户在浏览器输入的内容是:

```
http://192.168.1.100:8080/example1_1.jsp
```

也许没有为 Tomcat 服务器所在的机器设置过一个有效的 IP 地址,那么为了调试 JSP 页面,可以打开 Tomcat 服务器所在机器上的浏览器,在浏览器的地址栏中输入:

```
http://127.0.0.1:8080/example1_1.jsp
```

2. webapps 下的 Web 服务目录

Tomcat 服务器安装目录的 webapps 目录下的任何一个子目录都可以作为一个 Web 服务目录。可以在 webapps 下新建子目录,例如 ch1 子目录,那么 ch1 就成为一个 Web 服务目录。如果将 JSP 页面文件 example1_1. jsp 保存到 webapps 下的 Web 服务目录中,那么应当在浏览器的地址栏中输入 Tomcat 服务器的 IP 地址(或域名)、端口号、Web 服务目录和 JSP 页面的名字,比如,example1_1. jsp 保存到 ch1 中,输入的内容为:

```
http://127.0.0.1:8080/ch1/example1_1.jsp
```

将例 1-1 中的 example1_1. jsp 保存到 D:\apache-tomcat-8. 0. 3\webapps\ch1 中,用浏览器访问 example1_1. jsp 的效果如图 1-7 所示。

注意:安装 Tomcat 服务器后,webapps 下默认已有:examples、host-manager、manager 子目录;在 webapps 下新建一个 Web 服务目录后不必重新启动 Tomcat 服务器。

3. 新建 Web 服务目录

可以将 Tomcat 服务器所在计算机的某个目录(非 webapps 下的子目录)设置成一个 Web 服务目录,并为该 Web 服务目录指定虚拟目录,即隐藏 Web 服务目录的实际位置,用户只能通过虚拟目录访问 Web 服务目录中的 JSP 页面。

通过修改 Tomcat 服务器安装目录下 conf 文件夹中的 server. xml 文件来设置新的 Web 服务目录。假设要将 D:\MyBook\zhang 以及 C:\wang 作为 Web 服务目录,并让用户分别使用 apple 和 cloud 虚拟目录访问 Web 服务目录 D:\MyBook\zhang 和 C:\wang 下的 JSP 页面。首先用记事本打开 conf 文件夹中的主配置文件:server. xml,找到出现 </Host>的部分(接近 server. xml 文件尾部处)。然后在</Host>的前面加入:

```
<Context path = "/apple" docBase = "D:\MyBook\zhang" debug = "0" reloadable = "true" />
<Context path = "/cloud" docBase = "C:\wang" debug = "0" reloadable = "true" />
```

注意:xml 文件是区分大小写的,不可以将<Context>写成<context>。

主配置文件 server.xml 修改后，必须重新启动 Tomcat 服务器。这样，就可以将 JSP 页面存放到 D:\MyBook\zhang 或 C:\wang 中，用户可以通过虚拟目录 apple 或 cloud 访问 JSP 页面，例如，example1_1.jsp 保存到 D:\MyBook\zhang 或 C:\wang 中，在浏览器地址栏中输入：

```
http://127.0.0.1:8080/apple/example1_1.jsp
```

或

```
http://127.0.0.1:8080/cloud/example1_1.jsp
```

4. 相对目录

Web 服务目录下的目录称为该 Web 服务目录下的相对 Web 服务目录。例如，可以在 Web 服务目录 D:\MyBook\zhang（虚拟目录是 apple）下再建立一个子目录 image，将 example1_1.jsp 文件保存到 image 中。那么可以在浏览器的地址栏中输入：

```
http://127.0.0.1:8080/apple/image/example1_1.jsp
```

来访问 example1_1.jsp。

1.4 JSP 运行原理

当服务器上的一个 JSP 页面被第一次请求执行时，服务器上的 JSP 引擎首先将 JSP 页面文件转译成一个 java 文件，并编译这个 java 文件生成字节码文件，然后执行字节码文件响应用户的请求。而当这个 JSP 页面再次被请求执行时，JSP 引擎将直接执行字节码文件来响应用户。根据 Tomcat 执行 JSP 页面的这一特点，Web 程序设计完毕后，可以由管理者首次访问 JSP 页面，这样一来，后续用户再访问该 JSP 页面时，JSP 引擎将直接执行已经驻留在内存中的字节码文件来响应用户，提高了 JSP 页面的访问速度，这也正是 JSP 技术比 ASP 技术速度快的一个主要原因。

JSP 引擎对每个 JSP 页面都对应地产生一个字节码文件，以下阐述 Tomcat 服务器执行 JSP 页面所对应的字节码文件的原理，即给出 JSP 页面的运行原理：

(1) 把 JSP 页面中的 HTML 标记（页面的静态部分）发送给用户的浏览器，由浏览器中的 HTML 解释器负责解释执行 HTML 标记。

(2) 负责处理 JSP 标记，并将有关的处理结果发送到用户的浏览器。

(3) 执行“<%”和“%>”之间的 Java 程序片（JSP 页面中的动态部分），并把执行结果交给用户的浏览器显示。

(4) 当多个用户请求一个 JSP 页面时，Tomcat 服务器为每个用户启动一个线程，该线程负责执行常驻内存的字节码文件来响应相应用户的请求。这些线程由 Tomcat 服务器来管理，将 CPU 的使用权在各个线程之间快速切换，以保证每个线程都有机会执行字节码文件，这与传统的 CGI 为每个用户启动一个进程相比较，效率要高得多。

注意：如果对 JSP 页面进行了修改、保存，那么 Tomcat 服务器会生成新的字节码。

下面是 JSP 引擎生成的 example1_1.jsp 的 java 文件：example1_005f1_jsp.java，文件把 JSP 引擎交给用户端负责显示的内容做了注释（见/ *** /）。如果 JSP 页面保存在 Root

目录中,可以在 Tomcat 服务器下的: work\Catalina\localhost_\org\apache\jsp 目录中找到 Tomcat 服务器生成的 JSP 页面对应的 Java 文件以及编译 Java 文件得到的字节码文件。

example1_005f1_jsp. java

```
package org.apache.jsp;
import javax.servlet.*;
import javax.servlet.http.*;
import javax.servlet.jsp.*;
public final class example1_005f1_jsp extends org.apache.jasper.runtime.HttpJspBase
  implements org.apache.jasper.runtime.JspSourceDependent {
  private static final JspFactory _jspxFactory = JspFactory.getDefaultFactory();
  private static java.util.List _jspx_dependants;
  private javax.el.ExpressionFactory _el_expressionfactory;
  private org.apache.AnnotationProcessor _jsp_annotationprocessor;
  public Object getDependants() {
    return _jspx_dependants;
  }
  public void _jspInit() {
  _el_expressionfactory = _jspxFactory.getJspApplicationContext
  (getServletConfig().getServletContext()).getExpressionFactory();
  public void _jspDestroy() {
  }
  public void _jspService(HttpServletRequest request, HttpServletResponse response)
        throws java.io.IOException, ServletException {
    PageContext pageContext = null;
    HttpSession session = null;
    ServletContext application = null;
    ServletConfig config = null;
    JspWriter out = null;
    Object page = this;
    JspWriter _jspx_out = null;
    PageContext _jspx_page_context = null;
    try {
      response.setContentType("text/html;charset = GB2312");
      pageContext =
      _jspxFactory.getPageContext(this, request, response,null, true, 8192, true);
      _jspx_page_context = pageContext;
      application = pageContext.getServletContext();
      config = pageContext.getServletConfig();
      session = pageContext.getSession();
      out = pageContext.getOut();
      _jspx_out = out;
 /***/out.write("\r\n");
 /***/out.write("<HTML><BODY BGCOLOR = yellow><FONT Size = 3>\r\n");
 /***/out.write("<P>这是一个简单的 JSP 页面\r\n");
 /***/out.write(" ");
      int i, sum = 0;
      for(i = 1;i <= 100;i++){
         if(i % 2 == 0)
           sum = sum + i;
```

```
        }
/ *** /out.write("\r\n");
/ *** /out.write("<P> 1 到 100 的偶数之和是：");
/ *** /out.print(sum );
/ *** /out.write("\r\n");
/ *** /out.write("</FONT></BODY></HTML>\r\n");
      } catch (Throwable t) {
        if (! (t instanceof SkipPageException)){
          out = _jspx_out;
          if (out != null && out.getBufferSize() != 0)
            try { out.clearBuffer(); } catch (java.io.IOException e) {}
          if (_jspx_page_context != null) _jspx_page_context.handlePageException(t);
        }
      } finally {
        _jspxFactory.releasePageContext(_jspx_page_context);
      }
    }
}
```

1.5 实验：编写、保存、运行 JSP 页面

1. 实验目的

本实验的目的是让学生掌握怎样设置 Web 服务目录，怎样修改 Tomcat 服务器的端口号，怎样访问 Web 服务目录下的 JSP 页面。

2. 实验步骤

(1) 安装 Tomcat 服务器。

将下载的 apache-tomcat-8.0.3.zip 解压到硬盘某个分区，比如 C。

(2) 设置 Web 服务目录。

在硬盘分区 C 下新建一个 Web 服务目录，名字为 student。

打开 Tomcat 安装目录中 conf 文件夹里的 server.xml 文件，找到出现</Host>的部分(server.xml 文件尾部)。然后在</Host>的前面加入：

```
<Context path = "/friend" docBase = "c:/student" debug = "0" reloadable = "true" />
```

将 student 设置为 Web 服务目录，并为该 Web 服务目录指定名字为 friend 的虚拟目录。

(3) 修改端口号。

在 server.xml 文件中找到修改端口号部分，将端口号修改为 9999。

(4) 启动 Tomcat 服务器。

如果已经启动，必须关闭 Tomcat 服务器，并重新启动。

(5) 编写 JSP 页面。

用文本编辑器编写一个简单的 JSP 页面 number.jsp(见后面的参考代码)，该 JSP 页面可以显示 1 至 100 之间的完数。将 JSP 页面保存到 Web 服务目录 student 中。

(6) 访问 JSP 页面。

用浏览器访问 Web 服务目录 student 中的 JSP 页面 number.jsp，在浏览器地址栏

输入:

```
http://127.0.0.1:9999/friend/number.jsp
```

3. 参考代码

number.jsp

```
<%@ page contentType="text/html;charset=GB2312" %>
<HTML>
<BODY BGCOLOR=yellow>
<FONT Size=3>
   如果一个正整数刚好等于它的真因子之和,这样的正整数为完数,
<br>例如,6=1+2+3,因此 6 就是一个完数。
<br>1 到 1000 内的完数有:
   <% int i,j,sum;
        for(i=1,sum=0;i<=1000;i++){
           for(j=1;j<i;j++){
              if(i%j==0)
                 sum=sum+j;
           }
           if(sum==i)
              out.print(" "+i);
        }
   %>
</FONT></BODY></HTML>
```

习　题　1

1. 怎样启动和关闭 Tomcat 服务器?

2. 请在 C:\下建立一个名字为 book 的目录,并将该目录设置成一个 Web 服务目录,然后编写一个简单的 JSP 页面,保存到该目录中,让用户使用虚拟目录 red 访问该 JSP 页面。

3. 怎样访问 Web 服务目录子目录中的 JSP 页面?

4. 如果想修改 Tomcat 服务器的端口号,应当修改哪个文件?能否将端口号修改为 80。

第2章 JSP页面与JSP标记

本章导读

主要内容

- JSP页面的基本结构
- 变量和方法的声明
- Java程序片
- 表达式
- JSP中的注释
- JSP指令标记
- JSP动作标记

难点

- Java程序片的运行原理
- include指令标记与include动作标记

关键实践

- 编写一个JSP页面，让该JSP页面包含5种基本的元素
- 编写含有JSP指令标记的JSP页面
- 编写含有JSP动作标记的JSP页面

2.1 JSP页面的基本结构

在传统的HTML页面文件中加入Java程序片和JSP标记就构成了一个JSP页面文件。一个JSP页面可由5种元素组合而成：

- 普通的HTML标记。
- JSP标记，如指令标记、动作标记。
- 变量和方法的声明。
- Java程序片。
- Java表达式。

当服务器上的一个JSP页面被第一次请求执行时，服务器上的JSP引擎首先将JSP页面文件转译成一个Java文件，再将这个Java文件编译生成字节码文件，然后通过执行字节码文件响应用户的请求，执行原理是：

(1) 把JSP页面中普通的HTML标记交给用户的浏览器执行显示。

(2) JSP标记、数据和方法声明、Java程序片由服务器负责执行，将需要显示的结果发送给用户的浏览器。

(3) Java表达式由服务器负责计算，并将结果转化为字符串，然后交给用户的浏览器负责显示。

Tomcat服务器的webapps目录的子目录都可以作为一个Web服务目录，在webapps目录下新建一个Web服务目录：ch2，除非特别约定，本章例子中的JSP页面均保存在ch2中。

例2-1中，example2_1.jsp页面包含了5种元素，页面效果如图2-1所示。

图2-1　包含了5种元素的JSP页面

例2-1

example2_1.jsp

```
<%@ page contentType = "text/html; charset = GB2312" %> <!-- jsp 指令标记 -->
<%@ page import = "java.util.Date" %>                  <!-- jsp 指令标记 -->
    <%!
        Date date;                                       // 数据声明
        int sum;
        public int getFactorSum(int n) {                 // 方法声明
            for(int i = 1; i < n; i++ ) {
                if(n % i == 0)
                    sum = sum + i;
            }
            return sum;
        }
    %>
<HTML><BODY bgcolor = cyan>                              <!-- html 标记 -->
<FONT Size = 4><P>程序片创建 Date 对象：
  <% date = new Date();                                  //Java 程序片
       out.println("<br>" + date + "<br>");
       int m = 100;
  %>
  <% = m %>                                              <!-- Java 表达式 -->
   的因子之和是(不包括<% = m%>)：
  <% = getFactorSum(m) %>                                <!-- Java 表达式 -->
</FONT></BODY></HTML>
```

2.2　变量和方法的声明

在“＜％！”和“％＞”标记符号之间声明变量和方法。

2.2.1　声明变量

在“＜％！”和“％＞”标记符之间声明变量，即在“＜％！”和“％＞”之间放置Java的变量声明语句，变量的类型可以是Java语言允许的任何数据类型。将“＜％！”和“％＞”之间声明的变量称为JSP页面的成员变量。例如：

```
<%!
      int x,y=100,z;
      String tom=null,jerry="love JSP";
      Date date;
%>
```

“<%!”和“%>”之间声明的变量在整个JSP页面内都有效，与标记符号<%!、%>所在的位置无关，但习惯将标记符号<%!、%>写在Java程序片的前面。JSP引擎将JSP页面转译成Java文件时，将“<%!”和“%>”之间声明的变量作为类的成员变量，这些变量的内存空间直到服务器关闭才被释放。当多个用户请求一个JSP页面时，JSP引擎为每个用户启动一个线程，这些线程由JSP引擎服务器来管理。这些线程共享JSP页面的成员变量，因此任何一个用户对JSP页面成员变量操作的结果，都会影响到其他用户。

例2-2利用成员变量被所有用户共享这一性质，实现了一个简单的计数器，页面效果如图2-2所示。

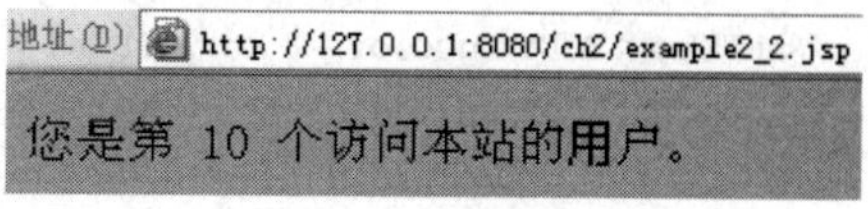

图2-2　JSP页面的成员变量

例2-2

example2_2.jsp

```
<%@ page contentType="text/html; charset=GB2312" %>
<HTML><BODY BGCOLOR=cyan>
<FONT Size=4>
   <%!   int i=0;
    %>
   <%    i++;
    %>
<P>你是第 <%=i%> 个访问本站的用户。
</FONT></BODY></HTML>
```

由于成员变量的有效范围与标记符号<%!、%>所在的位置无关，因此例2-2中的example2_2.jsp等价于下面的JSP页面(将<%!、%>写在JSP页面的后面)：

```
<%@ page contentType="text/html; charset=GB2312" %>
<HTML><BODY BGCOLOR=cyan><FONT Size=4>
  <% i++;
  %>
<P>你是第 <%=i%> 个访问本站的用户。
  <%! int i=0;
  %>
</FONT></BODY></HTML>
```

2.2.2　声明方法

在“<%!”和“%>”之间声明方法，该方法在整个JSP页面有效(与标记符号<%!、%>所在的位置无关)，但是该方法内定义的变量只在该方法内有效。这些方法将在Java程序片中被调用。当方法被调用时，方法内定义的变量被分配内存，调用完毕即可释放所占的内存。在例 2-3 中,example2_3.jsp 在"<%!"和"%>"之间声明了两个方法:getArea(double a)和 getLength(double a),在程序片中调用这两个方法,分别计算圆的面积和周长。example2_3.jsp 页面效果如图 2-3 所示。

```
地址(D) http://127.0.0.1:8080/ch2/example2_3.jsp

调用getArea方法计算半径是100.0的圆的面积: 31415.926535897932
调用getLength方法计算半径是50.0的圆的周长: 314.1592653589793
```

图 2-3 JSP 页面中方法的声明与调用

例 2-3

example2_3.jsp

```
<%@ page contentType = "text/html; charset = GB2312" %>
<HTML><BODY bgcolor = cyan><FONT size = 4>
   <%!   final double PI = Math.PI;
          double r;
          double getArea(double a){
             return PI * a * a;
          }
          double getLength(double a) {
              return 2 * PI * a;
          }
    %>
   <%     r = 100;
          out.println("调用 getArea 方法计算半径是" + r + "的圆的面积:");
          double area = getArea(r);
          out.println(area);
          r = 50;
          out.println("<BR>调用 getLength 方法计算半径是" + r + "的圆的周长:");
          double length = getLength(r);
          out.println(length);
    %>
</FONT></BODY></HTML>
```

2.3 Java 程序片

可以在"<%"和"%>"之间插入 Java 程序片。一个 JSP 页面可以有许多 Java 程序片,这些 Java 程序片将被 JSP 引擎按顺序执行。在 Java 程序片中声明的变量称作 JSP 页面的局部变量。局部变量的有效范围与其声明的位置有关,即局部变量在 JSP 页面后继的所有 Java 程序片以及 Java 表达式部分内都有效。JSP 引擎将 JSP 页面转译成 Java 文件时,将各个 Java 程序片的这些变量作为类中某个方法的变量,即局部变量。

现在已经知道,当多个用户请求一个 JSP 页面时,JSP 引擎为每个用户启动一个线程,该线程负责执行字节码文件响应用户的请求。因此当多个用户访问一个 JSP 页面时,该 JSP 页面中的 Java 程序片将被运行多次,分别运行在不同的线程中。

Java 程序片的执行有如下特点(如图 2-4 所示):

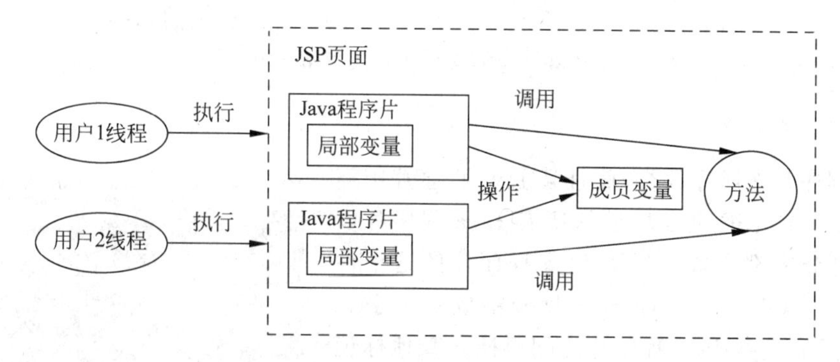

图 2-4　程序片的执行

1. 操作 JSP 页面的成员变量

Java 程序片中操作的成员变量是各个线程共享的变量，任何一个线程对 JSP 页面成员变量操作的结果，都会影响到其他线程。

2. 调用 JSP 页面的方法

Java 程序片中可以出现方法调用语句，该方法必须是 JSP 页面的方法（在"<%!"和"%>"之间声明方法）。

3. 声明操作局部变量

当一个线程享用 CPU 资源时，JSP 引擎让该线程执行 Java 程序片，这时，Java 程序片中的局部变量被分配内存空间，当轮到另一个线程享用 CPU 资源时，JSP 引擎让这个线程执行 Java 程序片，那么，Java 程序片中的局部变量会再次被分配内存空间。也就是说，Java 程序片已经被执行了两次，分别运行在不同的线程中，即运行在不同的时间片内。运行在不同线程中的 Java 程序片的局部变量互不干扰，即一个用户改变 Java 程序片中的局部变量的值不会影响其他用户的 Java 程序片中的局部变量。当一个线程将 Java 程序片执行完毕，运行在该线程中的 Java 程序片的局部变量释放所占的内存。

根据 Java 程序片的上述特点，对于某些特殊情形必须给予特别注意。例如，如果一个用户在执行 Java 程序片时调用 JSP 页面的方法操作成员变量时，可能不希望其他用户也调用该方法操作成员变量，以免对其产生不利的影响（成员变量被所有的用户共享），那么就应该将操作成员变量的方法用 synchronized 关键字修饰。当一个线程在执行 Java 程序片期间调用 synchronized 方法时，其他线程想在 Java 程序片中调用这个 synchronized 方法时就必须等待，直到调用 synchronized 方法的线程将该方法执行完毕。

在例 2-4 中，通过 synchronized 方法操作一个成员变量来实现一个简单的计数器。

例 2-4

example2_4.jsp

```
<%@ page contentType = "text/html; Charset = GB2312" %>
<HTML><BODY>
    <%!  int count = 0;                          //被用户共享的 count
         synchronized void setCount() {          //synchronized 修饰的方法
             count++;
         }
    %>
```

```
  <%    setCount();                          //程序片中调用同步方法
           out.println("你是第" + count + "个访问本站的用户");
    %>
</BODY></HTML>
```


Java 程序片按顺序执行，而且某 Java 程序片中声明的局部变量在 JSP 页面后继的所有 Java 程序片以及表达式部分内都有效。利用 Java 程序片的这个性质，有时候可以将一个 Java 程序片分割成几个 Java 程序片，然后在这些 Java 程序片之间再插入其他标记元素。例 2-5 通过将程序片分割成几部分，来验证用户输入的 E-mail 地址中是否含有非法的字符，页面效果如图 2-5 所示。

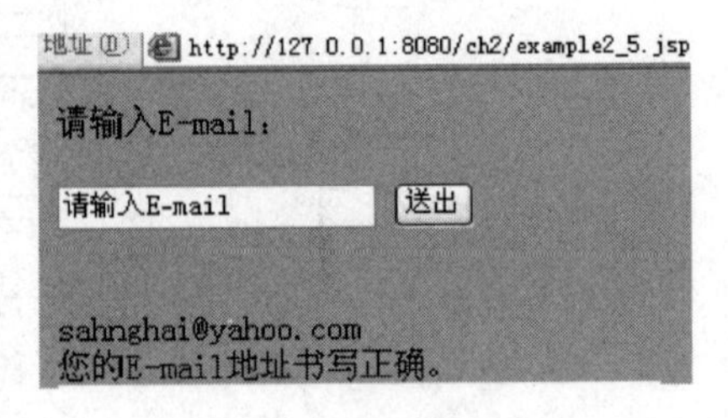

图 2-5 分割程序片

例 2-5

example2_5.jsp

```
<%@ page contentType = "text/html; charset = GB2312" %>
<HTML><BODY bgcolor = cyan><FONT Size = 3>
 <P>请输入 E-mail:<BR>
  <FORM action = "" method = get name = form>
       <INPUT type = "text" name = "client" value = "请输入 E-mail">
       <INPUT TYPE = "submit" value = "送出" name = submit>
  </FORM>
  <% String str = request.getParameter("client");
       if(str!= null){
         int index = str.indexOf("@");
         if(index == -1){
  %>       <BR>你的 E-mail 地址中没有@。
  <%    }
         else{
            int space = str.indexOf(" ");
            if(space!= -1){
  %>              <BR>你的 E-mail 地址含有非法的空格。
  <%          }
            else{
                 int start = str.indexOf("@");
                 int end = str.lastIndexOf("@");
                 if(start!= end){
  %>                  <BR>你的 E-mail 地址有两个以上的符号:@。
  <%                }
                 else{
                      out.print("<BR>" + str);
  %>                  <BR>你的 E-mail 地址书写正确。
  <%                }
            }
         }
     }
  %>
</FONT></BODY></HTML>
```

2.4 表　达　式

可以在"＜％＝"和"％＞"之间插入一个表达式，例如：＜％＝ x＋y ％＞，不可以在"＜％＝"和"％＞"之间插入语句。例如：＜％＝ x＝100；％＞是错误的。需要特别注意的是，"＜％＝"是一个完整的符号，"＜％"和"＝"之间不要有空格。表达式的值由服务器负责计算，并将计算结果以字符串形式发送至用户端显示。如果表达式无法求值，Tomcat 引擎将给出编译错误。例 2-6 计算了表达式的值，页面效果如图 2-6 所示。

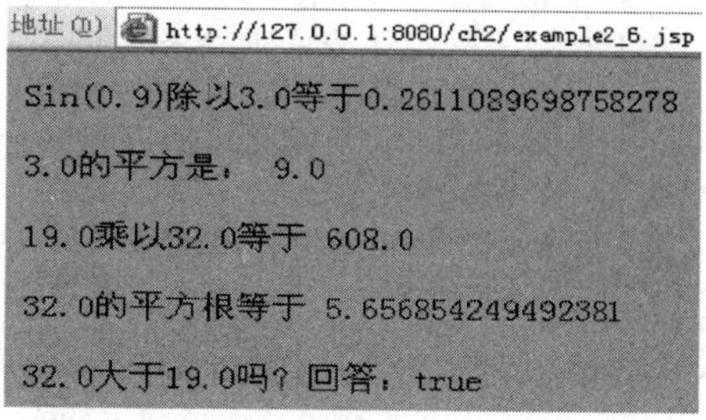

地址(D) http://127.0.0.1:8080/ch2/example2_6.jsp

Sin(0.9)除以3.0等于0.2611089698758278

3.0的平方是： 9.0

19.0乘以32.0等于 608.0

32.0的平方根等于 5.656854249492381

32.0大于19.0吗？回答：true

图 2-6　计算表达式的值

例 2-6

example2_6.jsp

```
<%@ page contentType = "text/html; charset = GB2312" %>
<HTML><BODY bgcolor = cyan><FONT Size = 3>
<% double x = 0.9,y = 3;
 %>
<P> Sin(<% = x %>)除以<% = y %>等于<% = Math.sin(x)/y %>
  <p><% = y %>的平方是：<% = Math.pow(y,2) %>
<%   x = 19;
     y = 32;
 %>
<P><% = x %>乘以<% = y %>等于 <% = x * y %>
<P> <% = y %>的平方根等于 <% = Math.sqrt(y) %>
<P><% = y %>大于<% = x %>吗？ 回答：<% = y > x %>
</FONT></BODY></HTML>
```

2.5　JSP 中的注释

注释可以增强 JSP 文件的可读性，并易于 JSP 文件的维护。JSP 中的注释可分为两种。

1. HTML 注释

在标记符号"＜!--"和"--＞"之间加入注释内容。例如：

```
<!-- 下面是一个提交密码的表单 -->
```

JSP 引擎把 HTML 注释交给用户，用户通过浏览器查看 JSP 的源文件时，能够看到 HTML 注释。

2. JSP 注释

在标记符号"＜％--"和"--％＞"之间加入注释内容。例如：

```
<%-- 利用 for 循环输出用户的全部信息 --%>
```

JSP 引擎忽略 JSP 注释，即在编译 JSP 页面时忽略 JSP 注释。

例 2-7 中的 JSP 页面使用了 HTML 注释和 JSP 注释。

例 2-7

example2_7.jsp

```
<%@ page contentType = "text/html; charset = GB2312" %>
<HTML><BODY>
 <P> 请输入三角形的三条边 a,b,c 的长度：
<!-- 以下是 HTML 表单,向服务器发送三角形的三条边的长度 -->
<!-- 要特别注意 action = ""的引号中不要有空格 -->
  <FORM action = "" method = post name = form>
        <P>请输入三角形边 a 的长度：<INPUT type = "text" name = "a">
        <P>请输入三角形边 b 的长度：<INPUT type = "text" name = "b">
        <P>请输入三角形边 c 的长度：<INPUT type = "text" name = "c">
        <INPUT TYPE = "submit" value = "送出" name = submit>
  </FORM>
 <%-- 获取用户提交的数据 --%>
  <%   String string_a = request.getParameter("a"),
               string_b = request.getParameter("b"),
               string_c = request.getParameter("c");
            double a = 0,b = 0,c = 0;
   %>
  <%-- 判断字符串是否是空对象,如果是空对象就初始化 --%>
      <% if(string_a == null){
              string_a = "0";
              string_b = "0";
              string_c = "0";
           }
        %>
   <%-- 求出边长,并计算面积 --%>
      <% try{ a = Double.valueOf(string_a).doubleValue();
               b = Double.valueOf(string_b).doubleValue();
               c = Double.valueOf(string_c).doubleValue();
               if(a + b > c&&a + c > b&&b + c > a){
                     double p = (a + b + c)/2.0;
                     double mianji = Math.sqrt(p * (p - a) * (p - b) * (p - c));
                     out.print("<BR>" + "三角形面积：" + mianji);
               }
               else
                   out.print("<BR>" + "你输入的三边不能构成一个三角形");
           }
           catch(NumberFormatException e){
               out.print("<BR>" + "请输入数字字符");
           }
        %>
 </BODY></HTML>
```

2.6 JSP 指令标记

2.6.1 page 指令

page 指令用来定义整个 JSP 页面的一些属性和这些属性的值。page 指令标记可以指定如下属性的值：

```
contentType、import、language、session、buffer、auotFlush、isThreadSafe、pageEncoding
```

例如，可以用 page 指令指定 JSP 页面的 contentType 属性的值是 text/html；charset=GB2312，这样，JSP 页面就可以显示标准汉语。例如：

```
<%@ page contentType="text/html; charset=GB2312" %>
```

page 指令对整个页面有效，与其书写的位置无关，但习惯把 page 指令写在 JSP 页面的最前面。page 指令的格式：

```
<%@ page 属性 1="属性 1 的值" 属性 2="属性 2 的值" …… %>
```

属性值需用单引号或双引号括起来，如果为一个属性指定几个值的话，这些值用逗号分隔。例如：

```
<%@ page import="java.util.*","java.io.*" %>
```

可以用一个 page 指令指定多个属性的值，例如：

```
<%@ page 属性 1="属性 1 的值" 属性 2="属性 2 的值" …… %>
```

也可以使用多个 page 指令分别为每个属性指定值。例如：

```
<%@ page 属性 1="属性 1 的值" %>
<%@ page 属性 2="属性 2 的值" %>
      ⋮
<%@ page 属性 n="属性 n 的值" %>
```

当为 import 属性指定多个属性值时，JSP 引擎把 JSP 页面转译成的 Java 文件中会有如下的 import 语句：

```
import java.util.*;
import java.io.*;
import java.sql.*;
```

在一个 JSP 页面中，也可以使用多个 page 指令来指定属性及其值。需要注意的是：可以使用多个 page 指令指定 import 属性几个值，但其他属性只能使用 page 指令指定一个值。例如：

```
<%@ page contentType="text/html; charset=GB2312" %>
<%@ page import="java.util.*" %>
<%@ page import="java.util.*","java.awt.*" %>
```

不允许两次使用 page 指令给 contentType 属性指定不同的属性值，下列用法是错误的：

```
<%@ page contentType="text/html; charset=GB2312" %>
<%@ page contentType="application/msword" %>
```

以下将分别讲述 page 指令可以指定的属性及其作用。

1. language 属性

定义 JSP 页面使用的脚本语言，该属性的值目前只能取 java。

为 language 属性指定值的格式：

```
<%@ page language="java" %>
```

language 属性的默认值是 java,即如果在 JSP 页面中没有使用 page 指令指定该属性的值的话,那么 JSP 页面默认有如下 page 指令:

```
<%@ page language="java" %>
```

2. import 属性

该属性的作用是为 JSP 页面引入 Java 核心包中的类,这样就可以在 JSP 页面的程序片部分、变量及方法声明部分、表达式部分使用核心包中的类。可以为 import 属性指定多个值,该属性的值可以是 Java 某核心包中的所有类或一个具体的类。例如:

```
<%@ page import="java.io.*","java.util.Date" %>
```

JSP 页面默认 import 属性已经有如下的值:

```
"java.lang.*"、"javax.servlet.*"、"javax.servlet.jsp.*"、"javax.servlet.http.*"
```

3. contentType 属性

contentType 属性值确定 JSP 页面响应的 MIME(Multipurpose Internet Mail Extention)类型和 JSP 页面字符的编码。属性值的一般形式是 MIME 类型;charset=编码。例如:

```
<%@ page contentType="text/html; charset=GB2312" %>
```

如果不使用 page 指令为 contentType 指定一个值,那么 contentType 属性的默认值是 text/html;charset=ISO-8859-1。

当用户请求一个 JSP 页面时,Tomcat 服务器负责解释执行 JSP 页面,并将某些信息发送到用户的浏览器,以便用户浏览这些信息。Tomcat 服务器同时负责通知用户的浏览器使用怎样的方式来处理所接收到的信息,这就要求 JSP 页面必须设置响应的 MIME 类型和 JSP 页面字符的编码。例如,如果希望用户的浏览器启用 HTML 解析器来解析执行所接收到的信息(即所谓的网页形式),就可以如下设置 contentType 属性的值:

```
<%@ page contentType="text/html; charset=GB2312" %>
```

如果希望用户的浏览器启用本地的 MS-Word 应用程序来解析执行收到的信息,就可以如下设置 contentType 属性的值:

```
<%@ page contentType="application/msword" %>
```

JSP 页面使用 page 指令只能为 contentType 指定一个值,不允许两次使用 page 指令给 contentType 属性指定不同的属性值,下列用法是错误的:

```
<%@ page contentType="text/html; charset=GB2312" %>
<%@ page contentType="application/msword" %>
```

可以使用 page 指令为 contentType 属性指定的值有:text/html、text/plain、image/gif、image/x-xbitmap、image/jpeg、image/pjpeg、application/x-shockwave-flash、application/vnd.ms-powerpoint、application/vnd.ms-excel、application/msword 等。

如果用户的浏览器不支持某种 MIME 类型,那么用户的浏览器就无法用相应的手段处理所接收到的信息。例如,使用 page 指令设置 contentType 属性的值是 application/msword,

如果用户浏览器所驻留的计算机没有安装 MS-Word 应用程序，那么浏览器就无法处理所接收到的信息。

例 2-8 中有两个 JSP 页面，其中的 first. jsp 页面使用 page 指令设置 contentType 属性的值是 text/html；charset＝GB2312，当用户请求 first. jsp 页面时，用户的浏览器启用 HTML 解析器来解析执行收到的信息；second. jsp 页面使用 page 指令设置 contentType 属性的值是 application/msword，当用户请求 second. jsp 页面时，用户的浏览器将启动本地的 MS-Word 应用程序来解析执行收到的信息，页面效果如图 2-7 所示。

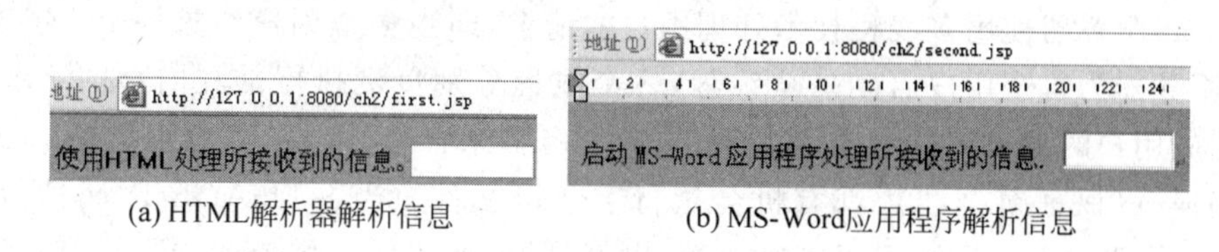

(a) HTML解析器解析信息　　(b) MS-Word应用程序解析信息

图 2-7　两个页面的效果图

例 2-8

first. jsp

```
<%@ page contentType = "text/html; Charset = GB2312" %>
<HTML><BODY BGCOLOR = cyan>
<FONT Size = 3>
<P>使用 HTML 处理所接收到的信息。
   <input type = text size = 10>
</FONT></BODY></HTML>
```

second. jsp

```
<%@ page contentType = "application/msword" %>
<HTML><BODY BGCOLOR = cyan>
<FONT Size = 8>
<P>启动 MS-Word 应用程序处理所接收到的信息。
 <input type = text size = 10>
</FONT></BODY></HTML>
```

4. session 属性

用于设置是否需要使用内置的 session 对象。

session 属性的属性值可以是 true 或 false，默认的属性值是 true。

5. buffer 属性

内置输出流对象 out 负责将服务器的某些信息或运行结果发送到用户端显示，buffer 属性用来指定 out 设置的缓冲区的大小或不使用缓冲区。buffer 属性可取值 none 来设置 out 不使用缓冲区。buffer 属性的默认值是 8KB。例如：

```
<%@ page buffer = "24KB" %>
```

6. autoFlush 属性

指定 out 的缓冲区被填满时，缓冲区是否自动刷新。autoFlush 可以取值 true 或 false。autoFlush 属性的默认值是 true。当 autoFlush 属性取值 false 时，如果 out 的缓冲区被填满,就会出现缓存溢出异常。当 buffer 的值是 none 时,autoFlush 的属性值就不能设置成 false。

7. isThreadSafe 属性

isThreadSafe 的属性值取 true 或 false,用来设置 JSP 页面是否可多线程访问。当 isThreadSafe 属性值设置为 true 时,JSP 页面能同时响应多个用户的请求;当 isThreadSafe 属性值设置成 false 时,JSP 页面同一时刻只能处理响应一个用户的请求,其他用户需排队等待。isThreadSafe 属性的默认值是 true。

Tomcat 服务器使用多线程技术处理用户的请求,即当多个用户请求一个 JSP 页面时,Tomcat 服务器为每个用户启动一个线程,每个线程分别负责执行常驻内存的字节码文件来响应相应用户的请求。这些线程由 Tomcat 服务器来管理,将 CPU 的使用权在各个线程间快速切换,以保证每个线程都有机会执行字节码文件。当 isThreadSafe 属性值为 true 时,CPU 的使用权在各个线程间快速切换,也就是说,即使一个用户的线程没有执行完毕,CPU 的使用权也可能要切换给其他的线程,如此轮流,直到各个线程执行完毕;当 JSP 使用 page 指令将 isThreadSafe 属性值设置成 false 时,该 JSP 页面同一时刻只能处理响应一个用户的请求,其他用户需排队等待,也就是说,CPU 要保证一个线程将 JSP 页面执行完毕才会把 CPU 使用权切换给其他线程。

下列 JSP 页面 computer. jsp 将 isThreadSafe 属性的值设置为 false。

computer. jsp

```
<%@ page contentType = "text/html; charset = GB2312" %>
<%@ page isThreadSafe = "false" %>
<HTML><BODY>
  <%!   int i = 1;          //被所有用户共享
   %>
  <%
      for(int k = 1; k <= 100; k++){
          out.println(i);
          i++;
      }
   %>
</BODY></HTML>
```

首先注意到,上述 computer. jsp 页面中的成员变量 *i* 被所有用户共享。假设有两个用户访问上述 computer. jsp 页面,那么当第一个用户访问该页面时,Tomcat 引擎要保证该用户的线程执行完该页面,才会让第 2 个用户线程执行该 JSP 页面。因此,第一个用户看到的页面运行效果是输出 1 至 100 的整数,第 2 个用户看到的页面运行效果是输出 101 至 200 的整数。但是,如果 computer. jsp 将 isThreadSafe 属性的值设置为 true,Tomcat 引擎就会将 CPU 的使用权在各个线程间快速切换,这样一来,就可出现第 1 个用户的线程在没有执行完 JSP 页面中的程序片中的循环时,例如,当循环到第 50 次时,Tomcat 引擎就可能将 CPU 的使用权切换给第 2 个用户的线程,那么第 2 个用户看到的第一个整数将是 51。

8. info 属性

info 属性的属性值是一个字符串,其目的是为 JSP 页面准备一个常用且可能需要经常修改的字符串。例如:

```
<%@ page info= "we are students" %>
```

可以在JSP页面中使用方法：

```
getServletInfo();
```

获取info属性的属性值。

注意：当JSP页面被转译成Java文件时，转译成的类是Servlet的一个子类，所以在JSP页面中可以使用Servlet类的方法：getServletInfo()。

2.6.2 include 指令标记

如果需要在JSP页面的某处整体插入一个文件，就可以考虑使用这个指令标记。该指令标记语法如下：

```
<%@ include file="文件的 URL" %>
```

该指令标记的作用是在JSP页面出现该指令的位置处，静态插入一个文件。被该指令插入的文件必须是可访问和可使用的，如果该文件和当前JSP页面在同一Web服务目录中，那么"文件的URL"就是文件的名字；如果该文件在JSP页面所在的Web服务目录的一个子目录中，例如fileDir子目录中，那么"文件的URL"就是"fileDir/文件的名字"。

静态插入就是当前JSP页面和插入的文件合并成一个新的JSP页面，然后JSP引擎再将这个新的JSP页面转译成Java文件。因此，一个JSP页面使用该指令插入文件后，必须保证新合并成的JSP页面符合JSP语法规则，即能够成为一个JSP页面文件。例如，JSP页面A.jsp已经使用page指令为contentType属性设置了值：

```
<%@ page contentType="text/html; charset=GB2312" %>
```

如果A.jsp使用include指令标记插入一个JSP页面：B.jsp，而B.jsp页面使用page指令为contentType属性设置的值是：

```
<%@ page contentType="application/msword" %>
```

那么，合并后的JSP页面就两次使用page指令为contentType属性设置了不同的属性值，导致出现语法错误，因为JSP页面中的page指令只能为contentType指定一个值。

Tomcat 5.0版本以后的服务器每次都要检查include指令标记插入的文件是否被修改过，因此，JSP页面成功静态插入一个文件后，如果对插入的文件进行了修改，那么Tomcat服务器会重新编译JSP页面，即将当前的JSP页面和修改后的文件合并成一个JSP页面，然后Tomcat服务器再将这个新的JSP页面转译成Java类文件。

使用include指令可以实现代码的复用。例如，每个JSP页面上都可能需要一个导航条，以便用户在各个JSP页面之间方便地切换，那么每个JSP页面都可以使用include指令在页面的适当位置整体插入一个相同的文件。

需要特别注意的是，在Tomcat 4版本中，被插入的文件不允许使用page指令指定contentType属性的值，但是，在Tomcat 5版本中，被插入的文件允许使用page指令指定contentType属性的值，但指定的值必须和插入该文件的JSP页面中的page指令指定的

contentType 属性的值相同。

例 2-9 中的 example2_9.jsp 页面使用 include 指令标记静态插入一个文本文件 Hello.txt,该文本文件的内容如下:

```
<%@ page contentType="text/html; charset=GB2312" %>
很高兴认识你们!
nice to meet you.
```

Hello.txt 文件和当前 example2_9.jsp 页面在同一 Web 服务目录中,页面效果如图 2-8 所示。

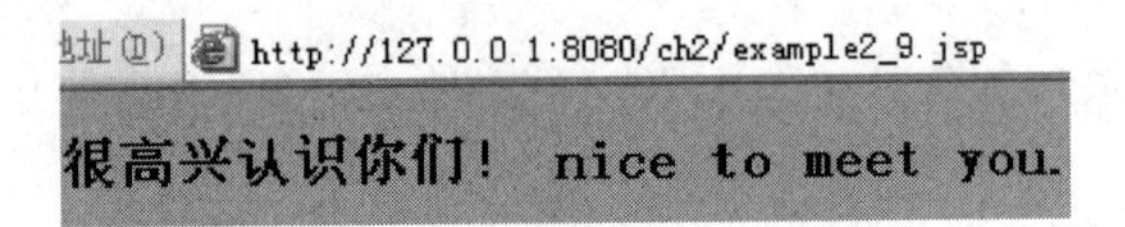

图 2-8 使用 include 指令标记(a)

例 2-9

example2_9.jsp

```
<%@ page contentType="text/html; charset=GB2312" %>
<HTML><BODY bgcolor=cyan>
<H3><%@ include file="Hello.txt" %>
</H3>
</BODY></HTML>
```

上述 example2_9.jsp 等价于下面的 JSP 文件:example2_9_1.jsp。

example2_9_1.jsp

```
<%@ page contentType="text/html; charset=GB2312" %>
<html><BODY>
<H3>很高兴认识你们!
nice to meet you.
</H3>
</BODY></HTML>
```

例 2-10 中的 JSP 页面 example2_10.jsp 使用 include 指令标记静态插入一个 JSP 文件 computer.jsp。computer.jsp 和 example2_10.jsp 均保存在 Web 服务目录 ch2 中,example2_10.jsp 页面的效果如图 2-9 所示。

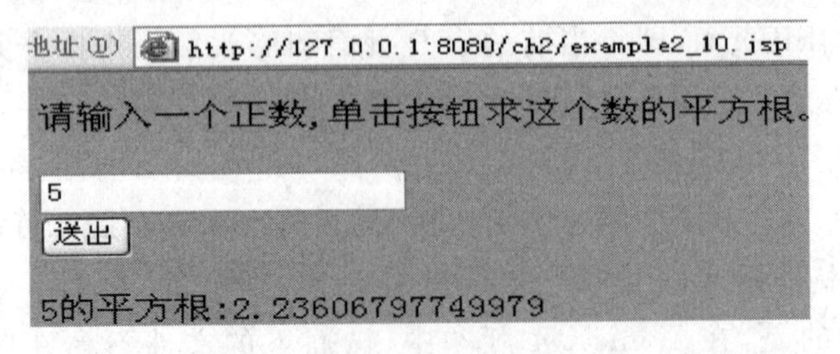

图 2-9 使用 include 指令标记(b)

例 2-10

example2_10.jsp

```
<%@ page contentType = "text/html; charset = GB2312" %>
<HTML><BODY Bgcolor = cyan><FONT size = 3>
<P>请输入一个正数,单击按钮求这个数的平方根。
<%@ include file = "computer.jsp" %>
</FONT></BODY></HTML>
```

computer.jsp

```
<%@ page contentType = "text/html; charset = GB2312" %>
<FORM action = "" method = post name = form>
<INPUT type = "text" name = "ok">
<BR><INPUT TYPE = "submit" value = "送出" name = submit>
</FORM>
<% String a = request.getParameter("ok");
    if(a == null){
        a = "1";
    }
    try{   double number = Integer.parseInt(a);
            out.print(a + "的平方根：" + Math.sqrt(number));
    }
    catch(NumberFormatException e){
            out.print("<BR>" + "请输入数字字符");
    }
 %>
```

上述 example2_10.jsp 等同于下述 JSP 文件：example2_10_1.jsp。

example2_10_1.jsp

```
<%@ page contentType = "text/html; charset = GB2312" %>
<HTML><BODY Bgcolor = cyan><FONT size = 1>
<P>请输入一个正数,单击按钮求这个数的平方根。
<FORM action = "" method = post name = form>
<INPUT type = "text" name = "ok">
<BR><INPUT TYPE = "submit" value = "送出" name = submit>
</FORM>
  <% String a = request.getParameter("ok");
      if(a == null){
         a = "1";
      }
      try{   double number = Integer.parseInt(a);
              out.print(a + "的平方根：" + Math.sqrt(number));
      }
      catch(NumberFormatException e){
            out.print("<BR>" + "请输入数字字符");
      }
  %>
</FONT></BODY></HTML>
```

2.7 JSP 动作标记

动作标记是一种特殊的标记,它影响 JSP 运行时的功能。

2.7.1 include 动作标记

include 动作标记语法格式为:

```
<jsp:include page= "文件的 URL"/>
```

或

```
<jsp:include page= "文件的 URL">
      param 子标记
</jsp:include>
```

需要注意的是,当 include 动作标记不需要 param 子标记时,必须使用上述第一种形式。

include 动作标记告诉 JSP 页面动态加载一个文件。与静态插入文件的 include 指令标记不同,当 JSP 引擎把 JSP 页面转译成 Java 文件时,不把 JSP 页面中动作指令 include 所指定的文件与原 JSP 页面合并一个新的 JSP 页面,而是告诉 Java 解释器,这个文件在 JSP 运行时(JSP 页面对应的字节码文件被加载执行)才被处理。如果包含的文件是普通的文本文件,就将文件的内容发送到用户端,由用户端负责显示;如果包含的文件是 JSP 文件,JSP 引擎就执行这个文件,然后将执行的结果发送到用户端,并由用户端负责显示这些结果。

使用 include 动作标记也可以实现代码的复用,尽管 include 动作标记和 include 指令标记的作用都是处理所需要的文件,但是处理方式和处理时间上是不同的。include 指令标记是在编译阶段就处理所需要的文件,被处理的文件在逻辑和语法上依赖于当前 JSP 页面,其优点是页面的执行速度快;而 include 动作标记是在 JSP 页面运行时才处理文件,被处理的文件在逻辑和语法上独立于当前 JSP 页面,其优点是可以使用 param 子标记更加灵活地处理所需要的文件(见后面的 param 标记),缺点是执行速度要慢一些。

书写 include 动作标记<jsp:include ……/>时要注意:jsp、:、include 三者之间不要有空格。

例 2-11 中的 example2_11.jsp 页面动态加载两个文件:imageCar.html 和 car.txt。example2_11.jsp 页面保存在 Web 服务目录 ch2 中。example2_11.jsp 页面要动态加载的 imageCar.html 文件以及 imageCar.html 文件所使用的图像文件 car.jpg 均保存在 ch2 中;example2_11.jsp 页面要动态加载的 car.txt 文件保存在 ch2 的子目录 Myfile 中。

在浏览器的地址栏输入:http://127.0.0.1:8080/ch2/example2_11.jsp,页面的效果如图 2-10 所示。

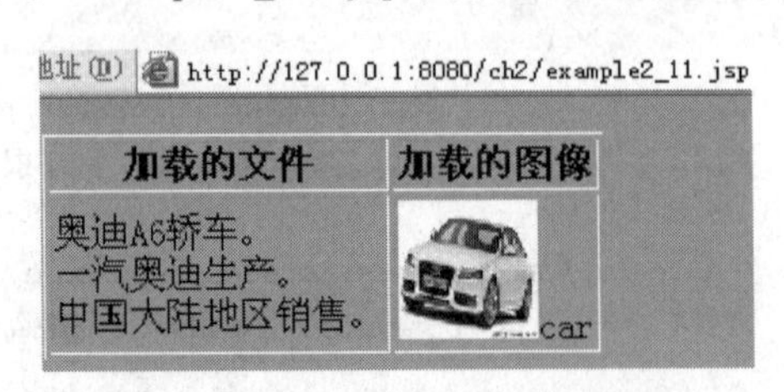

图 2-10 使用 include 动作标记

例 2-11

car. txt

```
奥迪 A6 轿车。
<BR>一汽奥迪生产。
<BR>中国大陆地区销售。
```

imageCar. html

```
<image src="car.jpg" width=60 height=60>car</iamge>
```

example2_11. jsp

```
<%@ page contentType="text/html; charset=GB2312" %>
<HTML><BODY BGCOLOR=Cyan><FONT Size=2>
 <table border=1>
 <tr><th>加载的文件</th>
     <th>加载的图像</th>
 </tr>
 <tr><td><jsp:include page="Myfile/car.txt" /></td>
     <td><jsp:include page="imageCar.html" /></td>
 </tr>
 </table>
</FONT></BODY></HTML>
```

2.7.2 param 动作标记

param 标记以“名字-值”对的形式为其他标记提供附加信息，这个标记与 jsp：include、jsp：forward、jsp：plugin 标记一起使用。

param 动作标记：

```
<jsp:param name="属性名字" value="属性的值" />
```

param 标记不能独立使用，需作为 jsp：include、jsp：forward、jsp：plugin 标记的子标记来使用。

当该标记与 jsp：include 标记一起使用时，可以将 param 标记中的值传递到 include 指令要加载的文件中去。也就是说，JSP 页面在使用 include 动作标记加载文件时，可以动态地向所加载文件传递数据。相对使用 include 指令标记，JSP 页面通过使用 include 动作标记能更为灵活地处理所需要的文件。

例 2-12 中，example2_12. jsp 页面使用 include 动作标记动态加载文件 tom. jsp，当 tom. jsp 文件被加载时获取 example2_12. jsp 页面中 include 动作标记的 param 子标记中 name 属性的值(tom. jsp 文件使用 Tomcat 服务器提供的 request 内置对象获取 param 子标记中 name 属性的值，有关内置对象见后面的第 4 章)。

example2_12. jsp 和 tom. jsp 页面保存在 Web 服务目录 ch2 中。在浏览器的地址栏输入：http：//127. 0. 0. 1：8080/ch2/example2_12. jsp，页面的效果如图 2-11 所示。

地址(D) http://127.0.0.1:8080/ch2/example2_12.jsp

加载文件效果：

从1到300的连续和是： 45150

图 2-11 动态加载文件计算连续和

例 2-12

example2_12.jsp

```
<%@ page contentType = "text/html; charset = GB2312" %>
<HTML><BODY>
 <P>加载文件效果:
   <jsp:include page = "tom.jsp">
       <jsp:param name = "computer" value = "300" />
   </jsp:include>
</BODY></HTML>
```

tom.jsp

```
<%@ page contentType = "text/html; charset = GB2312" %>
<HTML><BODY>
  <% String str = request.getParameter("computer"); //获取 example2_12.jsp 中 param 子标记中
      int n = Integer.parseInt(str);                    //name 属性的值
      int sum = 0;
      for(int i = 1; i <= n; i++ )
        sum = sum + i;
  %>
<P>从 1 到<% = n %>的连续和是:<% = sum %>
</BODY></HTML>
```

2.7.3 forward 动作标记

forward 动作标记的语法格式:

```
<jsp:forward page = "要转向的页面" />
```

或

```
<jsp:forward page = "要转向的页面" >
    param 子标记
</jsp:forward>
```

该指令的作用是:从该指令处停止当前页面的继续执行,转向执行该指令中 page 属性指定的 JSP 页面,但是在浏览器的地址栏中并不显示 forward 指令所转向的 JSP 页面的 URL 表示,浏览器的地址栏仍然显示 forward 转向前的 JSP 页面的 URL 表示。

需要注意的是,当 forward 动作标记不需要 param 子标记时,必须使用上述第一种形式。forward 标记可以使用 param 动作标记作为子标记,以便向所转向的页面传送信息。forward 动作标记指定的所转向的 JSP 文件可以使用 Tomcat 服务器提供的 request 内置对象获取 param 子标记中 name 属性的值。

例 2-13 中的 example2_13.jsp 页面使用 forward 动作标记转向 come.jsp 页面,并向 come.jsp 页面传递一个数值。example2_13.jsp 和 come.jsp 页面保存在 Web 服务目录 ch2 中。在浏览器的地址栏输入:http://127.0.0.1:8080/ch2/example2_13.jsp,页面的效果如图 2-12 所示。

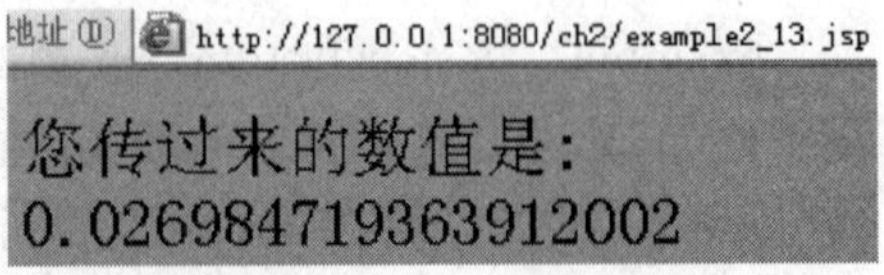

图 2-12 向转向的页面传递数据

例 2-13

example2_13.jsp

```
<%@ page contentType = "text/html; charset = GB2312" %>
<HTML><BODY>
 <% double i = Math.random();
  %>
    <jsp:forward page = "come.jsp" >
        <jsp:param name = "number" value = "<% = i %>" />
    </jsp:forward>
</BODY></HTML>
```

come.jsp

```
<%@ page contentType = "text/html; charset = GB2312" %>
<HTML><BODY bgcolor = cyan><FONT Size = 5>
<%   String str = request.getParameter("number");
        double n = Double.parseDouble(str);
 %>
        <P>你传过来的数值是:<BR>
        <% = n %>
</FONT></BODY></HTML>
```

2.7.4 plugin 动作标记

在页面中使用普通的 HTML 标记<applet ……></applet>可以让用户下载运行一个 Java applet 小程序,但并不是所有用户的浏览器都支持 Java applet 小程序。如果 Java applet 小程序使用了 JDK1.2 以后的类,那么,有些浏览器并不支持这个 Java 小应用程序,而使用 plugin 动作标记可以保证用户能执行小应用程序。

该动作标记指示 JSP 页面加载 Java plugin 插件。该插件由用户负责下载,并使用该插件来运行 Java applet 小程序。

plugin 动作标记:

```
<jsp:plugin type = "applet" code = "小程序的字节码文件"
      jreversion = "Java 虚拟机版本号" width = "小程序宽度值" height = "小程序高度值" >
      <jsp:fallback>
          提示信息:用来提示用户的浏览器是否支持插件下载
    </jsp:fallback>
</jsp:plugin>
```

假设有一个 Java applet 小程序,主类字节码文件是 B.class,该文件存放在 Web 服务目录 ch2 中,含有 plugin 动作标记的 JSP 文件 example2_14.jsp 也存放在 ch2 中。

例 2-14

example2_14.jsp

```
<%@ page contentType = "text/html; charset = GB2312" %>
<HTML><BODY>
  <jsp:plugin type = "applet"code = "B.class" jreversion = "1.2" width = "200" height  = "260" >
     <jsp:fallback>
```

```
            Plugin tag OBJECT or EMBED not supported by browser.
        </jsp:fallback>
    </jsp:plugin>
    </BODY></HTML>
```


当用户访问上述 JSP 页面时,将导致登录 Sun 公司的网站下载 Java plugin 插件,出现用户选择是否下载插件的界面。用户下载插件完毕后,接受许可协议,就可以根据向导一步一步地安装插件。安装完毕后,小程序就开始用 Java 的虚拟机(不再使用浏览器自带的虚拟机)加载执行 Java applet 小程序。以后用户再访问带有 plugin 动作标记的 JSP 页面就能直接执行该页面中包含的 Java applet 小程序。

2.7.5 useBean 动作标记

该标记用来创建并使用一个 JavaBean,是非常重要的一个动作标记,将在第 7 章详细讨论它的用法和作用。Sun 公司倡导:用 HTML 完成 JSP 页面的静态部分,用 JavaBean 完成动态部分,实现真正意义上的静态和动态之分割。

2.8 实验 1:JSP 页面的基本结构

1. 相关知识点

一个 JSP 页面可由普通的 HTML 标记、JSP 标记、成员变量和方法的声明、Java 程序片和 Java 表达式组成。JSP 引擎把 JSP 页面中的 HTML 标记交给用户的浏览器执行显示;JSP 引擎负责处理 JSP 标记、变量和方法声明;JSP 引擎负责运行 Java 程序片、计算 Java 表达式,并将需要显示的结果发送给用户的浏览器。

JSP 页面中的成员变量是被所有用户共享的变量。Java 程序片可以操作成员变量,任何一个用户对 JSP 页面成员变量操作的结果,都会影响到其他用户。如果多个用户访问一个 JSP 页面,那么该页面中的 Java 程序片就会被执行多次,分别运行在不同的线程中,即运行在不同的时间片内。运行在不同线程中的 Java 程序片的局部变量互不干扰,即一个用户改变 Java 程序片中的局部变量的值不会影响其他用户的 Java 程序片中的局部变量。

2. 实验目的

本实验的目的是让学生掌握怎样在 JSP 页面中使用成员变量,怎样使用 Java 程序片、Java 表达式。

3. 实验要求

编写两个 JSP 页面,名字分别为 inputName.jsp 和 people.jsp。

1) inputName.jsp 的具体要求

该页面有一个表单,用户通过该表单输入自己的姓名并提交给 people.jsp 页面,inputName.jsp 的效果如图 2-13(a)所示。

2) people.jsp 的具体要求

(1) JSP 页面有名字为 personList、类型是 StringBuffer 以及名字是 count、类型为 int 的成员变量。

(2) JSP 页面有 public void judge ()方法。该方法负责创建 personList 对象,当 count

的值是 0 时，judge ()方法创建 personList 对象。

(3) JSP 页面有 public void addPerson(String p)的方法，该方法将参数 p 指定的字符串尾加到成员变量 personList，同时将 count 作自增运算。

(4) JSP 页面在程序片中获取 inputName. jsp 页面提交的姓名，然后调用 judge()创建 personList 对象，调用 addPerson 方法将用户的姓名尾加到成员变量 personList 中。

(5) 如果 inputName. jsp 页面没有提交姓名，或姓名含有的字符个数大于 10，就使用 <jsp:forward page="要转向的页面" />标记将用户转到 inputName. jsp 页面。

(6) 通过 Java 表达式输出 person 和 count 的值。

people. jsp 的效果如图 2-13(b)所示。

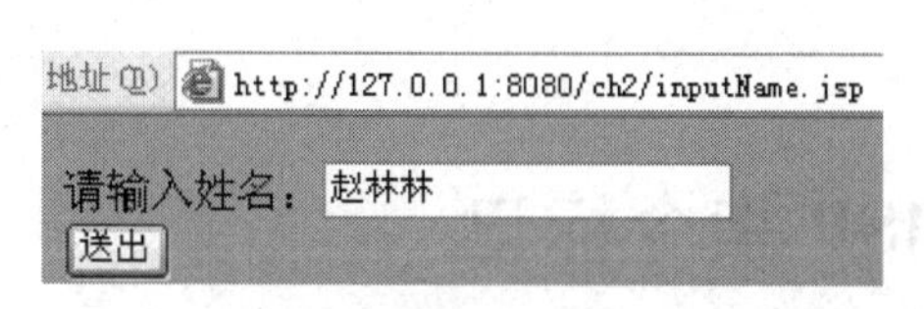

(a) inputName.jsp页面的效果

(b) people.jsp页面的效果

图 2-13　两个页面的效果图

4. 参考代码

代码仅供参考，学生可按着实验要求，参考本代码编写代码。

JSP 页面参考代码如下：

inputName. jsp

```
<%@ page contentType = "text/html; charset = GB2312" %>
<HTML><BODY bgcolor = cyan><FONT Size = 3>
  <FORM action = "people.jsp" method = get name = form>
    请输入姓名：<INPUT type = "text" name = "name">
    <BR><INPUT TYPE = "submit" value = "送出" name = submit>
  </FORM>
</BODY></HTML>
```

people. jsp

```
<%@ page contentType = "text/html; charset = GB2312" %>
<HTML><BODY BGCOLOR = yellow><FONT Size = 3>
 <%! int count;
        StringBuffer personList;
        public void judge(){
             if(count == 0)
                  personList = new StringBuffer();
        }
        public void addPerson(String p){
             if(count == 0)
                  personList.append(p);
             else
                personList.append("," + p);
             count++;
    }
```

```
 %>
<%   String name = request.getParameter("name");
     byte bb[] = name.getBytes("iso - 8859 - 1");
     name = new String(bb);
     if(name.length() == 0||name.length()> 10){
  %>   <jsp:forward page = "inputName.jsp" />
<%   }
     judge();
     addPerson(name);
 %>
<BR> 目前共有<% = count %>人浏览了该页面,他们的名字是:
<BR><% = personList %>
</FONT></BODY></HTML>
```

2.9 实验 2:JSP 指令标记

1. 相关知识点

page 指令:

```
<%@ page 属性 1 = "属性 1 的值" 属性 2 = "属性 2 的值" …… %>
```

用来定义整个 JSP 页面的一些属性和这些属性的值。比较常用的两个属性是 contentType 和 import。page 指令只能为 contentType 指定一个值,但可以为 import 属性指定多个值。

include 指令标记:

```
<%@ include file = "文件的 URL " %>
```

的作用是在 JSP 页面出现该指令的位置处静态插入一个文件。被插入的文件必须是可访问和可使用的,如果该文件和当前 JSP 页面在同一 Web 服务目录中,那么"文件的 URL"就是文件的名字;如果该文件在 JSP 页面所在的 Web 服务目录的一个子目录中,例如 fileDir 子目录中,那么"文件的 URL"就是"fileDir/文件的名字"。include 指令标记是在编译阶段就处理所需要的文件,被处理的文件在逻辑和语法上依赖于当前 JSP 页面,其优点是页面的执行速度快。

2. 实验目的

本实验的目的是让学生掌握怎样在 JSP 页面中使用 page 指令设置 contentType 的值;使用 include 指令在 JSP 页面中静态插入一个文件的内容。

3. 实验要求

编写 3 个 JSP 页面:first. jsp、second. jsp 和 third. jsp。另外,要求用"记事本"编写一个 TXT 文件 hello. txt。hello. txt 的每行有若干个英文单词,这些单词之间用空格分隔,每行之间用"
"分隔,如下所示:

hello. txt

```
package apple void back public
<BR>
private throw class hello welcome
```

1）first.jsp 的具体要求

first.jsp 使用 page 指令设置 contentType 属性的值是 text/plain，使用 include 指令静态插入 hello.txt 文件。

2）second.jsp 的具体要求

second.jsp 使用 page 指令设置 contentType 属性的值是 application/vnd.ms-powerpoint，使用 include 指令静态插入 hello.txt 文件。

3）third.jsp 的具体要求

third.jsp 使用 page 指令设置 contentType 属性的值是 application/msword，使用 include 指令静态插入 hello.txt 文件。

4. 参考代码

代码仅供参考，学生可按着实验要求，参考本代码编写代码。

JSP 页面参考代码如下：

first.jsp

```
<%@ page contentType = "text/plain" %>
<HTML><BODY><FONT Size = 4 color = blue>
    <%@include file = "hello.txt" %>
 </FONT></BODY></HTML>
```

second.jsp

```
<%@ page contentType = "application/vnd.ms-powerpoint" %>
<HTML><BODY><FONT Size = 2 color = blue>
    <%@include file = "hello.txt" %>
 </FONT></BODY></HTML>
```

third.jsp

```
<%@ page contentType = "application/msword" %>
<HTML><BODY><FONT Size = 4 color = blue>
    <%@include file = "hello.txt" %>
</FONT></BODY></HTML>
```

2.10 实验 3：JSP 动作标记

1. 相关知识点

include 动作标记：

```
<jsp:include page = "文件的 URL"/>
```

是在 JSP 页面运行时才处理加载的文件，被加载的文件在逻辑和语法上独立于当前 JSP 页面。include 动作标记可以使用 param 子标记向被加载的 JSP 文件传递信息。

forward 动作标记：

```
<jsp:forward page = "要转向的页面" />
```

的作用是从该指令处停止当前页面的继续执行，而转向执行 page 属性指定的 JSP 页面。

forward 标记可以使用 param 动作标记作为子标记,以便向要转向的 JSP 页面传送信息。

2. 实验目的

本实验的目的是让学生掌握怎样在 JSP 页面中使用 include 标记动态加载文件；使用 forward 动作标记实现页面的转向。

3. 实验要求

编写 4 个 JSP 页面：one. jsp、two. jsp、three. jsp 和 error. jsp。one. jsp、two. jsp 和 three. jsp 页面都含有一个导航条,以便让用户方便地单击超链接访问这 3 个页面,要求这 3 个页面通过使用 include 动作标记动态加载导航条文件 head. txt。导航条文件 head. txt 的内容如下所示：

head. txt

```
<%@ page contentType = "text/html; charset = GB2312" %>
 <table cellSpacing = "1" cellPadding = "1" width = "60 %" align = "center" border = "0" >
   <tr valign = "bottom">
   <td><A href = "one.jsp"><FONT Size = 3 >one.jsp 页面</FONT></A></td>
   <td><A href = "two.jsp"><FONT Size = 3 >two.jsp 页面</FONT></A></td>
   <td><A href = "three.jsp"><FONT Size = 3 >three.jsp 页面</FONT></A></td>
  </tr>
  </FONT>
</table>
```

1) one. jsp 的具体要求

要求 one. jsp 页面有一个表单,用户使用该表单可以输入一个 1～100 之间的整数,并提交给本页面；如果输入的整数在 50 至 100 之间(不包括 50)就转向 three. jsp,如果在 1 至 50 之间就转向 two. jsp；如果输入不符合要求的整数,例如 120,就转向 error. jsp。要求 forward 标记在实现页面转向时,使用 param 子标记将整数传递到转向的 two. jsp 或 three. jsp 页面,将有关输入错误的信息传递到转向的 error. jsp 页面。one. jsp 页面的效果如图 2-14(a)所示。

2) two. jsp、three. jsp 和 error. jsp 的具体要求

要求 two. jsp 和 three. jsp 能输出 one. jsp 传递过来的值,并显示一幅图像,该图像的宽和高刚好是 one. jsp 页面传递过来的值。error 页面能显示有关错误信息和一幅警告出错的图像。two. jsp、three. jsp 和 error. jsp 页面的效果如图 2-14 所示。

4. 参考代码

代码仅供参考,学生可按着实验要求,参考本代码编写代码。

JSP 页面参考代码如下：

one. jsp

```
<%@ page contentType = "text/html; charset = GB2312" %>
<HEAD>
   <jsp:include page = "head.txt"/>
</HEAD>
<HTML><BODY bgcolor = yellow>
 <FORM action = "" method = get name = form>
   请输入 1 至 100 之间的整数：<INPUT type = "text" name = "number">
```

(a) one.jsp页面的效果

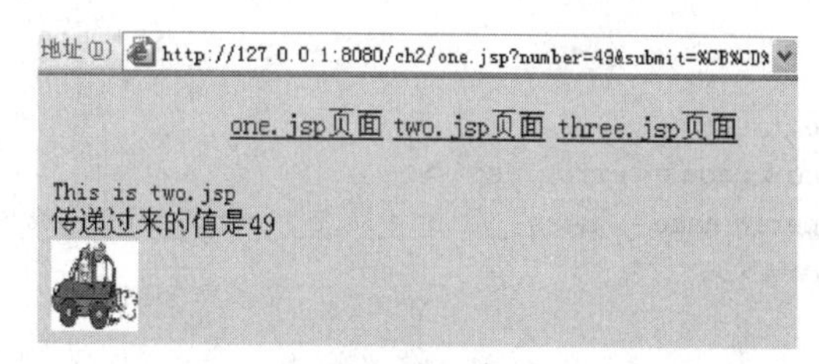

(b) two.jsp页面的效果

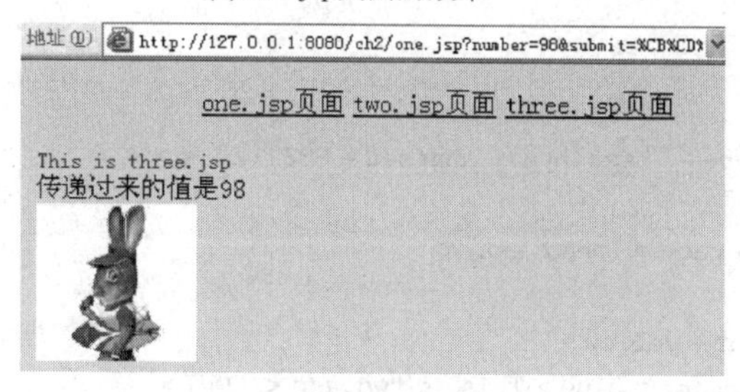

(c) three.jsp页面的效果

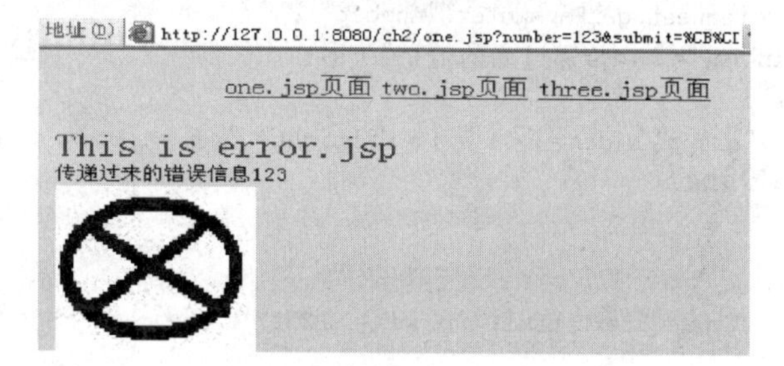

(d) error.jsp页面的效果

图 2-14　4 个页面的效果图

```
    <BR><INPUT TYPE = "submit" value = "送出" name = submit>
   </FORM>
<% String num = request.getParameter("number");
      if(num == null)
         num = "0";
      try{ int n = Integer.parseInt(num);
           if(n >= 1&&n <= 50){
%>              <jsp:forward page = "two.jsp" >
                     <jsp:param name = "number" value = "<% = n %>" />
                </jsp:forward>
<%         }
           else if(n > 50&&n <= 100){
%>           <jsp:forward page = "three.jsp" >
```

```
                <jsp:param name="number" value="<%=n%>" />
            </jsp:forward>
<%          }
            else if(n>100){
%>             <jsp:forward page="error.jsp" >
                <jsp:param name="mess" value="<%=n%>" />
             </jsp:forward>
<%          }
   }
   catch(Exception e){
 %>      <jsp:forward page="error.jsp" >
              <jsp:param name="mess" value="<%=e.toString()%>" />
         </jsp:forward>
<% }
 %>
</BODY></HTML>
```

two.jsp

```
<%@ page contentType="text/html;charset=GB2312" %>
<HEAD>
      <jsp:include page="head.txt"/>
</HEAD>
<HTML><BODY bgcolor=yellow>
 <P><FONT size=2 color=blue>This is two.jsp </FONT>
 <FONT size=3>
  <%   String s=request.getParameter("number");
         out.println("<BR>传递过来的值是"+s);
   %>
 <BR><img src="a.jpg" width="<%=s%>" height="<%=s%>" ></img>
</FONT></BODY></HTML>
```

three.jsp

```
<%@ page contentType="text/html;charset=GB2312" %>
<HEAD>
   <jsp:include page="head.txt"/>
</HEAD>
<HTML><BODY bgcolor=yellow>
 <P><FONT size=2 color=red>This is three.jsp </FONT>
  <FONT size=3>
  <%   String s=request.getParameter("number");
         out.println("<BR>传递过来的值是"+s);
   %>
 <BR><img src="b.jpg" width="<%=s%>" height="<%=s%>" ></img>
 </FONT></BODY></HTML>
```

error.jsp

```
<%@ page contentType="text/html;charset=GB2312" %>
<HEAD>
     <jsp:include page="head.txt"/>
</HEAD>
```

```
<HTML><BODY bgcolor = yellow>
 <P><FONT size = 5 color = red> This is error.jsp </FONT>
  <FONT size = 2>
  <%   String s = request.getParameter("mess");
         out.println("<BR>传递过来的错误信息" + s);
   %>
<BR><img src = "error.jpg" width = "120" height = "120" ></img>
</FONT> </BODY></HTML>
```

习 题 2

1. “<%!”和“%>”之间声明的变量与“<%”和“%>”之间声明的变量有何不同?

2. 如果有两个用户访问一个 JSP 页面，该页面中的 Java 程序片将被执行几次?

3. 是否允许一个 JSP 页面同时含有如下两条 page 指令：

```
<%@ page contentType = "text/html; charset = GB2312" %>
<%@ page contentType = "application/msword" %>
```

是否允许 JSP 页面同时含有如下两条 page 指令：

```
<%@ page import = "java.util.*" %>
<%@ page import = "java.sql.*" %>
```

4. 假设有两个用户访问下列 JSP 页面 hello.jsp，请问第一个访问和第二个访问 hello.jsp 页面的用户所看到的页面的效果有何不同?

hello.jsp

```
<%@ page contentType = "text/html; charset = GB2312" %>
<%@ page isThreadSafe = "false" %>
<HTML><BODY>
<%!   int sum = 10;
        void add(int m){
          sum = sum + m;
        }
%>
<%   int n = 600;
       add(n);
%>
       <% = sum %>
</BODY></HTML>
```

5. 请编写一个简单的 JSP 页面，显示大写英文字母表。

6. 请简单叙述 include 指令标记和 include 动作标记的不同。

7. 编写两个 JSP 页面：main.jsp 和 lader.jsp，将两个 JSP 页面保存在同一 Web 服务目录中。main.jsp 使用 include 动作标记动态加载 lader.jsp 页面。lader.jsp 页面可以计算并显示梯形的面积。当 lader.jsp 被加载时获取 main.jsp 页面中 include 动作标记的 param 子标记提供的梯形的上底、下底和高的值。

第3章 Tag文件与Tag标记

本章导读

主要内容

- Tag文件的结构
- Tag文件的存储目录
- Tag标记
- Tag文件中的常用指令
- Tag标记的嵌套

难点

- 掌握Tag文件中的attribute指令
- 掌握Tag文件中的variable指令

关键实践

- 使用标记体
- 使用attribute指令和variable指令

一个Web应用中的许多JSP页面可能需要使用某些相同的信息,如都需要使用相同的导航栏、标题等。如果能将许多页面都需要的共同的信息形成一种特殊文件,而且各个JSP页面都可以使用这种特殊的文件,那么这样的特殊文件就是可复用的代码。代码复用是软件设计的一个重要方面,是衡量软件可维护性的重要指标之一。

前面学习了include指令标记和include动作标记,使用这两个标记可以实现代码的复用。但是,在某些情况下,使用include指令标记和include动作标记有一定的缺点。例如,如果include指令标记或动作标记要处理的文件是一个JSP文件,那么用户可以在浏览器的地址栏中直接输入该JSP文件所在Web服务目录来访问这个JSP文件,这可能不是设计者所希望发生的,因为该JSP文件也许仅仅是个导航条,仅仅供其他JSP文件使用include指令标记或动作标记来嵌入或动态加载的,而不是让用户直接访问的。另外,include指令标记和include动作标记允许所要处理的文件存放在Web服务目录中的任意子目录中,不仅显得杂乱无章,而且使得include标记和所处理文件的所在目录的结构形成了耦合,不利于Web应用的维护。

本章将学习一种特殊的文本文件:Tag文件。Tag文件和JSP文件很类似,可以被JSP页面动态加载调用,但是用户不能通过该Tag文件所在Web服务目录直接访问这个Tag文件。

使用Tag文件具有以下重要的两点好处:

(1) 在设计 Web 应用时，可以通过编写 Tag 文件实现代码复用。

(2) 可将 JSP 页面中的关于数据处理的代码放在一个 Tag 文件中，让 JSP 页面只负责显示数据，即通过使用 Tag 文件将数据的处理和显示相分离，有利于 Web 应用的维护。

Tomcat 服务器的 webapps 目录的子目录都可以作为一个 Web 服务目录，在 webapps 目录下新建一个 Web 服务目录 ch3，除非特别约定，本章例子中的 JSP 页面均保存在 ch3 中。

3.1 Tag 文件的结构

Tag 文件是扩展名为.tag 的文本文件，其结构几乎和 JSP 文件相同。一个 Tag 文件中可以有普通的 HTML 标记符、某些特殊的指令标记(见 3.3 节)、成员变量和方法的声明、Java 程序片和 Java 表达式。

为了便于在名字上明显地区分 Tag 文件和 JSP 文件，本章中的 Tag 文件的名字的首写字母为大写，JSP 文件的名字的首写字母为小写。以下是两个简单的 Tab 文件：AddSum.tag 和 EvenSum.tag。AddSum.tag 负责计算 1～100 内的全部奇数之和；EvenSum.tag 负责计算 1～100 内的全部偶数之和。

AddSum.tag

```
<%@ tag pageEncoding = "gb2312" %>
<P>这是一个 Tab 文件,负责计算 1～100 内的奇数之和：
  <% int sum = 0,i = 1;
       for(i = 1; i <= 100; i++ ){
          if(i % 2 == 1)
            sum = sum + i;
       }
       out.println(sum);
  %>
```

EvenSum.tag

```
<%@ tag pageEncoding = "gb2312" %>
<P>这是一个 Tab 文件,负责计算 1～100 内的偶数之和：
  <% int sum = 0,i = 1;
       for(i = 1; i <= 100; i++ ){
          if(i % 2 == 0)
            sum = sum + i;
       }
       out.println(sum);
  %>
```

3.2 Tag 文件的存储目录

Tag 文件可以实现代码的复用，即 Tag 文件可以被许多 JSP 页面使用。为了能让一个 Web 应用中的 JSP 页面使用某一个 Tag 文件，必须把这个 Tag 文件存放到 Tomcat 服务器指定的目录中，也就是说，如果某个 Web 服务目录下的 JSP 页面准备调用一个 Tag 文件，那么必须在该 Web 服务目录下建立如下的目录结构：

```
Web 服务目录\WEB - INF\tags
```

例如:

```
ch3\WEB - INF\tags
```


其中的 WEB-INF 和 tags 都是固定的目录名称,而 tags 下的子目录的名称可由用户给定。

一个 Tag 文件必须保存到 tags 目录或其下的子目录中。把 3.1 节中的 AddSum.tag 保存到

```
ch3\WEB - INF\tags
```

目录中,将 EvenSum.tag 保存到

```
ch3\WEB - INF\tags\geng
```

目录中。

注意:Tag 文件必须使用 ANSI 编码保存。

3.3 Tag 标记

3.3.1 Tag 标记与 Tag 文件

某个 Web 服务目录下的 Tag 文件只能由该 Web 服务目录(包括该 Web 服务目录的子目录)中的 JSP 页面调用,JSP 页面必须通过 Tag 标记来调用一个 Tag 文件。

Tag 标记是伴随着 Tag 文件一同诞生的,即编写了一个 Tag 文件并保存到特定目录中后,也就自定义出了一个标记,该标记的格式为:

```
<Tag 文件名字 />
```

或

```
<Tag 文件名字 >
    标记体
</ Tag 文件名字>
```

一个 Tag 文件对应着一个标记,该标记被习惯地称为 Tag 标记,将存放在同一目录中的若干个 Tag 文件所对应的 Tag 标记的全体称为一个自定义标记库或简称为标记库。

3.3.2 Tag 标记的使用

一个 JSP 页面通过使用 Tag 标记来调用一个 Tag 文件。Web 服务目录下的一个 JSP 页面在使用 Tag 标记来调用一个 Tag 文件之前,必须首先使用 taglib 指令标记引入该 Web 服务目录下的标记库,只有这样,JSP 页面才可以使用 Tag 标记调用相应的 Tag 文件。

taglib 指令的格式如下:

```
<%@ taglib tagdir = "自定义标记库的位置" prefix = "前缀">
```

一个 JSP 页面可以使用几个 taglib 指令标记引入若干个标记库。例如:

```
<%@ taglib tagdir = "/WEB - INF/tags" prefix = "beijing" %>
```

```
<%@ taglib tagdir="/WEB-INF/tags/geng" prefix="dalian" %>
```

引入标记库后,JSP页面就可以使用带前缀的Tag标记调用相应的Tag文件,其中的前缀由taglib指令中的prefix属性指定。例如:

```
<beijing:AddSum/>
<dalian:EvenSum/>
```

注意:通过前缀可以有效地区分不同标记库中具有相同名字的标记文件。

JSP引擎处理JSP页面中的Tag标记的原理如下:

(1) 如果该Tag标记对应的Tag文件是首次被JSP页面调用,那么JSP引擎会将Tag文件转译成一个Java文件,并编译这个Java文件生成字节码文件,然后执行这个字节码文件来实现Tag文件的动态处理,将有关的结果发送到用户端(这和执行JSP页面的原理类似)。

(2) 如果该Tag文件已经被编译为字节码文件,那么JSP引擎将直接执行这个字节码文件来实现Tag文件的动态处理。

(3) 如果对Tag文件进行了修改,那么JSP引擎会重新将Tag文件转译成一个Java文件,并编译这个Java文件生成字节码文件,然后执行这个字节码文件来实现Tag文件的动态处理。

现在用一个例子来说明怎样在JSP页面中调用Tag文件。例3-1中的JSP页面保存在Web服务目录ch3中,该JSP页面所调用的Tag文件是3.1节中提到的AddSum.tag和EvenSum.tag。example3_1.jsp的效果如图3-1所示。

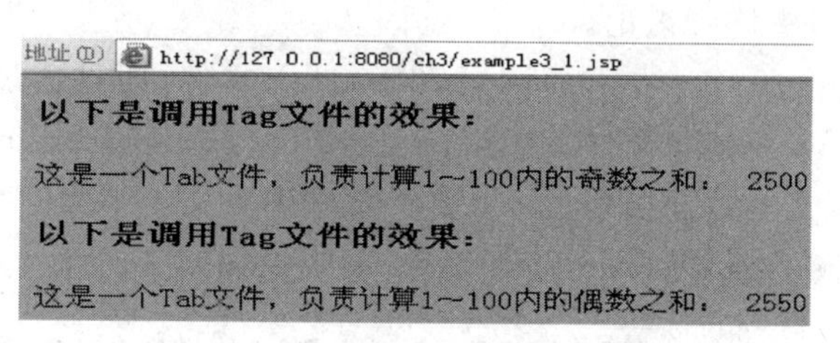

地址(D) http://127.0.0.1:8080/ch3/example3_1.jsp
以下是调用Tag文件的效果:
这是一个Tab文件,负责计算1~100内的奇数之和: 2500
以下是调用Tag文件的效果:
这是一个Tab文件,负责计算1~100内的偶数之和: 2550

图3-1 在JSP页面中使用Tag标记

例3-1

example3_1.jsp

```
<%@ page contentType="text/html;charset=GB2312" %>
<%@ taglib tagdir="/WEB-INF/tags" prefix="beijing" %>
<%@ taglib tagdir="/WEB-INF/tags/geng" prefix="dalian" %>
<html><body bgcolor=cyan>
  <h3>以下是调用Tag文件的效果:</h3>
      <beijing:AddSum />
  <h3>以下是调用Tag文件的效果:</h3>
      <dalian:EvenSum />
</body></html>
```

注意:Tag文件在"<%!"和"%>"标记符号之间声明的变量,和JSP在"<%!"和"%>"之间声明的变量类似,其有效范围是整个Tag文件。但是有一点非常不同,每当Tag文件对应的字节码被执行完毕后,这些变量即可释放所占有的内存空间。

3.3.3 Tag 标记的标记体

一个 Tag 文件会对应一个 Tag 标记,并让 JSP 页面使用这个 Tag 标记动态执行该 Tag 文件。Tag 标记的格式为:

```
<Tag文件名字 />
```

或

```
<Tag文件名字>
    标记体
</Tag文件名字>
```

默认情况下,JSP 可以使用没有标记体的 Tag 标记,也可以使用带有标记体的 Tag 标记调用一个 Tag 文件。那么,Tag 标记的标记体起着怎样的作用呢?

当 JSP 页面调用一个 Tag 文件时可能希望动态地向该 Tag 文件传递信息,那么就可以使用带有标记体的 Tag 标记来执行一个 Tag 文件,Tag 标记中的"标记体"就会传递给相应的 Tag 文件,这个 Tag 文件通过使用

```
<jsp:doBody />
```

标记处理 JSP 页面传递过来的"标记体"。

那么,标记体可以是怎样的形式呢?默认的情况下,标记体可以是一些文本数据,有关细节会在后续的 3.4 节中详细讨论。

在例 3-2 中,example3_2.jsp 页面调用 Show.tag 文件,而且通过标记体向 Show.tag 文件传递文本数据。Show.tag 文件使用<jsp:doBody/>处理 example3_2.jsp 页面传递过来的文本数据,将该文本数据循环显示 3 次并逐次增大文本字体的字号。Show.tag 保存在 ch3\WEB-INF\tags 目录中,example3_2.jsp 的效果如图 3-2 所示。

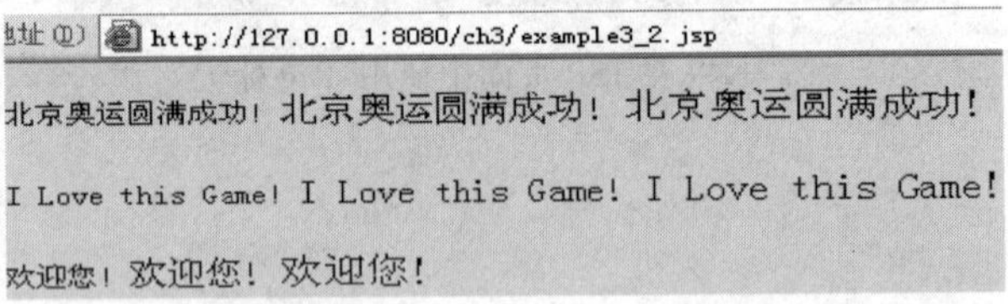

图 3-2 Tag 标记的标记体

例 3-2

example3_2.jsp

```
<%@ page contentType="text/html; charset=GB2312" %>
<%@ taglib prefix="look" tagdir="/WEB-INF/tags" %>
<HTML>
   <look:Show>
       北京奥运圆满成功!                    <%-- 标记体 --%>
   </look:Show>
   <look:Show>
```

```
        I Love this Game!                   <%-- 标记体 --%>
    </look:Show>
    <look:Show>
        欢迎您!                              <%-- 标记体 --%>
    </look:Show>
</HTML>
```

Show.tag

```
<body bgcolor=yellow><P>
<%   int size=1;
       for(int i=1; i<=3; i++){
           size=size+1;
 %>       <font size=<%=size%>>
               <jsp:doBody />
           </font>
<% }
 %>
</P></body>
```

3.4 Tag文件中的常用指令

与JSP文件类似,Tag文件中也有一些常用指令,这些指令将影响Tag文件的行为。Tag文件中经常使用的指令有tag、variable、include、attribute和taglib。

以下将分别讲解上述指令在Tag文件中的作用和用法。

3.4.1 tag指令

Tag文件中的tag指令类似于JSP文件中的page指令。Tag文件通过使用tag指令可以指定某些属性的值,以便从总体上影响Tag文件的处理和表示。tag指令的语法如下:

```
<%@ tag 属性1="属性值" 属性2="属性值" ……属性n="属性值"%>
```

在一个Tag文件中可以使用多个tag指令,因此经常使用多个tag指令为属性指定需要的值:

```
<%@ tag 属性1="属性值"%>
<%@ tag 属性2="属性值"%>
……
<%@ tag 属性n="属性值"%>
```

tag指令可以操作的属性有body-content、language、import和pageEncoding。以下将分别讲解怎样设置这些属性的值。

1. body-content属性

标记的格式为:

```
<Tag文件名字 />
```

或

```
<Tag文件名字>
    标记体
</Tag文件名字>
```


JSP文件通过使用Tag标记调用相应的Tag文件，那么JSP文件到底应该使用Tag标记的哪种格式来调用Tag文件呢？答案是：一个Tag文件通过tag指令指定body-content属性的值可以决定Tag标记的使用格式。也就是说，body-content属性的值可以确定JSP页面使用Tag标记时是否可以有标记体，如果允许有标记体，该属性会给出标记体内容的类型。

body-content属性的值有empty、tagdependent、scriptless，默认值是scriptless。

如果body-content属性的值是empty，那么JSP页面必须使用没有标记体的Tag标记：

```
<Tag文件名字 />
```

来调用相应的Tag文件。

如果body-content属性的值是tagdependent或scriptless，那么JSP页面可以使用无标记体或有标记体的Tag标记：

```
<Tag文件名字>
    标记体
</Tag文件名字>
```

来调用相应的Tag文件。如果属性值是scriptless，那么标记体中不能有Java程序片；如果属性值是tagdependent，那么Tag文件将标记体的内容按纯文本处理。

Tag标记中的标记体由相应的Tag文件负责处理，因此，当JSP页面使用有标记体的Tag标记调用一个Tag文件时，可以通过"标记体"向该Tag文件动态地传递文本数据或必要的JSP指令。

Tag文件通过使用标记：

```
<jsp:doBody />
```

来获得JSP页面传递过来的"标记体"，并进行处理，也就是说，在一个Tag文件中，<jsp:doBody />标记被替换成对"标记体"进行处理后所得到的结果，有关<jsp:doBody/>的用法可参见例3-2。

2. language属性

language属性的值指定Tag文件使用的脚本语言，目前只能取值java，其默认值就是java，因此在编写Tag文件时，没有必要使用tag指令指定language属性的值。

3. import属性

import属性的作用是为Tag文件引入Java核心包中的类，这样就可以在Tag文件的程序片部分、变量及方法声明部分、表达式部分使用Java核心包中的类。import属性可以取多个值，import属性已经有如下值：java.lang.*、javax.servlet.*、javax.servlet.jsp.*和javax.servlet.http.*。

4. pageEncoding

该属性的值指定Tag文件的字符编码，其默认值是ISO-8859-1。

3.4.2 include 指令

在 Tag 文件中也有和 JSP 文件类似的 include 指令标记，其使用方法和作用与 JSP 文件中的 include 指令标记类似。

3.4.3 attribute 指令

Tag 文件充当着可复用代码的角色，如果一个 Tag 文件允许使用它的 JSP 页面向该 Tag 文件传递数据，就使得 Tag 文件的功能更为强大。在 Tag 文件中通过使用 attribute 指令，可以让使用它的 JSP 页面向该 Tag 文件传递需要的数据。attribute 指令的格式如下：

```
<%@ attribute name="对象名字" required="true"|"false" type="对象的类型" %>
```

attribute 指令中的 name 属性是必需的，该属性的值是一个对象的名字。JSP 页面在调用 Tag 文件时，可向 name 属性指定的对象传递一个引用。type 指定对象的类型。例如：type= " java. util. Date"。需要特别注意的是，对象的类型必须带有包名。例如，不可以将 java. util. Date 简写为 Date。如果 attribute 指令中没有使用 type 指定对象的类型，那对象的类型是 java. lang. String 类型。

JSP 页面使用 Tag 标记向所调用的 Tag 文件中 name 属性指定的对象传递一个引用，方式如下：

```
<前缀：Tag 文件名字 对象名字="对象的引用" />
```

或

```
<前缀：Tag 文件名字 对象名字= "对象的引用">
    标记体
</前缀：Tag 文件名字>
```

例如，一个 Tag 文件：MyTag. tag 中有如下的 attribute 指令：

```
<%@ attribute name="length" required="true" %>
```

那么 JSP 页面就可以如下使用 Tag 标记（假设标记的前缀为 computer）调用 MyTag. tag：

```
<computer：MyTag length="1000" />
```

或

```
<computer：MyTag length="1000" >
    我向 Tag 文件中传递的值是 1000
<computer：MyTag />
```

再举一例，一个 Tag 文件 YourTag. tag 中已有如下的 attribute 指令：

```
<%@ attribute name="result" required="true" type="java.lang.Double" %>
```

那么 JSP 页面可以使用 Tag 标记（假设标记的前缀为 computer）调用 YourTag. tag，将一个 java. lang. Double 类型对象的引用传递给 YourTag. tag 文件中的 result 对象：

```
<computer：YourTag result="<% = new Double(666.999) %>" />
```

attribute 指令中的 required 属性也是可选的,如果省略 required 属性,那么 required 的默认值是 false。当指定 required 的值是 true 时,调用该 Tag 文件的 JSP 页面必须向该 Tag 文件中 attribute 指令中的 name 属性指定的对象传递一个引用,即当 required 的值是 true 时,如果使用"<前缀: Tag 文件名字 />"调用 Tag 文件就会出现错误。当指定 required 的值是 false 时,调用该 Tag 文件的 JSP 可以向该 Tag 文件中 attribute 指令中的 name 属性指定的对象传递或不传递对象的引用。

注意:在 Tag 文件中不可以再定义和 attribute 指令中的 name 属性指定的对象具有相同名字的变量,否则将隐藏 attribute 指令中的对象,使其失效。

在例 3-3 中,Triangle. tag 存放在 ch3\WEB-INF\tags 目录中,该 Tag 文件负责计算、显示三角形的面积。example3_3. jsp 使用 Tag 标记调用 Triangle. tag 文件,并且向 Triangle. tag 文件传递三角形三边的长度。example3_3. jsp 的效果如图 3-3 所示。

地址(D) http://127.0.0.1:8080/ch3/example3_3.jsp

以下是调用Tag文件的效果:

这是一个Tag文件,负责计算三角形的面积。
JSP页面传递过来的三条边: 5,6,7
三角形的面积:14.696938456699069

图 3-3 调用 Tag 文件计算面积

例 3-3

example3_3. jsp

```
<%@ page contentType = "text/html; charset = GB2312" %>
<%@ taglib tagdir = "/WEB-INF/tags" prefix = "computer" %>
<HTML><BODY>
   <H3>以下是调用 Tag 文件的效果:</H3>
   <computer:Triangle sideA = "5" sideB = "6" sideC = "7"/>
</BODY></HTML>
```

Triangle. tag

```
<%@ tag pageEncoding = "gb2312" %>
<h4>这是一个 Tag 文件,负责计算三角形的面积。
<%@ attribute name = "sideA" required = "true" %>
<%@ attribute name = "sideB" required = "true" %>
<%@ attribute name = "sideC" required = "true" %>
 <%!   public String getArea(double a,double b,double c){
           if(a + b > c&&a + c > b&&c + b > a) {
              double p = (a + b + c)/2.0;
              double area = Math.sqrt(p * (p - a) * (p - b) * (p - c));
              return "<BR>三角形的面积:" + area;
           }
           else
              return("<BR>" + a + "," + b + "," + c + "不能构成一个三角形,无法计算面积。");
       }
 %>
 <%  out.println("<BR>JSP页面传递过来的三条边:" + sideA + "," + sideB + "," + sideC");
      double a = Double.parseDouble(sideA);
      double b = Double.parseDouble(sideB);
      double c = Double.parseDouble(sideC);
      out.println(getArea(a,b,c));
 %>
```

在例 3-4 中，JSP 页面只负责将学生的姓名和成绩分别存放到链表(java. util. LinkedLst 类型对象)中，然后将链表传递给 Sort. tag，Sort. tag 负责按从低到高的顺序显示学生的成绩。example3_4. jsp 的效果如图 3-4 所示。

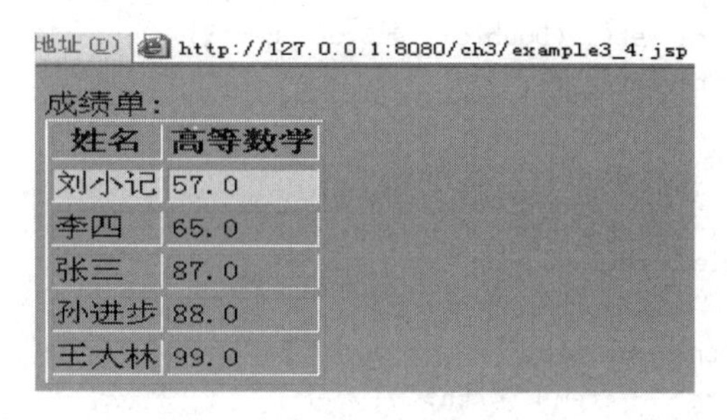

图 3-4 调用 Tag 文件排序成绩

例 3-4

example3_4. jsp

```
<%@ page contentType = "text/html; charset = GB2312" %>
<%@ page import = "java.util. * " %>
<%@ taglib tagdir = "/WEB - INF/tags" prefix = "show" %>
<HTML><BODY bgcolor = cyan>
   <%  LinkedList listName = new LinkedList();
        LinkedList listScore = new LinkedList();
        listName.add("张三");
        listScore.add(new Double(87));
        listName.add("李四");
        listScore.add(new Double(65));
        listName.add("刘小记");
        listScore.add(new Double(57));
        listName.add("王大林");
        listScore.add(new Double(99));
        listName.add("孙进步");
        listScore.add(new Double(88));
   %>
  <p>成绩单：
  <show:Sort title = "姓名" item = "高等数学"
             listName = "<% = listName %>" listScore = "<% = listScore %>"/>
</BODY></HTML>
```

Sort. tag

```
<%@ tag pageEncoding = "gb2312" %>
<%@ attribute name = "listName" required = "true" type = "java.util.LinkedList" %>
<%@ attribute name = "listScore" required = "true" type = "java.util.LinkedList" %>
<%@ attribute name = "title" required = "true" %>
<%@ attribute name = "item" required = "true" %>
<% for(int i = 0; i < listName.size(); i ++ ){
       for(int j = i + 1; j < listName.size(); j ++ ){
          double a = ((Double)listScore.get(i)).doubleValue();
          double b = ((Double)listScore.get(j)).doubleValue();
          if(b < a){
```

```
                String temp = (String)listName.get(i);
                Double r = (Double)listScore.get(i);
                listName.set(i,(String)listName.get(j));
                listName.set(j,temp);
                listScore.set(i,(Double)listScore.get(j));
                listScore.set(j,r);
            }
        }
    }
    out.print("< table border = 1 >");
    out.print("< tr >");
        out.print("< th >" + title + "</th >");
        out.print("< th >" + item + "</th >");
    out.print("< tr >");
    for(int k = 0; k < listName.size(); k++ ){
        out.print("< tr >");
        double score = (Double)listScore.get(k);
        String name = (String)listName.get(k);
        if(score < 60){
            out.print("< td bgcolor = yellow >" + name + "</td >");
            out.print("< td bgcolor = yellow >" + score + "</td >");
        }
        else{
            out.print("< td >" + name + "</td >");
            out.print("< td >" + score + "</td >");
        }
        out.print("< tr >");
    }
    out.print("</table >");
%>
```

3.4.4 variable 指令

Tag 文件通过使用 attribute 指令,可以使得调用该 Tag 文件的 JSP 页面动态地向其传递数据。在某些 Web 设计中,JSP 页面不仅希望向 Tag 文件传递数据,而且还希望 Tag 文件能返回某些数据给 JSP 页面。例如,许多 JSP 页面可能都需要调用某个 Tag 文件对某些数据进行基本的处理,但不希望 Tag 文件做进一步的特殊处理以及显示数据,因为各个 JSP 页面对数据进行进一步处理的细节以及对数据显示格式的要求是不同的。因此,JSP 页面希望 Tag 文件将数据的基本处理结果存放在某些对象中,并将这些对象返回给当前 JSP 页面,由 JSP 页面负责进一步处理和显示这些对象,这样做可以很好地实现代码的复用,将数据的基本处理和特殊数据以及显示相分离。

Tag 文件通过使用 variable 指令可以将 Tag 文件中的对象返回给调用该 Tag 文件的 JSP 页面。

1. variable 指令的格式

variable 指令的格式如下:

```
<%@ variable name-given = "对象名字" variable-class = "对象的类型" scope = "有效范围" %>
```

variable 指令中属性 name 的值用来指定对象的名字，名字必须符合标识符规定，即名字可以由字母、下划线、美元符号和数字组成，并且第一个字符不能是数字字符。

variable 指令中属性 variable-class 的值指定对象的类型。例如，可以是 java. lang. String、java. lang. Integer、java. lang. Float、java. lang. Double、java. util. Date 等类型。如果 variable 指令中没有使用 variable-class 指定对象的类型，那么对象的类型是 java. lang. String 类型。需要特别注意的是，对象的类型必须带有包名。例如，不可以将 java. util. Date 简写为 Date。

variable 指令中属性 scope 的值指定对象的有效范围，scope 的值可以取 AT_BEGIN、NESTED 和 AT_END。当 scope 的值是 AT_BEGIN 时，JSP 页面一旦开始使用 Tag 标记，就可以使用 variable 指令中给出的对象。例如，JSP 页面可以在 Tag 标记的标记体中或 Tag 标记结束后的各个部分中使用 variable 指令给出的对象。当 scope 的值是 NESTED 时，JSP 页面只可以在 Tag 标记的标记体使用 variable 指令给出的对象。

当 scope 的值是 AT_END 时，JSP 页面只有在 Tag 标记结束后，才可以使用 variable 指令中给出的对象，也就是说 JSP 页面不可以在 Tag 标记的标记体中使用该 variable 指令给出的对象，必须在调用完 Tag 标记后才可以使用 variable 指令给出的对象。

下面的 variable 指令给出的对象的名字是 time、类型为 java. util. Date、有效范围是 AT_END：

```
<%@ variable name-given="time" variable-class="java.util.Date" scope="AT_END" %>
```

2. 对象的返回

Tag 文件为了给 JSP 页面返回一个对象，就必须将对象的名字以及该对象的引用存储到 Tomcat 引擎提供的内置对象 jspContext 中。Tag 文件只有将对象的名字及其引用存储到 jspContext 中，JSP 页面才可以使用该对象。

jspContext 调用

```
setAttribute("对象的名字",对象的引用);
```

方法存储对象的名字以及该对象的引用。例如：

```
jspContext.setAttribute("time",new Date());
```

将名字是 time 的 Date 对象存储到 jspContext 中。

以下的 variable 指令：

```
<%@ variable name-given="time" variable-class="java.util.Date" scope="AT_END" %>
```

为 JSP 页面返回名字是 time 的 Date 对象。下列 Tag 文件 GiveDate. tag 使用了上述 variable 指令：

GiveTag. tag

```
<%@ tag import="java.util.*" %>
<%@ variable name-given="time" variable-class="java.util.Date" scope="AT_END" %>
  <%  jspContext.setAttribute("time",new Date());
  %>
```

需要特别注意的是，不能在 Tag 文件中的 Java 程序片中直接操作 variable 指令中的对

象,Tag 文件只能将对象的名字及其引用存储到 jspContext 中。例如,下列程序片中的代码是不允许的:

```
<%
    time = new Date();
%>
```

以下的 JSP 页面使用 Tag 标记调用上述 GiveTag. tag 文件。在下面的 JSP 页面 lookTime. jsp 中,在出现 Tag 标记

```
<showTime: GiveDate />
```

之后,该 JSP 页面就可以在程序片中或表达式部分使用 GiveDate. tag 文件为它返回的 time 对象。以下是使用 GiveDate. tag 文件的 lookTime. jsp 的完整代码:

lookTime. jsp

```
<%@ page contentType = "text/html; Charset = GB2312" %>
<%@ page import = "java.text.*" %>
<%@ taglib tagdir = "/WEB-INF/tags" prefix = "showTime" %>
<HTML><BODY bgcolor = cyan>
    <showTime: GiveDate />        <%-- Tag标记 --%>
 <h3>当前时间:</h3>
    <% = time %>                   <%-- 使用GiveDate.tag文件返回的time对象 --%>
</BODY></HTML>
```

注意:在 JSP 页面中不可以再定义与 Tag 文件返回的对象具有相同名字的变量,否则 Tag 文件无法将 variable 指令给出的对象返回给 JSP 页面(将出现编译错误)。

如果 Tag 文件同时使用 variable 指令和 attribute 指令,那么 variable 指令和 attribute 指令中的 name 指定的对象不能相同(将出现编译错误)。

在例 3-5 中,Tag 文件 GiveRoot. tag 负责求出一元二次方程的根。JSP 页面在调用 Tag 文件时,使用 attribute 指令将方程的系数传递给 Tag 文件;Tag 文件 GiveRoot. tag 使用 variable 指令返回一元二次方程的根给调用该 Tag 文件的 JSP 页面。例 3-5 中的 userOne. jsp 和 userTwo. jsp 都使用 Tag 标记调用 GiveRoot. tag,二者都可以得到 GiveRoot. tag 返回的方程的两个根,但是二者使用不同的方式来处理和显示方程的两个根。userOne. jsp 将方程的根保留最多 3 位小数,并计算方程的两个根之和,userTwo. jsp 将方程的根保留最多 5 位小数,并计算方程的两个根之积。userOne. jsp 和 userTwo. jsp 的效果如图 3-5 所示。

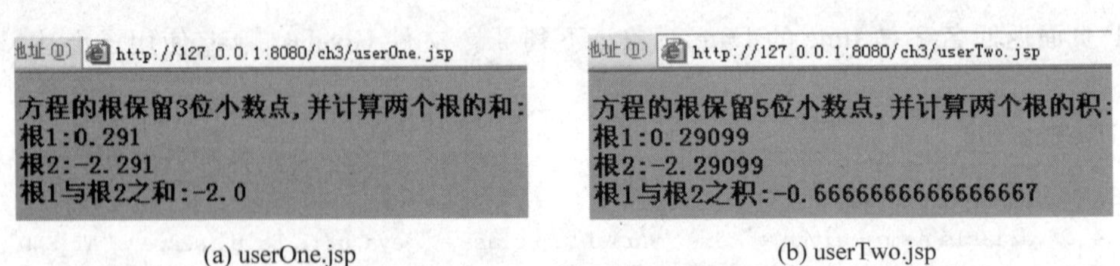

(a) userOne.jsp (b) userTwo.jsp

图 3-5 userOne. jsp 和 userTwo. jsp 的效果图

例 3-5

userOne. jsp

```
<%@ page contentType = "text/html; charset = GB2312" %>
<%@ page import = "java.text.*" %>
<%@ taglib tagdir = "/WEB-INF/tags" prefix = "computer" %>
<HTML><BODY bgcolor = cyan>
  <computer:GiveRoot coefficientA = "3" coefficientB = "6" coefficientC = "-2"/>
  <h4>方程的根保留 3 位小数点，并计算两个根的和：
  <% NumberFormat f = NumberFormat.getInstance();
     f.setMaximumFractionDigits(3);
     if(rootOne!= null&&rootTwo!= null){
        double r1 = rootOne.doubleValue(); //rootOne 是 GetRoot.tag 文件返回的 Double 型对象
        double r2 = rootTwo.doubleValue(); //rootTwo 是 GetRoot.tag 文件返回的 Double 型对象
        String s1 = f.format(r1);
        String s2 = f.format(r2);
        out.println("<br>根 1: " + s1);
        out.println("<br>根 2: " + s2);
        double sum = r1 + r2;
        out.println("<br>根 1 与根 2 之和: " + sum);
     }
     else{
        out.println("<br>方程没有实根");
     }
  %>
</BODY></HTML>
```

userTwo. jsp

```
<%@ page contentType = "text/html; charset = GB2312" %>
<%@ page import = "java.text.*" %>
<%@ taglib tagdir = "/WEB-INF/tags" prefix = "computer" %>
<HTML><BODY bgcolor = cyan>
<computer:GiveRoot coefficientA = "3" coefficientB = "6" coefficientC = "-2"/>
<h4>方程的根保留 5 位小数点，并计算两个根的积：
 <% NumberFormat f = NumberFormat.getInstance();
    f.setMaximumFractionDigits(5);
    if(rootOne!= null&&rootTwo!= null){
        double r1 = rootOne.doubleValue(); //rootOne 是 GetRoot.tag 文件返回的 Double 型对象
        double r2 = rootTwo.doubleValue(); //rootTwo 是 GetRoot.tag 文件返回的 Double 型对象
        String s1 = f.format(r1);
        String s2 = f.format(r2);
        out.println("<br>根 1: " + s1);
        out.println("<br>根 2: " + s2);
        double ji = r1 * r2;
        out.println("<br>根 1 与根 2 之积: " + ji);
    }
    else{
        out.println("<br>方程没有实根");
    }
  %>
</BODY></HTML>
```

GiveRoot.tag

```
<%@ tag pageEncoding="gb2312" %>
<%@ tag import="java.util.*" %>
<%@ attribute name="coefficientA" required="true" %>
<%@ attribute name="coefficientB" required="true" %>
<%@ attribute name="coefficientC" required="true" %>
<%@ variable name-given="rootOne" variable-class="java.lang.Double"
 scope="AT_END" %>
<%@ variable name-given="rootTwo" variable-class="java.lang.Double" scope="AT_END" %>
  <%  double disk,root1,root2;
      double a=Double.parseDouble(coefficientA);
      double b=Double.parseDouble(coefficientB);
      double c=Double.parseDouble(coefficientC);
      disk=b*b-4*a*c;
      if(disk>=0&&a!=0){
        root1=(-b+Math.sqrt(disk))/(2*a);
        root2=(-b-Math.sqrt(disk))/(2*a);
        jspContext.setAttribute("rootOne",new Double(root1)); //返回对象 rootOne
        jspContext.setAttribute("rootTwo",new Double(root2)); //返回对象 rootTwo
      }
      else{
        jspContext.setAttribute("rootOne",null);
        jspContext.setAttribute("rootTwo",null);
    }
  %>
```

3.4.5 taglib 指令

和 JSP 页面使用 Tag 文件一样，一个 Tag 文件也可以使用 Tag 标记来调用其他 Tag 文件。和 JSP 页面使用 Tag 标记的要求相同，Tag 文件必须使用<taglib>指令引入该 Web 服务目录下的标记库，才可以使用 Tag 标记来调用相应的 Tag 文件。<taglib>指令的格式如下：

```
<%@ taglib tagdir="自定义标记库的位置" prefix="前缀">
```

一个 Tag 文件也可以使用几个 taglib 指令标记引入若干个标记库。例如：

```
<%@ taglib tagdir="/WEB-INF/tags" prefix="beijing" %>
<%@ taglib tagdir="/WEB-INF/tags/tagsTwo" prefix="dalian" %>
```

在例 3-6 中，FirstTag.tag 文件使用 Tag 标记调用 SecondTag.tag 文件。SecondTag.tag 文件负责从四组数中随机取出 m 个($m \leqslant 52$)，这四组数相同，都是由 1～13 组成的 13 个整数。实际上，SecondTag.tag 就是模拟从一副扑克牌中(不包括大王和小王)随机抽取 m 张牌，其中 m 的值由 FirstTag.tag 文件提供。FirstTag.tag 将 SecondTag.tag 返回的 m 个随机数从小到大排列，并计算出它们的和。example3_6.jsp 使用 Tag 标记调用 FirstTag.tag，example3_6.jsp 的效果如图 3-6 所示。

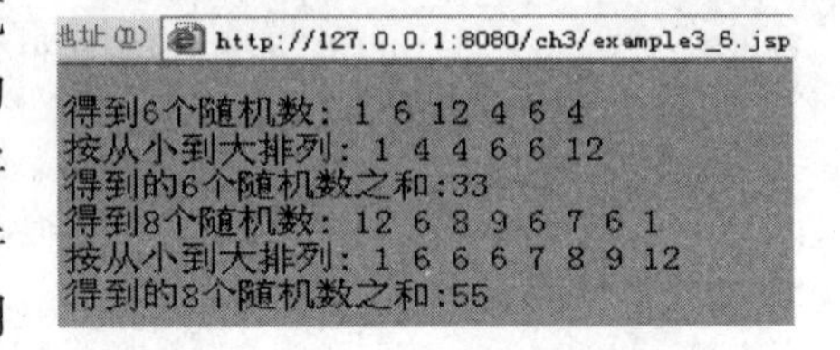

图 3-6　调用 Tag 文件得到随机数

例 3-6

example3_6.jsp

```
<%@ page contentType="text/html;charset=GB2312" %>
<%@ page import="java.util.*" %>
<%@ taglib tagdir="/WEB-INF/tags" prefix="first" %>
<HTML><BODY bgcolor=cyan>
  <first:FirstTag number="6"/>
  <first:FirstTag number="8" />
</BODY></HTML>
```

FirstTag.tag

```
<%@ tag pageEncoding="gb2312" %>
<%@ tag import="java.util.*" %>
<%@ taglib tagdir="/WEB-INF/tags" prefix="getNumber" %>
<%@ attribute name="number" required="true" %>
   <getNumber:SecondTag number="<%=number%>"/>
   <% out.println("得到"+number+"个随机数:");
        for(int i=0;i<listNumber.size();i++) //listNumber 是 SecondTag 文件返回的对象
            out.print((Integer)listNumber.get(i)+" ");
        out.println("<br>按从小到大排列:");
        for(int i=0;i<listNumber.size();i++){
                for(int j=i+1;j<listNumber.size();j++){
                    int a=((Integer)listNumber.get(i)).intValue();
                    int b=((Integer)listNumber.get(j)).intValue();
                    if(b<a){
                      Integer temp=(Integer)listNumber.get(i);
                      listNumber.set(i,(Integer)listNumber.get(j));
                      listNumber.set(j,temp);
                    }
                }
        }
        for(int i=0;i<listNumber.size();i++)
              out.print((Integer)listNumber.get(i)+" ");
        int sum=0;
        for(int i=0;i<listNumber.size();i++)
             sum=sum+((Integer)listNumber.get(i)).intValue();
        out.println("<br>得到的"+number+"个随机数之和:"+sum+"<br>");
%>
```

SecondTag.tag

```
<%@ tag pageEncoding="gb2312" %>
<%@ tag import="java.util.*" %>
<%@ attribute name="number" required="true" %>
<%@ variable name-given="listNumber"
               variable-class="java.util.LinkedList" scope="AT_END" %>
<%    int count=Integer.parseInt(number);
        LinkedList listBox=new LinkedList(),
                  listNeeded=new LinkedList();
        for(int k=1;k<=4;k++)
              for(int i=1;i<=13;i++)
                  listBox.add(new Integer(i));
```

```
            while(count > 0) {
                 int m = (int)(Math.random() * listBox.size());
                 Integer integer = (Integer)listBox.get(m);
                 listNeeded.add(integer);
                 listBox.remove(m);
                 count -- ;
            }
            jspContext.setAttribute("listNumber",listNeeded); //返回 listNumber
%>
```

3.5　Tag 标记的嵌套

使用 Tag 标记时，可以带有标记体，标记体还可以是一个 Tag 标记，这就实现了 Tag 标记的嵌套。Tag 标记中的标记体由 Tag 文件的<jsp:doBody/>标记负责处理，而在 Tag 文件中，<jsp:doBody />标记被替换成对“标记体”进行处理后所得到的结果。

在例 3-7 中，JSP 页面使用 Tag 标记嵌套显示一个表格，example3_7.jsp 的效果如图 3-7 所示。

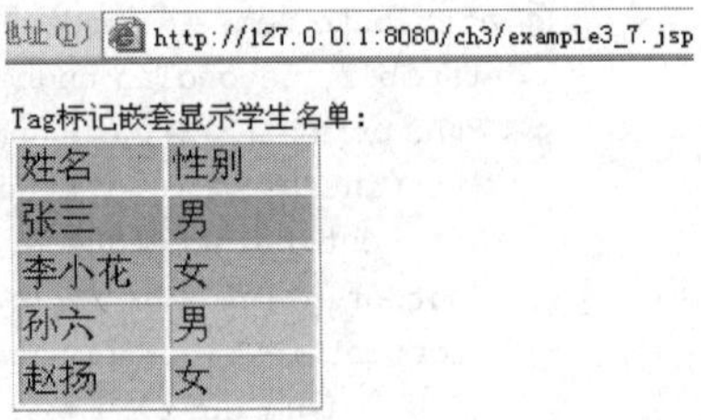

地址(D) http://127.0.0.1:8080/ch3/example3_7.jsp

Tag标记嵌套显示学生名单：

姓名	性别
张三	男
李小花	女
孙六	男
赵扬	女

图 3-7　Tag 标记的嵌套

例 3-7

example3_7.jsp

```
<%@ page contentType = "text/html; charset = GB2312" %>
<%@ taglib tagdir = "/WEB-INF/tags" prefix = "ok" %>
<html><body>
<p>
<Font size = 2>Tag 标记嵌套显示学生名单：</Font>
<table border = 1>
   <ok:Biaoge color = "#a9f002" name = "姓名" sex = "性别">
      <ok:Biaoge color = "cyan" name = "张三" sex = "男"/>
      <ok:Biaoge color = "#afc0ff" name = "李小花" sex = "女"/>
      <ok:Biaoge color = "pink" name = "孙六" sex = "男"/>
      <ok:Biaoge color = "#ffaaef" name = "赵扬" sex = "女"/>
   </ok:Biaoge>
</table>
</body></html>
```

Biaoge.tag

```
<%@ tag pageEncoding = "gb2312" %>
<%@ attribute name = "color" %>
<%@ attribute name = "name" %>
<%@ attribute name = "sex" %>
 <tr bgcolor = "<% = color %>">
      <td width = 60><% = name %></td>
      <td width = 60><% = sex %></td>
 </tr>
 <jsp:doBody/>
```

3.6　实验1：使用标记体

1. 相关知识点

Tag文件是扩展名为.tag的文本文件，其结构几乎和JSP文件相同，一个Tag文件中可以有普通的HTML标记符、某些特殊的指令标记、成员变量和方法的声明、Java程序片和Java表达式。JSP页面使用Tag标记动态执行一个Tag文件。当JSP页面调用一个Tag文件时可能希望动态地向该Tag文件传递信息，那么就可以使用带有标记体的Tag标记来执行一个Tag文件，Tag标记中的"标记体"就会传递给相应的Tag文件。标记体由Tag文件的<jsp：doBody/>标记负责处理，即<jsp：doBody />标记被替换成对"标记体"处理后所得到的结果。

2. 实验目的

本实验的目的是让学生灵活掌握在Tag标记中使用标记体。

3. 实验要求

编写一个JSP页面：putImage.jsp和一个Tag文件Image.tag。JSP页面通过调用Tag文件来显示若干幅图像，通过使用标记体将HTML图像标记传递给被调用的Tag文件。

1）putImage.jsp的具体要求

要求putImage.jsp页面使用带标记体Tag标记调用Tag文件来显示一幅图像，即标记体是"显示图像的HTML标记"，如下所示：

```
<pic：Image>
      <image src = "a.jpg" />
</pic：Image>
```

putImage.jsp的页面效果如图3-8所示。

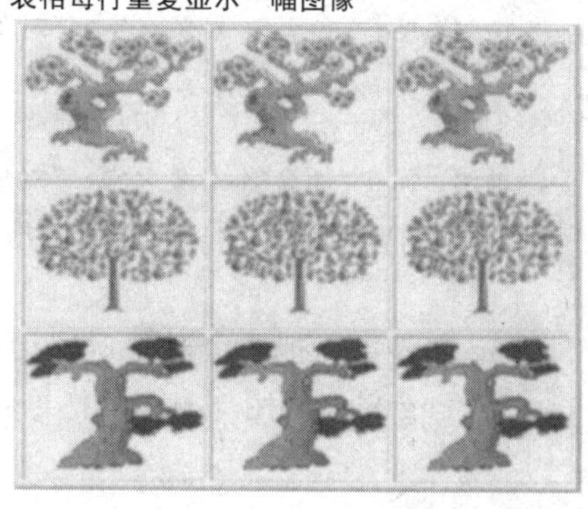

图3-8　putImage.jsp的页面效果

2）Image.tag的具体要求

Image.tag使用<jsp：doBody/>处理标记体，并将图像显示在表格的单元中，要求表格每行有三个单元，这三个单元重复显示一幅图像。

4. 参考代码

代码仅供参考，学生可按着实验要求，参考本代码编写代码。

JSP页面参考代码如下：

putImage.jsp

```
<%@ page contentType = "text/html; charset = GB2312" %>
<%@ taglib tagdir = "/WEB-INF/tags" prefix = "pic" %>
<html><body>
  <Font size = 2 color = blue>表格每行重复显示一幅图像</font>
  <table border = 2>
     <pic：Image>
        <image src = "a.jpg" width = 80 height = 60/>
```

```
    </pic:Image>
    <pic:Image>
        <image src="b.jpg" width=80 height=60/>
    </pic:Image>
    <pic:Image>
        <image src="c.jpg" width=80 height=60/>
    </pic:Image>
  </table>
  </body></html>
```

Tag 文件参考代码如下：

Image.Tag

```
<tr>
      <td><jsp:doBody/></td>
      <td><jsp:doBody/></td>
      <td><jsp:doBody/></td>
</tr>
```

3.7 实验 2：使用 attribute 指令和 variable 指令

1. 相关知识点

一个 Tag 文件中通过使用 attribute 指令：

```
<%@ attribute name="对象名字" required="true"|"false" type="对象的类型" %>
```

使得 JSP 页面调用 Tag 文件时，可以向该 Tag 文件中的对象传递一个引用，方式如下：

```
<前缀:Tag 文件名字 对象名字="对象的引用" />
```

或

```
<前缀:Tag 文件名字 对象名字="对象的引用">
     标记体
</前缀:Tag 文件名字>
```

Tag 文件可以把一个对象返回给调用它的 JSP 页面。步骤如下：

(1) Tag 文件使用 variable 指令：

```
<%@ variable name-given="对象名字" variable-class="对象的类型" scope="有效范围" %>
```

给出返回的对象的名字、类型和有效范围。

(2) 将返回的对象的名字和引用存储在内置对象 jspContext 中：

```
jspContext.setAttribute("对象名字",对象的引用);
```

2. 实验目的

本实验的目的是让学生灵活掌握在 Tag 标记中使用 attribute 指令和 variable 指令。

3. 实验要求

编写一个 Tag 文件 GetArea.tag，负责计算三角形或梯形的面积，并将计算结果返回给调用该 Tag 文件的 JSP 页面。编写一个 JSP 页面 inputAndShow.jsp，该页面负责向 Tag 文件提交三角形三边的长度或梯形的上底、下底和高，并负责显示 Tag 文件返回的相应面积。

1）inputAndShow.jsp 的具体要求

inputAndShow.jsp 提供一个表单。用户可以在表单中分别输入三个数值，并选择这三个数值代表三角形三边的长度或梯形的上底、下底和高，然后提交给当前页面。inputAndShow.jsp 通过 Tag 标记调用 GetArea.tag 文件，并向该 Tag 文件 GetArea.tag 传递三角形三边的长度或梯形的上底、下底和高。inputAndShow.jsp 负责显示 Tag 文件 GetArea.tag 返回的面积以及相关信息。inputAndShow.jsp 的效果如图 3-9 所示。

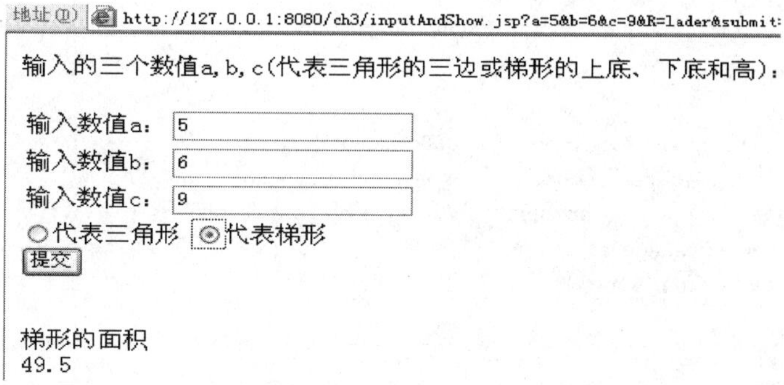

图 3-9　inputAndShow.jsp 的效果

2）GetArea.tag 的具体要求

要求 Tag 文件 GetArea.tag 使用 attribute 指令得到 JSP 页面传递过来三角形三边的长度或梯形的上底、下底和高，使用 variable 指令返回相应的面积以及字符串信息："三角形的面积"或"梯形的面积"。

4. 参考代码

代码仅供参考，学生可按着实验要求，参考本代码编写代码。

JSP 页面参考代码如下：

inputAndShow.jsp

```
<%@ page contentType = "text/html; charset = GB2312" %>
<%@ taglib tagdir = "/WEB-INF/tags" prefix = "computer" %>
<HTML>
输入的三个数值 a,b,c(代表三角形的三边或梯形的上底、下底和高)：
<BODY color = cyan>
   <FORM action = "" method = get name = form>
   <table>
   <tr><td>输入数值 a：</td>
         <td><INPUT type = "text" name = "a"></td>
   </tr>
  <tr><td>输入数值 b：</td>
         <td><INPUT type = "text" name = "b"></td>
   </tr>
  <tr><td>输入数值 c：</td>
         <td><INPUT type = "text" name = "c"></td>
  </tr>
```

```
  </table>
  <INPUT type="radio" name="R" value="triangle">代表三角形
  <INPUT type="radio" name="R" value="lader">代表梯形
  <br><INPUT TYPE="submit" value="提交" name=submit>
 </FORM>
 <%   String a=request.getParameter("a");
       String b=request.getParameter("b");
       String c=request.getParameter("c");
       String cd=request.getParameter("R");
       if(a==null||b==null||c==null){
          a="0";
          b="0";
          c="0";
          cd="0";
       }
       if(a.length()>0&&b.length()>0&&c.length()>0){
  %>  <computer:GetArea numberA="<%=a%>" numberB="<%=b%>"
                              numberC="<%=c%>" condition="<%=cd%>"/>
      <br><%=message%>
      <br><%=area %>
  <% }
   %>
</BODY></HTML>
```

Tag 文件参考代码如下：

GetArea. Tag

```
<%@ tag pageEncoding="gb2312"%>
<%@ attribute name="numberA" required="true" %>
<%@ attribute name="numberB" required="true" %>
<%@ attribute name="numberC" required="true" %>
<%@ attribute name="condition" required="true" %>
<%@ variable name-given="area" variable-class="java.lang.Double" scope="AT_END" %>
<%@ variable name-given="message" scope="AT_END" %>
<%!
    public double getTriangleArea(double a,double b,double c){
        if(a+b>c&&a+c>b&&c+b>a){
           double p=(a+b+c)/2.0;
           double area=Math.sqrt(p*(p-a)*(p-b)*(p-c));
           return area;
        }
        else
          return -1;
    }
    public double getLaderArea(double above,double bottom,double h){
         double area=(above+bottom)*h/2.0;
         return area;
    }
 %>
<% try{  double a=Double.parseDouble(numberA);
          double b=Double.parseDouble(numberB);
          double c=Double.parseDouble(numberC);
```

```
            double result = 0;
            if(condition.equals("triangle")){
               result = getTriangleArea(a,b,c);
               jspContext.setAttribute("area",new Double(result));
               jspContext.setAttribute("message","三角形的面积");
            }
            else if(condition.equals("lader")){
               result = getLaderArea(a,b,c);
               jspContext.setAttribute("area",new Double(result));
               jspContext.setAttribute("message","梯形的面积");
            }
        }
        catch(Exception e){
            jspContext.setAttribute("area",new Double(-1.0));
            jspContext.setAttribute("message",""+e.toString());
        }
    %>
```

习 题 3

1. 用户可以使用浏览器直接访问一个 Tag 文件吗?

2. Tag 文件应当存放在怎样的目录中?

3. Tag 文件中的 tag 指令可以设置哪些属性的值?

4. Tag 文件中的 attribute 指令有怎样的作用?

5. Tag 文件中的 variable 指令有怎样的作用?

6. 编写两个 Tag 文件 Rect.tag 和 Circle.tag。Rect.tag 负责计算并显示矩形的面积,Circle.tag 负责计算并显示圆的面积。编写一个 JSP 页面 lianxi6.jsp,该 JSP 页面使用 Tag 标记调用 Rect.tag 和 Circle.tag。调用 Rect.tag 时,向其传递矩形的两个边的长度;调用 Circle.tag 时,向其传递圆的半径。

7. 编写一个 Tag 文件 GetArea.tag 负责求出三角形的面积,并使用 variable 指令返回三角形的面积给调用该 Tag 文件的 JSP 页面。JSP 页面负责显示 Tag 文件返回的三角形的面积。JSP 在调用 Tag 文件时,使用 attribute 指令将三角形三边的长度传递给 Tag 文件。one.jsp 和 two.jsp 都使用 Tag 标记调用 GetArea.tag。one.jsp 将返回的三角形的面积保留最多 3 位小数,two.jsp 将返回的三角形的面积保留最多 6 位小数。

8. 参照例 3-7,编写一个 JSP 页面 lianxi8.jsp,该页面通过使用 Tag 标记的嵌套显示的效果如图 3-10 所示。

姓名	电话	email
张三	12345678	ss@163.com
李小花	9876543	cc@163.com
孙六	11223355	pp@163.com
吴老二	66553377	ee@163.com

图 3-10 嵌套显示的效果

第4章 JSP 内置对象

本章导读

主要内容

- request 对象
- response 对象
- session 对象
- out 对象
- application 对象

难点

- 使用 Tag 文件处理数据
- 理解 session 对象
- 使用 session 对象存储数据

关键实践

- 使用 request 对象
- 使用 response 对象
- 使用 session 对象

有些对象不用声明就可以在 JSP 页面的 Java 程序片和 Java 表达式中使用，这就是 JSP 的内置对象。JSP 的主要内置对象有 request、response、session、application、out，本章将给予详细的介绍。

request 和 response 对象是 JSP 的内置对象中较重要的两个，这两个对象提供了对服务器和浏览器通信方法的控制。在直接讨论这两个对象前，要先对 HTTP——World Wide Web 底层协议做简单介绍。

World Wide Web 是怎样运行的呢？在浏览器上输入一个正确的网址后，若一切顺利，网页就出现了。例如，在浏览器输入栏中输入 http://www.yahoo.com，Yahoo 网站的主页就出现在浏览器窗口中。这背后是什么在起作用呢？

HTTP 规定了信息在 Internet 上的传输方法，特别规定了浏览器与服务器的交互方法。使用浏览器从网站获取 HTML 页面时，就会使用 Hypertext Transfer Protocol (HTTP)，浏览器对服务器上的 HTML 页面发出请求，服务器收到请求后作出响应，所以 HTTP 的核心就是“请求和响应”。当然，浏览器可以使用 HTTP 协议请求服务器上的任何资源，这依赖于服务器所提供了相应的 Web 应用。

浏览器按着 HTTP 发出的请求有某种结构，按着 HTTP 请求包括一个请求行、头域和可能的信息体。最普通的请求类型是对页面的一个简单请求。例如：

```
GET/hello.htm HTTP/1.1
Host：www.sina.com.cn
```

这是对网站 www.sina.com.cn 上页面 hello.htm 的 HTTP 请求的例子。首行是请求行，规定了请求的方法、请求的资源及使用的 HTTP 的版本。

上例中，请求的方法是 GET 方法，此方法获取特定的资源。上例中 GET 方法用来获取名为 hello.htm 的网页。其他请求方法包括 POST、HEAD、DELETE、TRACE 及 PUT 方法等。

此例中的第二行是头(header)。Host 头规定了所请求的 hello.htm 文件所驻留的主机的 Internet 地址。此例中，请求的主机地址是 www.sina.com.cn。

一个典型请求通常包含许多头，称作请求的 HTTP 头。头提供了关于信息体的附加信息及请求的来源，其中有些头是标准的，有些头和特定的浏览器有关。

一个请求还可能包含信息体。例如，信息体可包含 HTML 表单的内容。在 HTML 表单上单击 Submit 按钮时，如果该表单使用 method＝POST 或 method＝GET 设置 method 属性的值，那么表单中输入的内容都被发送到服务器上，即该表单内容就由 POST 方法或 GET 方法在请求的信息体中发送。

服务器在收到请求时，返回 HTTP 响应。响应也有某种结构，每个响应都由状态行开始，可以包含几个头及可能的信息体，称作响应的 HTTP 头和响应信息体。这些头和信息体由服务器发送给用户的浏览器，信息体就是用户请求的网页的运行结果，对于 JSP 页面，就是网页的静态信息。状态行说明了正在使用的协议、状态代码及文本信息。例如，若服务器在相应请求时出错，则状态行返回错误及对错误的描述(如 HTTP/1.1 404 Object Not Found)。若服务器成功地响应了请求，返回包含 200 OK 的状态行。

Tomcat 服务器的 webapps 目录的子目录都可以作为一个 Web 服务目录，在 webapps 目录下新建一个 Web 服务目录 ch4，除非特别约定，本章例子中的 JSP 页面均保存在 ch4 中。

4.1 request 对象

HTTP 通信协议是用户与服务器之间一种请求与响应(request/response)的通信协议。在 JSP 中，内置对象 request 封装了用户请求页面时所提交的信息，那么该对象调用相应的方法可以获取封装的信息，即使用该对象可以获取用户提交的信息。

用户在请求 JSP 页面时，通常会使用 HTML 表单提交信息，表单的一般格式是：

```
<FORM method = get|post action = "提交信息的目的地页面">
    提交手段
</FORM>
```

其中<FORM>是表单标签，method 取值 get 或 post。get 方法和 post 方法的主要区别是：使用 get 方法提交的信息会在提交的过程中显示在浏览器的地址栏中，而使用 post 方法提交的信息不会显示在地址栏中。提交手段包括通过文本框、列表、文本区等。例如：

```
<FORM action = "tom.jsp" method = post >
    <INPUT type = "text" name = "boy" value = "ok" >
    <INPUT TYPE = "submit" value = "送出" name = "submit">
</FORM>
```

该表单使用 post 方法请求 tom.jsp 页面,并同时向页面 tom.jsp 提交信息,提交信息的手段是在文本框输入信息,其中默认信息是 ok。用户单击“送出”按钮请求 tom.jsp 页面,并向 tom.jsp 提交信息。tom.jsp 页面可以使用内置 request 对象获得用户提交的信息。例如,request 对象可以使用 getParameter(String s)方法获取该表单通过 text 提交的信息。例如:

```
request.getParameter("boy");
```

4.1.1 获取用户提交的信息

request 对象获取用户提交信息的最常用的方法是 getParameter(String s)。在例 4-1 中,example4_1.jsp 页面通过表单向 tree.jsp 页面提交信息 I am a student,tree.jsp 页面通过 request 对象获取表单提交的信息,包括 text 的值以及按钮的值。example4_1.jsp 页面和 tree.jsp 页面的效果如图 4-1 所示。

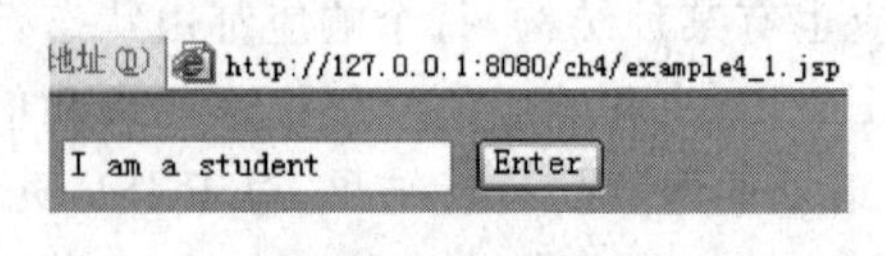

(a) 使用表单提交数据

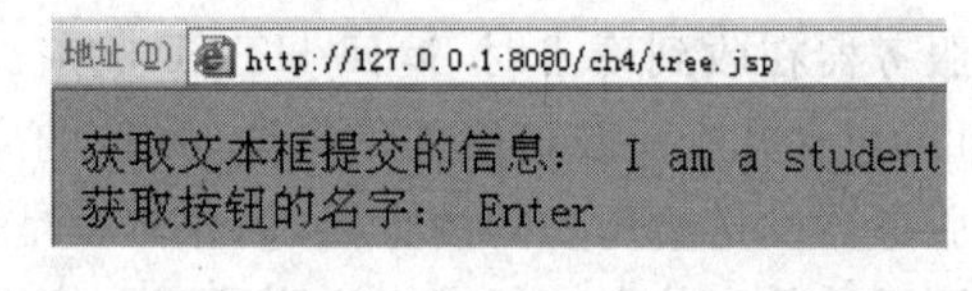

(b) 获取表单提交的数据

图 4-1 两个页面的效果图

例 4-1

example4_1.jsp

```
<%@ page contentType = "text/html; charset = GB2312" %>
<HTML><BODY bgcolor = cyan><FONT size = 3>
  <FORM action = "tree.jsp" method = post name = form>
      <INPUT type = "text" name = "boy">
      <INPUT TYPE = "submit" value = "Enter" name = "submit">
  </FORM>
</FONT></BODY></HTML>
```

tree.jsp

```
<%@ page contentType = "text/html; charset = GB2312" %>
<HTML><BODY bgcolor = cyan><FONT size = 4>
  获取文本框提交的信息:
  <%   String textContent = request.getParameter("boy");
  %>
  <% = textContent %>
<BR>获取按钮的名字:
  <%   String buttonName = request.getParameter("submit");
  %>
```

```
    <% = buttonName %>
</FONT></BODY></HTML>
```

在例 4-2 中，example4_2.jsp 页面通过表单向自己提交一个正数，example4_2.jsp 页面获取表单提交的正数，并计算这个数的平方根。表单中的 action 是所请求的页面，如果是当前页面，可以用双引号""代替当前页面，注意双引号中不能含有空格。example4_2.jsp 页面的效果如图 4-2 所示。

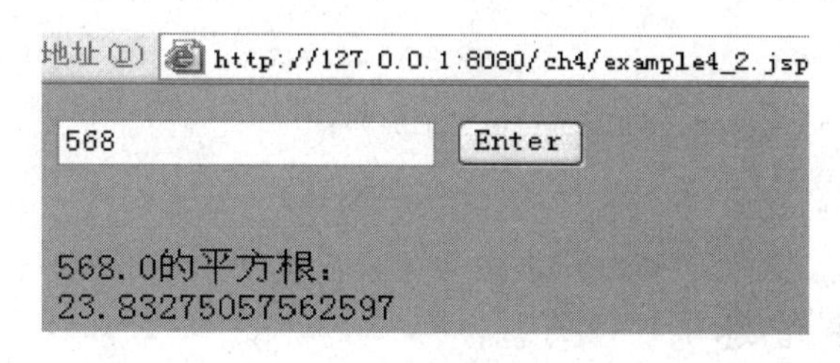

图 4-2 计算平方根

例 4-2

example4_2.jsp

```
<%@ page contentType = "text/html; charset = GB2312" %>
<HTML><BODY bgcolor = cyan><FONT size = 3>
    <FORM action = "" method = post name = form>
          <INPUT type = "text" name = "girl">
          <INPUT TYPE = "submit" value = "Enter" name = "submit">
    </FORM>
    <% String textContent = request.getParameter("girl");
         double number = 0,r = 0;
         if(textContent == null)
             textContent = "";
         try{   number = Double.parseDouble(textContent);
                 if(number >= 0){
                   r = Math.sqrt(number);
                   out.print("<BR>" + number + "的平方根：");
                   out.print("<BR>" + r);
                }
                else
                  out.print("<BR>" + "请输入一个正数");
         }
         catch(NumberFormatException e){
                out.print("<BR>" + "请输入数字字符");
         }
      %>
</FONT></BODY></HTML>
```

注意：使用 request 对象获取信息时要格外小心。在例 4-2 中：

```
String textContent = request.getParameter("girl");
```

获取提交的字符串 textContent，并且在下面的代码中使用了这个字符串对象 textContent：

```
number = Doule.parseDoubel(textContent);
```

那么,JSP 引擎在执行 JSP 页面对应的字节码文件时,会认为使用了空对象,因为在这个字节码被执行时(用户请求页面时),用户可能还没有提交数据,textContent 也就没有被创建。如果使用了空对象,即还没有创建对象,就使用了该对象,JSP 引擎就会提示出现了 NullPointerException 异常,当然如果不使用空对象就不会出现异常。

因此,可以像例 4-2 那样,为了避免在字节码文件运行时,JSP 引擎认为是使用了空对象,使用下面代码:

```
String textContent = request.getParameter("girl");
 if(textContent == null)
    textContent = "";
```

4.1.2 处理汉字信息

当用 request 对象获取用户提交的汉字字符时,会出现乱码问题,所以对含有汉字字符的信息必须进行特殊的处理。首先,将获取的字符串用 ISO-8859-1 进行编码,并将编码存放到一个字节数组中,然后再将这个数组转化为字符串对象即可,如下所示:

```
String str = request.getParameter("girl");
byte b[] = str.getBytes("ISO - 8859 - 1");
str = new String(b);
```

例 4-3 对例 4-1 按上述办法做了改动,并将按钮上的字变成汉字,文本框提交的默认信息是:“苹果:apple,12 斤,5 $”。

例 4-3

example4_3.jsp

```
<%@ page contentType = "text/html; charset = GB2312" %>
<HTML><BODY bgcolor = green><FONT size = 1>
   <FORM action = "apple.jsp" method = post name = form>
       <INPUT type = "text" name = "boy" value = "苹果:apple,12 斤,5 $">
       <INPUT TYPE = "submit" value = "提交" name = "submit">
   </FORM>
</FONT></BODY></HTML>
```

apple.jsp

```
<%@ page contentType = "text/html; charset = GB2312" %>
<MHML><BODY>
获取文本框提交的信息:
   <%   String textContent = request.getParameter("boy");
          byte b[] = textContent.getBytes("ISO - 8859 - 1");
          textContent = new String(b);
   %>
  <% = textContent %>
<BR>获取按钮的名字:
   <%   String buttonName = request.getParameter("submit");
          byte c[] = buttonName.getBytes("ISO - 8859 - 1");
          buttonName = new String(c);
```

```
    %>
    <%=buttonName%>
</BODY></HTML>
```

4.1.3 常用方法举例

当用户访问一个页面时，会提交一个 HTTP 请求给服务器的 JSP 引擎，这个请求包括一个请求行、HTTP 头和信息体，例如：

```
post/tree.jsp/HTTP.1.1
host: localhost: 8080
accept-encoding: gzip,deflate
```

其中首行叫请求行，规定了向请求的页面提交信息的方式，如 post、get 等方式以及请求的页面的文件名字和使用的通信协议。

第 2、3 行分别是两个头（Header），host、accept-encoding 是头名字，而 localhost：8080 以及 gzip，deflate 分别是 host 和 accept-encoding 头的值，其中 host 头的值就是所请求的 tree.jsp 的主机地址。上面的请求有两个头：host 和 accept-encoding，一个典型的请求通常包含很多的头，有些头是标准的，有些头和特定的浏览器有关。

一个请求还包含信息体，即 HTML 标记组成的部分，可能包括各式各样用于提交信息的表单等。例如：

```
<FORM action="tree.jsp" method=post name=form>
    <INPUT type="text" name="boy">
    <INPUT TYPE="submit" value="" name="submit">
</FORM>
```

可以使用 JSP 引擎的内置对象 request 对象来获取请求所提交的各类信息。说明如下：

- getProtocol()：获取用户向服务器提交信息所使用的通信协议。例如 http/1.1 等。
- getServletPath()：获取用户请求的 JSP 页面的所在目录。
- getContentLength()：获取用户提交的整个信息的长度。
- getMethod()：获取用户提交信息的方式。例如 post 或 get。
- getHeader(String s)：获取 HTTP 头文件中由参数 s 指定的头名字的值。一般来说 s 参数可取的头名有 accept、referer、accept-language、content-type、accept-encoding、user-agent、host、content-length、connection、cookie 等。例如，s 取值 user-agent 将获取用户的浏览器的版本号等信息。
- getHeaderNames()：获取头名字的一个枚举。
- getHeaders(String s)：获取头文件中指定头名字的全部值的一个枚举。
- getRemoteAddr()：获取用户的 IP 地址。
- getRemoteHost()：获取用户机的名称（如果获取不到，就获取 IP 地址）。
- getServerName()：获取服务器的名称。
- getServerPort()：获取服务器的端口号。
- getParameterNames()：获取用户提交的信息体部分中 name 参数值的一个枚举。

例 4-4 使用了 request 对象的一些常用方法。

例 4-4

example4_4. jsp(效果如图 4-3(a)所示)

```
<HTML><BODY bgcolor=cyan><FONT size=1>
<%@ page contentType="text/html; charset=GB2312" %>
   <FORM action="mess.jsp" method=post name=form>
        <INPUT type="text" name="boy">
        <INPUT TYPE="submit" value="enter" name="submit">
   </FORM>
</FONT></BODY></HTML>
```

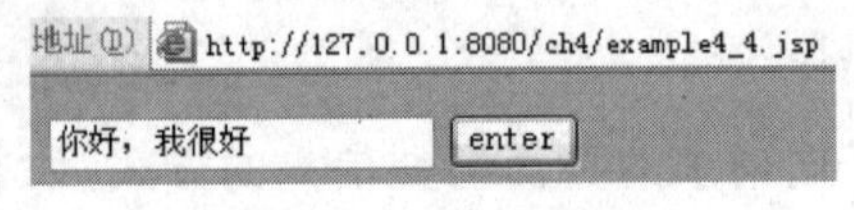

(a) 使用表单提交信息

```
用户使用的协议是: HTTP/1.1
获取接受用户提交信息的页面: /mess.jsp
接受用户提交信息的长度: 53
用户提交信息的方式: POST
获取HTTP头文件中User-Agent的值: Mozilla/4.0 (compatible; MSIE 6.0; Windows NT 5.1; SV1)
获取HTTP头文件中accept的值: image/gif, image/x-xbitmap, image/jpeg, image/pjpeg, application/x-shockwave-flash,
application/vnd.ms-excel, application/vnd.ms-powerpoint, application/msword, application/x-silverlight, */*
获取HTTP头文件中Host的值: 127.0.0.1:8080
获取HTTP头文件中accept-encoding的值: gzip, deflate
获取用户的IP地址: 127.0.0.1
获取用户机的名称: 127.0.0.1
获取服务器的名称: 127.0.0.1
获取服务器的端口号: 8080
获取用户端提交的所有参数的名字: boy submit
获取头名字的一个枚举: accept referer accept-language content-type accept-encoding user-agent host content-lengt
connection cache-control cookie
获取头文件中指定头名字的全部值的一个枚举: JSESSIONID=ACDB2E61D85E3C6E80F6D9367656FC41

文本框text提交的信息:
你好，我很好
按钮的名字:
enter
```

(b) 使用request对象得到请求所提交的信息

图 4-3 example4_4. jsp 和 mess. jsp 的效果图

mess. jsp(效果如图 4-3(b)所示)

```
<%@ page contentType="text/html; charset=GB2312" %>
<%@ page import="java.util.*" %>
<MHML><BODY bgcolor=white><Font size=3>
<BR>用户使用的协议是:
   <%  String protocol=request.getProtocol();
          out.println(protocol);
    %>
<BR>获取接收用户提交信息的页面:
   <%  String path=request.getServletPath();
          out.println(path);
    %>
<BR>接受用户提交信息的长度:
   <%  int length=request.getContentLength();
          out.println(length);
    %>
<BR>用户提交信息的方式:
   <%  String method=request.getMethod();
          out.println(method);
```

```
    %>
<BR>获取 HTTP 头文件中 User - Agent 的值:
   <%  String header1 = request.getHeader("User - Agent");
         out.println(header1);
    %>
<BR>获取 HTTP 头文件中 accept 的值:
   <%  String header2 = request.getHeader("accept");
         out.println(header2);
    %>
<BR>获取 HTTP 头文件中 Host 的值:
   <%  String header3 = request.getHeader("Host");
         out.println(header3);
    %>
<BR>获取 HTTP 头文件中 accept - encoding 的值:
   <%  String header4 = request.getHeader("accept - encoding");
         out.println(header4);
    %>
<BR>获取用户的 IP 地址:
   <%  String IP = request.getRemoteAddr();
         out.println(IP);
    %>
<BR>获取用户机的名称:
   <%  String clientName = request.getRemoteHost();
         out.println(clientName);
    %>
<BR>获取服务器的名称:
   <%  String serverName = request.getServerName();
         out.println(serverName);
    %>
<BR>获取服务器的端口号:
   <%  int serverPort = request.getServerPort();
         out.println(serverPort);
    %>
<BR>获取用户端提交的所有参数的名字:
   <%  Enumeration en = request.getParameterNames();
         while(en.hasMoreElements()){
            String s = (String)en.nextElement();
            out.println(s);
         }
       %>
<BR>获取头名字的一个枚举:
   <%  Enumeration enum_headed = request.getHeaderNames();
         while(enum_headed.hasMoreElements()){
            String s = (String)enum_headed.nextElement();
            out.println(s);
         }
    %>
<BR>获取头文件中指定头名字的全部值的一个枚举:
   <%  Enumeration enum_headedValues = request.getHeaders("cookie");
         while(enum_headedValues.hasMoreElements()){
            String s = (String)enum_headedValues.nextElement();
```

```
            out.println(s);
          }
      %>
    <BR>
      <P> 文本框 text 提交的信息:
      <%    String str = request.getParameter("boy");
              byte b[] = str.getBytes("ISO - 8859 - 1");
              str = new String(b);
      %>
     <BR><% = str %>
    <BR> 按钮的名字:
      <%    String buttonName = request.getParameter("submit");
              byte c[] = buttonName.getBytes("ISO - 8859 - 1");
              buttonName = new String(c);
      %>
    <BR><% = buttonName %>
    </FONT></BODY></HTML>
```

4.1.4 使用 Tag 文件处理有关数据

Tag 文件可以实现代码的复用,这些代码可能是许多 JSP 页面都需要的。因此我们应当尽量将数据的处理让 Tag 文件去完成。例如,JSP 页面使用 request 对象获取用户提交的数据,然后使用 Tag 标记调用 Tag 文件,并将必要的数据传递给 Tag 文件。Tag 文件负责处理数据,根据需要将处理结果显示给用户或返回给调用它的 JSP 页面。

在例 4-5 中,用户通过 example4_5.jsp 中的表单将三角形的三边的长度提交给 get.jsp,get.jsp 使用 Tag 标记调用 Tag 文件 Computer.tag,并将三角形三边的长度传递给 Computer.tag,Computer.tag 文件负责计算三角形的面积,并将计算结果返回给 get.jsp 页面。example4_5.jsp 和 get.jsp 的效果如图 4-4 所示。

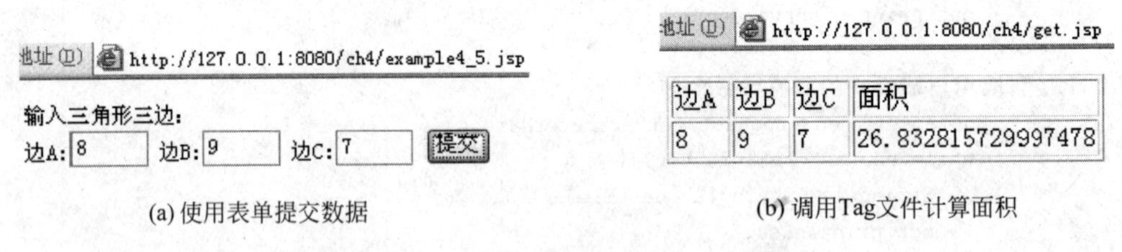

(a) 使用表单提交数据 (b) 调用Tag文件计算面积

图 4-4 example4_5.jsp 和 get.jsp 的效果图

例 4-5

example4_5.jsp

```
<%@ page contentType = "text/html; charset = GB2312"  %>
<HTML><BODY><Font size = 2>
<FORM action = "get.jsp" Method = "post" >
   <P>输入三角形三边:
   <BR>边 A: <Input type = text name = "sideA" value = 0 size = 5>
          边 B: <Input type = text name = "sideB" value = 0 size = 5>
          边 C: <Input type = text name = "sideC" value = 0 size = 5>
   <Input type = submit value = "提交">
```

```
</FORM>
</FONT></BODY></HTML>
```

get.jsp

```
<%@ page contentType="text/html; charset=GB2312" %>
<%@ taglib prefix="computer" tagdir="/WEB-INF/tags" %>
<%  String a=request.getParameter("sideA");
     String b=request.getParameter("sideB");
     String c=request.getParameter("sideC");
%>
<computer:Triangle a="<%=a%>" b="<%=b%>" c="<%=c%>"/>
<HTML><BODY>
<table border=1>
  <tr><td width=30>边 A</td>
        <td width=30>边 B</td>
        <td width=30>边 C</td>
        <td>面积</td>
  </tr>
   <tr><td><%=a%></td>
        <td><%=b%></td>
        <td><%=c%></td>
        <td><%=result%></td>    <%-- result 是 Tag 文件返回的对象 --%>
  </tr>
</table>
</BODY></HTML>
```

Triangle.tag

```
<%@ tag pageEncoding="GB2312" %>
<%@ attribute name="a" required="true" %>
<%@ attribute name="b" required="true" %>
<%@ attribute name="c" required="true" %>
<%@ variable name-given="result" scope="AT_END" %>
  <%  String mess="";
       try { double sideA=Double.parseDouble(a);
             double sideB=Double.parseDouble(b);
             double sideC=Double.parseDouble(c);
             if(sideA+sideB>sideC&&sideA+sideC>sideB&&sideC+sideB>sideA) {
                double p=(sideA+sideB+sideC)/2.0;
                double area=Math.sqrt(p*(p-sideA)*(p-sideB)*(p-sideC));
                mess=""+area;
             }
             else
                mess=="不能构成一个三角形,无法计算面积。";
       }
       catch(Exception e){
             mess ="无法计算面积。"+e;
       }
       jspContext.setAttribute("result",mess); //将 result 返回给 get.jsp
  %>
```

4.1.5 处理HTML标记

JSP页面可以含有HTML标记,当用户通过浏览器请求一个JSP页面时,Tomcat服务器将该JSP页面中的HTML标记直接发送到用户的浏览器,由用户的浏览器负责执行这些HTML标记。而JSP页面中的变量声明、程序片以及表达式由Tomcat服务器处理后,再将有关的结果用文本方式发送到用户端的浏览器。

HTML是Hypertext Marked Language的缩写,即超文本标记语言。目前的HTML有100多个标记,这些标记可以描述数据的显示格式,如果对HTML语言比较陌生,建议补充这方面的知识,本节对一些重要的HTML标记做一个简单的介绍。

1. <FORM>标记

<FORM>标记被习惯地称作表单,是非常重要的一个HTML标记,用户经常需要使用表单提交数据。

表单的一般格式是:

```
<FORM method = get| post action = "提交信息的目的地页面" name = "表单的名字">
    数据提交手段部分
</FORM>
```

其中<FORM……>……</FORM>是表单标记,其中的method属性取值get或post。get方法和post方法的主要区别是:使用get方法提交的信息会在提交的过程中显示在浏览器的地址栏中,而post方法提交的信息不会显示在地址栏中。提交手段包括:通过文本框、列表、文本区等。例如:

```
<FORM action = "tom.jsp" method = "post" >
   <INPUT type = "text" name = "boy" value = "ok" >
   <INPUT TYPE = "submit" value = "送出" name = "submit">
</FORM>
```

表单标记经常将下列标记作为表单的子标记,以便提供提交数据的手段,这些标记都以GUI形式出现,方便用户输入或选择数据。例如:

```
<INPUT ……>
<Select ……></Select>
<Option ……> </Option>
<TextArea ……> </TextArea>
```

2. <INPUT>标记

表单标记<FORM>将<INPUT>标记作为子标记来指定表单中数据的输入方式以及表单的提交按钮。<INPUT>标记中的type属性可以指定输入方式的GUI对象,name属性用来指定这个GUI对象的名称。<INPUT>标记的基本格式:

```
<INPUT type = "输入对象的GUI类型" name = "名字" >
```

服务器通过属性name指定的名字来获取"输入对象的GUI类型"中提交的数据。"输入对象的GUI类型"可以是:text(文本框)、checkbox(复选框)、submit(提交按钮)等。

1) 文本框(text)

当输入对象的GUI类型是text时,除了用name为text指定名字外,还可以为text指定其他一些值。例如：

```
<INPUT type="text" name="me" value="hi" size= "9" align= "left" maxlength="30">
```

其中，value 的值是 text 的初始值；size 是 text 对象的长度（单位是字符）；align 是 text 在浏览器窗体中的对齐方式；maxlength 指定 text 可输入的最多字符数目。request 对象通过 name 指定的名字来获取用户在文本框输入的字符串。如果用户没有在文本框中输入信息就单击表单中的提交按钮，request 对象调用 getParameter (String name)方法将获取由 value 指定的值；如果 value 未指定任何值，getParameter (String name)方法获取的字符串的长度为 0，即该字符串为：""。

2）单选框(radio)

当输入对象的 GUI 类型是 radio 时，除了用 name 为 radio 指定名字外，还可以为 radio 指定其他一些值。例如：

```
<INPUT type="radio" name="rad" value="red" align="top" checked="java">
```

其中，value 指定 radio 的值；align 是 radio 在浏览器窗体中的对齐方式；如果几个单选键的 name 取值相同，那么同一时刻只能有一个被选中。request 对象调用 getParameter (String name)方法，获取被选中的 radio 中 value 属性指定的值。checked 如果取值是一个非空的字符串，那么该单选框的初始状态就是选中状态。

3）复选框(checkbox)

当输入对象的 GUI 类型是 checkbox 时，除了用 name 为 checkbox 指定名字外，还可以为 checkbox 指定其他一些值。例如：

```
<INPUT type="checkbox" name="ch" value="pink" align="top" checked="java">
```

其中，value 指定 checkbox 的值；复选框与单选框的区别就是可以多选，即如果几个 checkbox 的 names 取值相同，那么同一时刻可有多个 checkbox 被选中，如果 checked 取值是一个非空的字符串，那么该复选框的初始状态就是选中状态。request 对象需调用 getParameterValues (String name)方法，不是调用 getParameter(String name)方法，获取被选中的多个 checkbox 中 value 属性指定的值。

4）口令框(password)

它是输入口令用的特殊文本框，输入的信息用“*”回显，防止他人偷看口令。

```
<INPUT type="password" name="me" size="12" maxlength="30">
```

服务器通过 name 指定的字符串获取 password 提交的值，在口令框中输入：“bird88_1”，那么 bird88_1 将被提交给服务器，口令框仅仅起着不让别人偷看的作用，不提供加密措施。

5）隐藏(hidden)

当<INPUT>中的属性 type 的值是 hidden 时，<INPUT>没有可见的输入界面，表单直接将<INPUT>中 value 属性的值提交给服务器。例如：

```
<INPUT type="hidden" name="h" value="123">
```

request 对象调用 getParameter(String name)方法，通过 name 的名字来获取由 value 指定的值。

6）提交按钮(submit)

为了能把表单的数据提交给服务器,一个表单至少要包含一个提交按钮。

```
< INPUT type = "submit" name = "me" value = "确定" size = "12">
```

用户单击提交按钮后,服务器就可以获取表单提交的各个数据。当然 request 对象调用 getParameter(String name)方法也可以获取提交按钮的值,getParameter (String name)方法通过 name 指定的名字来获取提交按钮本身由 value 指定的值。

7）重置按钮(reset)

重置按钮将表单中输入的数据清空,以便重新输入数据。

```
< INPUT type = "reset">
```

在例 4-6 中,用单选框和复选框实现一个网上小测试。用户在 example4_6. jsp 页面提供的表单中选中单选框或复选框,然后将选择提交给 answer. jsp 页面。example4_6. jsp 页面和 answer. jsp 页面的效果如图 4-5 所示。

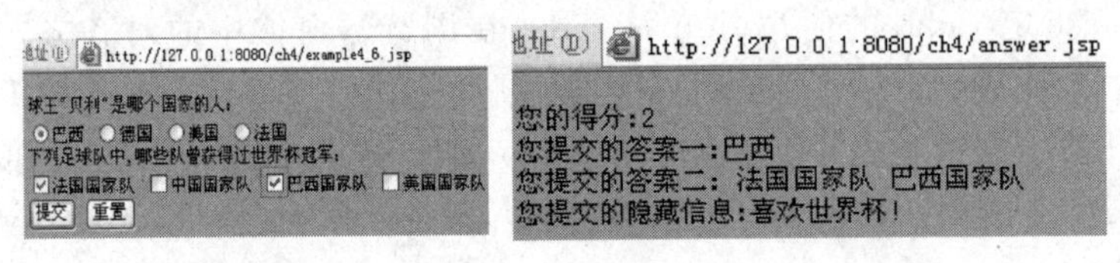

(a) 单选框和复选框　　　　(b) 获取单选框和复选框提交的数据

图 4-5　两个页面的效果图

例 4-6

example4_6. jsp

```
< HTML >< % @ page contentType = "text/html; charset = GB2312" % >
< BODY bgcolor = cyan >< FONT size = 2 >
 < FORM action = "answer. jsp" method = post name = form >
        球王"贝利"是哪个国家的人: < BR >
        < INPUT type = "radio" name = "R" value = "巴西">巴西
        < INPUT type = "radio" name = "R" value = "德国">德国
        < INPUT type = "radio" name = "R" value = "美国">美国
        < INPUT type = "radio" name = "R" value = "法国" checked = "ok">法国
        < BR>下列足球队中,哪些队曾获得过世界杯冠军: < BR >
        < INPUT type = "checkbox" name = "item" value = "法国国家队" >法国国家队
        < INPUT type = "checkbox" name = "item" value = "中国国家队" >中国国家队
        < INPUT type = "checkbox" name = "item" value = "巴西国家队" >巴西国家队
        < INPUT type = "checkbox" name = "item" value = "美国国家队" >美国国家队
        < INPUT TYPE = "hidden" value = "喜欢世界杯!" name = "secret">
        < BR >< INPUT TYPE = "submit" value = "提交" name = "submit">
        < INPUT TYPE = "reset" value = "重置">
   </FORM >
</FONT ></BODY ></HTML >
```

answer. jsp

```
< % @ page contentType = "text/html; Charset = GB2312" % >
```

```
<HTML><BODY bgcolor = cyan><Font size = 2>
   <% int score = 0;
        String countryName = request.getParameter("R");
        String itemNames[] = request.getParameterValues("item");
        String secretMess = request.getParameter("secret");
        if(countryName.equals("巴西"))
            score++;
        if(itemNames == null)
            out.print("没有选择球队<br>");
        else if(itemNames.length >= 2){
           if(itemNames[0].equals("法国国家队")&&itemNames[1].equals("巴西国家队"))
               score++;
        }
        out.print("你的得分：" + score);
        out.print("<br>你提交的答案一：" + countryName);
        out.print("<br>你提交的答案二：");
        if(itemNames! = null)
           for(int k = 0; k < itemNames.length; k++ )
              out.println(" " + itemNames[k]);
        out.println("<br> 你提交的隐藏信息：" + secretMess);
  %>
</FONT></BODY></HTML>
```

3. <Select>、<Option>格式

下拉式列表和滚动列表通过<Select>和<Option>标记来定义。基本格式为：

```
<Select>
    <Option>
    <Option>
    ……
</Select>
```

1）下拉列表

```
<Select name = "shulie" >
    <Option value = "cat">你选了小猫
    <Option value = "dog">你选了小狗
         ……
    <Option value = "600"> n = 600
</Select>
```

服务器通过 name 获取下拉列表中被选中的 Option 的值(参数 value 指定的值)。

2）滚动列表

在 Select 中指定 size 属性的值就变成滚动列表，size 的值是滚动列表的可见行的个数。

```
<Select name = "shulie" size = 2>
    <Option value = "1">计算 1 到 n 的连续和
    <Option value = "2">计算 1 到 n 的平方和
  <Option value = "3">计算 1 到 n 的立方和
</Select>
```

服务器通过 name 获取滚动列表中被选中的 Option 的值(参数 value 指定的值)。

在例 4-7 中,用户在 example4_7.jsp 页面提供的表单中,通过滚动列表选择计算求和的方式、通过下拉列表选择计算求和的项数,然后将选择提交给 sum.jsp。example4_7.jsp 页面和 sum.jsp 页面的效果如图 4-6 所示。

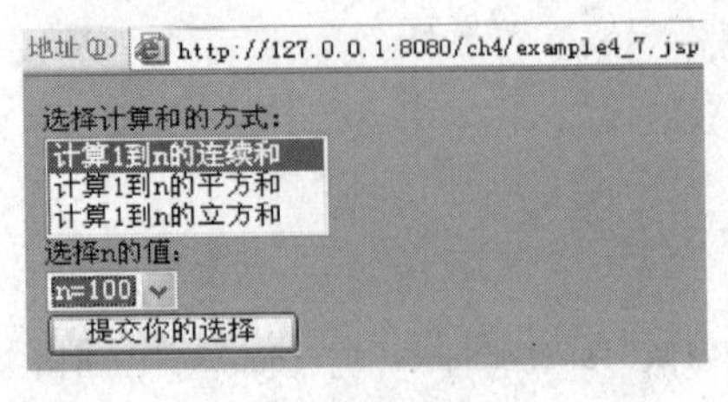

(a) 下拉列表和滚动列表

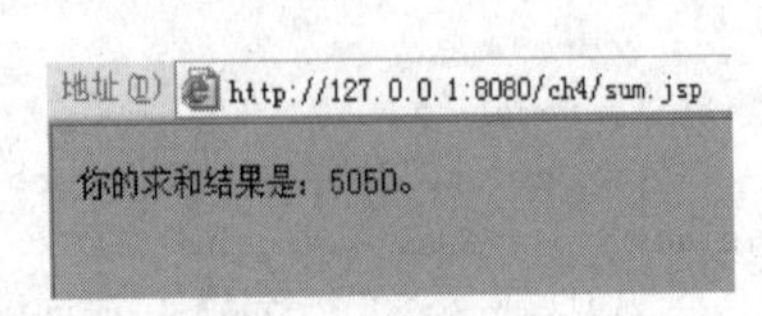

(b) 计算连续和

图 4-6 两个页面的效果图

例 4-7

example4_7.jsp

```
<HTML><%@ page contentType = "text/html; charset = GB2312" %>
<BODY bgcolor = cyan><FONT size = 2>
  <FORM action = "sum.jsp" method = post name = form>
      选择计算和的方式:<br>
      <Select name = "sum" size = 3>
         <Option Selected value = "1">计算 1 到 n 的连续和
         <Option value = "2">计算 1 到 n 的平方和
         <Option value = "3">计算 1 到 n 的立方和
      </Select>
      <br>选择 n 的值:<br>
      <Select name = "n" >
         <Option value = "10"> n = 10
         <Option value = "20"> n = 20
         <Option value = "30"> n = 30
         <Option value = "40"> n = 40
         <Option value = "50"> n = 50
         <Option value = "100"> n = 100
      </Select>
      <br><INPUT TYPE = "submit" value = "提交你的选择" name = "submit">
   </FORM>
</FONT></BODY></HTML>
```

sum.jsp

```
<HTML><%@ page contentType = "text/html; charset = GB2312" %>
<BODY bgcolor = cyan><FONT size = 2>
 <% long sum = 0;
    String s1 = request.getParameter("sum");
    String s2 = request.getParameter("n");
    if(s2.equals(""))
       s2 = "0";
    if(s1.equals("1")){
       int n = Integer.parseInt(s2);
```

```
            for(int i = 1; i <= n; i++ )
                sum = sum + i;
        }
        else if(s1.equals("2")){
            int n = Integer.parseInt(s2);
            for(int i = 1; i <= n; i++ )
                sum = sum + i * i;
        }
        else if(s1.equals("3")){
            int n = Integer.parseInt(s2);
            for(int i = 1; i <= n; i++ )
                sum = sum + i * i * i;
        }
    %>
<P>你的求和结果是：<% = sum %>。
</FONT></BODY></HTML>
```

4. <TextArea>格式

<TextArea> 标记在表单中指定一个能输入多行文本的文本区域。

```
<TextArea name = "ilovethisgame" Rows = "4" Cols = "20"></TextArea>
```

5. 表格

表格以行列形式显示数据，不提供输入数据功能。经常将某些数据或GUI放置在表格的单元格中，以使得界面更加简练、美观。表格由<table>、</table>标记定义。一般格式：

```
<table>
   <tr width = "该行的宽度">
       <th width = "单元格的宽度" >单元格中的数据</th>
              ……
       <td width = "单元格的宽度" >单元格中的数据</td>
   </tr>
    ……
</table>
```

其中 <tr> …… </tr>定义表格的一个行，<th>或<td>标记定义这一行中的表格单元，二者的区别是：<th>定义的单元加重显示，<td>称作普通单元，不加重显示。一行中的加重单元和普通单元可以交替出现，也可以全是加重单元或普通单元。<table border=1>中增加选项 border 可指明该表格是否带有边框。

在例 4-8 中对例 4-6 进行了改动，把表单的 GUI 显示在表格单元格中，example4_8.jsp 页面的效果如图 4-7 所示。

图 4-7　使用表格

例 4-8

example4_8.jsp

```
<HTML><%@ page contentType="text/html; charset=GB2312" %>
<BODY bgcolor=cyan><FONT size=1>
<FORM action="answer.jsp" method=post name=form>
  <table border=1>
    <tr>
      <th>"贝利"是哪个国家的人</th>
      <th>曾获得过世界杯冠军的球队</th>
    </tr>
    <tr>
      <td>
          <INPUT type="radio" name="R" value="巴西">巴西
          <INPUT type="radio" name="R" value="德国">德国
          <INPUT type="radio" name="R" value="美国">美国
          <INPUT type="radio" name="R" value="法国" checked="ok">法国
      </td>
      <td>
          <INPUT type="checkbox" name="item" value="法国国家队">法国国家队
          <INPUT type="checkbox" name="item" value="中国国家队">中国国家队
          <INPUT type="checkbox" name="item" value="巴西国家队">巴西国家队
          <INPUT type="checkbox" name="item" value="美国国家队">美国国家队
          <INPUT TYPE="hidden" value="喜欢世界杯!" name="secret">
      </td>
    </tr>
    <tr>
      <td><INPUT TYPE="submit" value="提交" name="submit"></td>
      <td><INPUT TYPE="reset" value="重置"></td>
    </tr>
  </table>
</FORM>
</FONT></BODY></HTML>
```

6. <img>与<image>标记

使用<img>或<image>标记都可以显示一幅图像，标记的基本格式为：

```
<img src="图像文件的 URL">描述文字</img>
<image src="图像文件的 URL">描述文字</image>
```

如果图像文件和当前页面在同一 Web 服务目录中，“图像文件的 URL”就是该图像文件的名字；如果图像文件在当前 Web 服务目录一个子目录中，例如 picture 子目录中，那么“图像文件的 URL”就是“picture/图像文件的名字”。

标记中可以使用 width 和 height 属性指定被显示的图像的宽和高，如果省略 width 和 height 属性，标记将按图像的原始宽度和高度来显示图像。

7. <embed>标记

使用<embed>标记可以播放音乐和视频，当浏览器执行该标记时，会把浏览器所在机器上的默认播放器嵌入到浏览器中，以便播放音乐或视频文件。<embed>标记的基本格式为：

```
<embed src = "音乐或视频文件的 URL" >描述文字</embed >
```

如果音乐或视频文件和当前页面在同一 Web 服务目录中，<embed>标记中 src 属性的值就是该文件的名字；如果视频文件在当前 Web 服务目录一个子目录中，例如 avi 子目录中，那么<embed>标记中 src 属性的值就是“avi/视频文件的名字”。

<embed>标记中经常使用的属性及取值如下：

autostart 取值 true 或 false，autostart 属性的值用来指定音乐或视频文件传送完毕后是否立刻播放，该属性的默认值是 false。

loop 取值为正整数，该属性的值用来指定音乐或视频文件重复播放的次数。

width、height 取值均为正整数，用 width 和 height 属性的值指定播放器的宽和高，如果省略 width 和 height 属性，将使用默认值。

例 4-9 中页面使用了<img>和<embed>标记。用户通过 example4_9.jsp 页面中的下拉列表选择一幅图像和一个视频文件或音乐文件，然后单击提交按钮将数据提交给 show.jsp 页面，该页面使用<image>标记显示图像，使用<embed>标记播放音乐和视频。其中图像文件和视频文件分别存放在当前 Web 服务目录 ch3 的子目录 picture 和 avi 中。example4_9.jsp 页面和 show.jsp 页面的效果如图 4-8 所示。

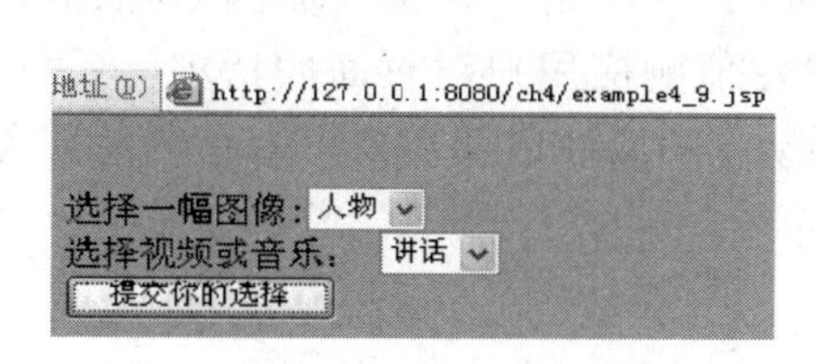

(a) 选择图像和视频

(b) 显示图像、播放视频

图 4-8 两个页面的效果图

例 4-9

example4_9.jsp

```
<%@ page contentType = "text/html; charset = GB2312" %>
<HTML><BODY bgcolor = cyan><Font size = 3>
  <FORM action = "show.jsp" method = post name = form>
   <br>选择一幅图像：<Select name = "image" >
          <Option value = "car.jpg">汽车
          <Option value = "person.jpg">人物
       </Select>
   <br>选择视频或音乐：
       <Select name = "video" >
          <Option value = "happy.wav">好汉歌
          <Option value = "moon.wmv">故乡
          <Option value = "clock.avi">表
       </Select>
   <br><INPUT TYPE = "submit" value = "提交你的选择" name = "submit">
   </FORM>
</FONT></BODY></HTML>
```

show.jsp

```
<%@ page contentType = "text/html; charset = GB2312" %>
<HTML><BODY bgcolor = cyan><Font size = 3>
 <% String s1 = request.getParameter("image");
    String s2 = request.getParameter("video");
  %>
 <image src = "picture/<% = s1 %>" width = 200 height = 160 >图像</image>
 <embed src = "avi/<% = s2 %>" width = 300 height = 180 >视频</embed>
</FONT></BODY></HTML>
```

4.2 response 对象

当用户访问一个服务器的页面时，会提交一个 HTTP 请求，服务器收到请求时，返回 HTTP 响应。响应和请求类似，也有某种结构。每个响应都由状态行开始，可以包含几个头及可能的信息体(网页的结果输出部分)。

上一节学习了用 request 对象获取用户请求提交的信息，与 request 对象相对应的对象是 response 对象。可以用 response 对象对用户的请求做出动态响应，向用户端发送数据。例如，当一个用户请求访问一个 JSP 页面时，该页面用 page 指令设置页面的 contentType 属性的值是 text/html，那么 JSP 引擎将按照这种属性值响应用户对页面的请求，将页面的静态部分返回给用户。如果想动态地改变 contentType 的属性值就需要用 response 对象改变页面的这个属性的值，做出动态的响应。

4.2.1 动态响应 contentType 属性

当用户请求访问一个 JSP 页面时，如果该页面用 page 指令设置页面的 contentType 属性的值是 text/html，那么 JSP 引擎将按着这种属性值做出响应，将页面的静态部分返回给用户。由于 page 指令只能为 contentType 属性指定一个值来决定响应的 MIME 类型，如果想动态地改变这个属性的值来响应用户，就需要使用 response 对象的 setContentType (String s)方法来改变 contentType 的属性值。

```
public void setContentType(String s);
```

方法动态设置响应的 MIME 类型，参数 s 可取：text/html、text/plain、image/gif、image/x-xbitmap、image/jpeg、image/pjpeg、application/x-shockwave-flash、application/vnd.ms-powerpoint、application/vnd.ms-excel、application/msword。

当用 setContentType(String s)方法动态改变了 contentType 属性的值，即响应的 MIME 类型，JSP 引擎就会按着新的 MIME 类型将 JSP 页面的输出结果返回给用户。

在例 4-10 中，当用户单击按钮，选择将当前页面保存为一个 Word 文档时，JSP 页面动态地改变 contentType 属性的值为 application/msword。这时，用户的浏览器会提示用户用 MS-Word 程序来显示或保存当前页面。

例 4-10

example4_10.jsp

```
<%@ page contentType = "text/html; charset = GB2312" %>
```

```
<HTML>
<BODY bgcolor=cyan><FONT size=1>
 <P>我正在学习 response 对象的
 <BR>setContentType 方法
 <P>将当前页面保存为 word 文档吗?
  <FORM action="" method="get" name=form>
     <INPUT TYPE="submit" value="yes" name="submit">
   </FORM>
 <%   String str=request.getParameter("submit");
      if(str==null)
         str="";
      if(str.equals("yes"))
         response.setContentType("application/msword; charset=GB2312");
  %>
</FONT></BODY></HTML>
```

在例 4-11 中,用 response 对象将 contentType 属性的值设为 image/jpeg,使得用户可以看到 Java 程序片所绘制的图形。

例 4-11

example4_11.jsp(效果如图 4-9 所示)

```
<%@ page contentType="text/html; charset=GB2312" %>
<%@ page import="java.awt.*" %>
<%@ page import="java.io.*" %>
<%@ page import="java.awt.image.*" %>
<%@ page import="java.awt.geom.*" %>
<%@ page import="com.sun.image.codec.jpeg.*" %>
<HTML><BODY><BR>观看旋转的椭圆
  <FORM action="" method="post" name=form>
     <INPUT TYPE="submit" value="观看" name="submit">
  </FORM>
<% String str=request.getParameter("submit");
    if(str!=null){
      response.setContentType("image/jpeg");                //改变 MIME 类型
      int width=260,height=260;
      BufferedImage image=new BufferedImage(width,height,BufferedImage.TYPE_INT_RGB);
      Graphics g = image.getGraphics();
      g.setColor(Color.white);
      g.fillRect(0,0,width,height);
      Graphics2D g_2d=(Graphics2D)g;
      Ellipse2D ellipse=new Ellipse2D.Double (20,50,120,50);
      g_2d.setColor(Color.blue);
      AffineTransform trans=new AffineTransform();
      for(int i=1; i<=24; i++){
        trans.rotate(15.0*Math.PI/180,75,75);
        g_2d.setTransform(trans);
        g_2d.draw(ellipse);
      }
      g.dispose();
      OutputStream outClient = response.getOutputStream();      //获取指向用户端的输出流
```

```
        JPEGImageEncoder encoder = JPEGCodec.createJPEGEncoder(outClient);
        encoder.encode(image);
    }
%>
</BODY></HTML>
```

地址(D) http://127.0.0.1:8080/ch4/example4_11.jsp

观看旋转的椭圆

观看

(a) 单击按钮之前的效果

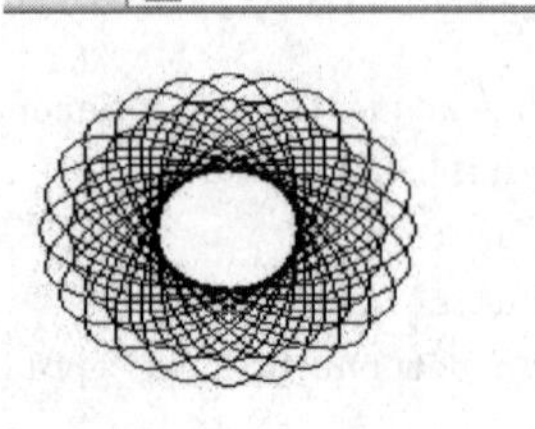

(b) 单击按钮之后的效果

图 4-9 单击按钮的效果

4.2.2 response 的 HTTP 文件头

当用户访问一个页面时,会提交一个 HTTP 头给服务器。这个请求包括一个请求行、HTTP 头和信息体,如下例:

```
post/example.jsp/HTTP.1.1
host: localhost: 8080
accept - encoding: gzip,deflate
```

第 2、3 行分别是两个头,host、accept-encoding 是头名字,而 localhost: 8080 以及 gzip,deflate 分别是它们的值。这里规定了 host 的值是 example.jsp 的地址。上面的请求有两个头: host 和 accept-encoding。一个典型的请求通常包含很多的头,有些头是标准的,有些头和特定的浏览器有关。

同样,响应也包括一些头。response 对象可以使用方法:

```
addHeader(String head,String value);
```

或使用方法:

```
setHeader(String head,String value)
```

动态添加新的响应头和头的值,并将这些头发送给用户的浏览器。如果添加的头已经存在,则先前的头被覆盖。

例 4-12 中,response 对象添加一个响应头 refresh,其头值是 5,那么用户收到这个头之后,5 秒钟后将再次刷新该页面,导致该网页每 5 秒刷新一次。

例 4-12

example4_12.jsp

```
<%@ page contentType = "text/html; charset = GB2312" %>
<%@ page import = "java.util.*" %>
<HTML><BODY bgcolor = cyan><FONT size = 4>
<P>现在的时间是: <BR>
```

```
<%  out.println("" + new Date());
      response.setHeader("Refresh","5");
  %>
</FONT></BODY></HTML>
```

4.2.3 response 重定向

在某些情况下，JSP 引擎在响应用户时，可能需要将用户重新引导至另一个页面。例如，如果用户输入的表单信息不完整，就会再次被引导到该表单的输入页面。

可以使用 response 对象的 sendRedirect(URL url)方法实现用户的重定向。

在例 4-13 中，用户在 example4_13.jsp 页面填写表单提交给 verify.jsp 页面，如果填写的表单不完整就会被重新定向到 example4_13.jsp 页面。

例 4-13

example4_13.jsp

```
<%@ page contentType = "text/html; charset = GB2312" %>
<HTML><BODY>
<P>填写姓名：<BR>
   <FORM action = "verify.jsp" method = "get" name = form>
     <INPUT TYPE = "text" name = "boy">
     <INPUT TYPE = "submit" value = "Enter">
   </FORM>
</BODY></HTML>
```

verify.jsp

```
<%@ page contentType = "text/html; charset = GB2312" %>
<HTML><BODY>
    <%  String str = null;
          str = request.getParameter("boy");
          if(str == null)
             str = "";
          byte b[] = str.getBytes("ISO-8859-1");
          str = new String(b);
          if(str.equals(""))
             response.sendRedirect("example4_13.jsp");
          else{
             out.print("欢迎你来到本网页!");
             out.print(str);
          }
    %>
</BODY></HTML>
```

4.2.4 response 的状态行

当 JSP 引擎对用户请求进行响应时，它发送的首行称作状态行。

状态行包括 3 位数字的状态代码和对状态代码的描述(称作原因短语)。下面列出了对 5 类状态代码的大概描述：

1yy(1 开头的 3 位数)：主要是实验性质的。

2yy(2 开头的 3 位数)：用来表明请求成功的。例如，状态代码 200 可以表明已成功取得了请求的页面。

3yy(3 开头的 3 位数)：用来表明在请求满足之前应采取进一步的行动。

4yy(4 开头的 3 位数)：当浏览器无法满足请求时，返回该状态代码。例如 404 表示请求的页面不存在。

5yy(5 开头的 3 位数)：用来表示服务器出现问题。例如，500 说明服务器内部发生错误。

一般不需要修改状态行，在出现问题时，JSP 引擎会自动响应，发送相应的状态代码。也可以使用 response 对象的 setStatus(int n)方法来增加状态行的内容。

在例 4-14 中，使用 setStatus(int n)设置响应的状态行来防止本网站的 JSP 页面被"盗链"。网站就是 JSP 引擎管理的一个 Web 服务目录，"盗链"就是指一个网站在其 JSP 页面中提供其他网站上资源的超链接。

在例 4-14 中，other.jsp 存放在 Web 服务目录 ch3 中(一个网站)；sameWebOne.jsp 和 sameWebTwo.jsp 存放在同一 Web 服务目录 ch4 中(一个网站)。other.jsp 和 sameWebOne.jsp 都提供了关于 sameWebTwo.jsp 的超链接。sameWebTwo.jsp 允许 sameWebOne.jsp 通过其中的超链接访问它，但不允许 other.jsp 通过其中的超链接访问它(因为 other.jsp 和 sameWebOne.jsp 不在同一个网站中)。

sameWebTwo.jsp 页面怎样防止自己被"盗链"呢？sameWebTwo.jsp 可以使用 request 对象调用方法：

```
String mess = request.getHeader("referer");
```

获取请求当前页面的其他网站上的 JSP 页面的 URL 表示。例如，如果请求当前 sameWebTwo.jsp 的页面是 other.jsp 上的超链接，那么 mess 就是：

```
http://127.0.0.1:8080/ch3/other.jsp
```

如果请求当前 sameWebTwo.jsp 的页面是 sameWebOne.jsp 上的超链接，那么 mess 就是：

```
http://127.0.0.1:8080/ch4/sameWebOne.jsp
```

这样一来，一个 JSP 页面就可以通过 getHeader 方法返回的信息决定是否允许其他页面通过超链接访问当前页面，如果不允许访问当前 JSP 页面，当前 JSP 页面就使用 response 对象的 setStatus(int n)方法增加 404 状态行，告知请求不成功，其他页面就会收到当前页面不可访问的信息。如果允许访问当前 JSP 页面，就不必修改状态行，使用默认的状态行 200(请求成功)即可。

图 4-10(a)是单击例 4-14 中 other.jsp 页面上的超链接的效果，图 4-10(b)是单击例 4-14 中 sameWebOne.jsp 页面上的超链接的效果。

例 4-14

other.jsp

```
<%@ page contentType = "text/html; charset = GB2312" %>
<HTML><BODY bgcolor = yellow><FONT size = 2>
```

地址(D) http://127.0.0.1:8080/ch4/sameWebTwo.jsp

找不到网页

你要查看的网页可能已被删除、名称已被更改，或者暂时不可用。

(a) 不允许other.jsp盗链sameWebTwo.jsp

地址(D) http://127.0.0.1:8080/ch4/sameWebTwo.jsp

欢迎访问本页面！
你来自:http://127.0.0.1:8080/ch4/sameWebOne.jsp

(b) 允许sameWebOne超链接到sameWebTwo.jsp

图 4-10　两个超链接的效果图

```
<P>我是 ch3 服务目录中的页面
<P>单击下面的超链接：<BR>
   <A HREF = "http://127.0.0.1:8080/ch4/sameWebTwo.jsp">访问 sameWebTwo.jsp
</FONT></BODY></HTML>
</HTML>
```

sameWebOne.jsp

```
<%@ page contentType = "text/html; charset = GB2312" %>
<HTML><BODY bgcolor = cyan><FONT size = 2>
<P>我是 ch4 服务目录中的页面
<br>单击下面的超链接：<BR>
   <A HREF = "http://127.0.0.1:8080/ch4/sameWebTwo.jsp">访问 sameWebTwo.jsp
</FONT></BODY></HTML>
```

sameWebTwo.jsp

```
<%@ page contentType = "text/html; charset = GB2312" %>
<HTML><BODY bgcolor = cyan><FONT size = 2>
<% String mess = request.getHeader("referer");
     if(mess == null){
        mess = "";
        response.setStatus(404);
     }
     if(!(mess.startsWith("http://127.0.0.1:8080/ch4"))) //不允许网站 ch4 以外的
              response.setStatus(404);                   //其他网站盗链本页面
%>
<P> 欢迎访问本页面！<BR>
     你来自：<% = mess %>
</FONT></BODY></HTML>
```

表 4-1 是状态代码表。

表 4-1　状态代码表

状 态 代 码	代 码 说 明
101	服务器正在升级协议
100	用户可以继续
201	请求成功且在服务器上创建了新的资源
202	请求已被接受但还没有处理完毕
200	请求成功
203	用户端给出的信息不是发自服务器的
204	请求成功，但没有新信息

续表

状态代码	代码说明
205	用户必须重置文档视图
206	服务器执行了部分 get 请求
300	请求的资源有多种表示法
301	资源已经被永久移动到新位置
302	资源已经被临时移动到新位置
303	应答可以在另外一个 URL 中找到
304	get 方式请求不可用
305	请求必须通过代理来访问
400	请求有语法错误
401	请求需要 HTTP 认证
403	取得了请求但拒绝服务
404	请求的资源不可用
405	请求所用的方法是不允许的
406	请求的资源只能用请求不能接受的内容特性来响应
407	用户必须得到认证
408	请求超时
409	发生冲突,请求不能完成
410	请求的资源已经不可用
411	请求需要一个定义的内容长度才能处理
413	请求太大,被拒绝
414	请求的 URL 太大
415	请求的格式被拒绝
500	服务器发生内部错误,不能服务
501	不支持请求的部分功能
502	从代理和网关接受了不合法的字符
503	HTTP 服务暂时不可用
504	服务器在等待代理服务器应答时发生超时
505	不支持请求的 HTTP 版本

4.3 session 对象

HTTP 是一种无状态协议。一个用户向服务器发出请求(request),然后服务器返回响应(response),链接就被关闭,在服务器端不保留链接的有关信息,因此当下一次链接时,服务器已没有以前的链接信息了,无法判断这一次链接和以前的链接是否属于同一用户。因此,必须使用会话记录有关链接的信息。

从一个用户打开浏览器链接到服务器的某个 Web 服务目录,到用户关闭浏览器离开服务器称作一个会话。当一个用户访问一个服务器时,可能会在某个 Web 服务目录中反复链接几个页面,反复刷新一个页面或不断地向一个页面提交信息等,服务器应当通过某种办法知道这是同一个用户,这就需要 session(会话)对象。

4.3.1 session 对象的 Id

当一个用户首次访问 Web 服务目录中的某个 JSP 页面时，JSP 引擎为其产生一个 session 对象。为用户产生的 session 对象调用相应的方法可以存储用户在访问各个页面期间提交的各种信息，比如姓名、号码等信息。为用户产生的 session 对象中含有一个 String 类型的 Id 号，JSP 引擎同时将这个 Id 号发送到用户端，存放在用户的浏览器的 Cookie 中。这样一来，session 对象和用户之间就建立起一一对应的关系，即每个用户都对应着一个 session 对象（该用户的会话），不同用户的 session 对象互不相同，即具有不同的 Id 号码。通过介绍已经知道，JSP 引擎为每个用户启动一个线程，也就是说，JSP 引擎为每个线程分配不同的 session 对象。当用户再访问链接该 Web 服务目录的其他页面时，或从该 Web 服务目录连接到其他 Web 服务器再回到该 Web 服务目录时，JSP 引擎不再分配给用户新的 session 对象，而是使用完全相同的一个，直到服务器关闭或这个 session 对象达到了最大生存时间，服务器端将销毁该用户的 session 对象，即和用户的会话对应关系消失，用户再访问该服务目录时，服务器为该用户再创建一个新的 session 对象。

注意：同一个用户在不同的服务目录（即不同网站）中的 session 是互不相同的。

在下面的例 4-15 中，用户在服务器的 Web 服务目录 ch4 中的 3 个页面 first. jsp、second 和 third. jsp 之间进行链接，3 个页面的 session 对象是完全相同的。其中，first. jsp 存放在 ch4 中，second. jsp 存放在 ch4 的子目录 two 中，third. jsp 存放在 ch4 的子目录 three 中。用户首先访问 first. jsp 页面，从这个页面再链接到 second. jsp 页面，然后从 second. jsp 再链接到 third. jsp 页面。first. jsp、second 和 third. jsp 的效果如图 4-11(a)，4-11(b)，4-11(c)所示。

(a) 用户访问first.jsp时session的ID

(b) 用户访问second.jsp时session的ID

(c) 用户访问third.jsp时session的ID

图 4-11　3 个 session 页面效果

例 4-15

first. jsp

```
<%@ page contentType="text/html;charset=GB2312" %>
<HTML><BODY bgcolor=yellow>
   我是 first.jsp 页面,输入你的姓名链接到 second.jsp
   <% String id=session.getId();
       out.println("<br>你的 session 对象的 ID 是:<br>"+id);
   %>
  <FORM action="two/second.jsp" method=post name=form>
        <INPUT type="text" name="boy">
        <INPUT TYPE="submit" value="送出" name=submit>
  </FORM>
</BODY></HTML>
```

second. jsp

```
<%@ page contentType="text/html;charset=GB2312" %>
<HTML><BODY>
   我是 second.jsp 页面
   <% String id=session.getId();
       out.println("你的 session 对象的 ID 是:<br>"+id);
   %>
<BR>单击超链接,链接到 third.jsp 的页面。
<BR><A HREF="/ch4/three/third.jsp">欢迎去 third.jsp 页面!</A>
</BODY></HTML>
```

third. jsp

```
<%@ page contentType="text/html;charset=GB2312" %>
<HTML><BODY bgcolor=cyan>
   我是 third.jsp 页面
   <% String id=session.getId();
       out.println("你的 session 对象的 ID 是:<br>"+id);
   %>
<BR>单击超链接,链接到 first.jsp 的页面。
<BR><A HREF="/ch4/first.jsp">欢迎去 first.jsp!</A>
</BODY></HTML>
```

4.3.2 session 对象与 URL 重写

session 对象能否和用户建立起一一对应关系依赖于用户的浏览器是否支持 Cookie。如果用户端不支持 Cookie,那么用户在不同网页之间的 session 对象可能是互不相同的,因为如果服务器无法将 ID 存放到用户端,就不能建立 session 对象和用户的一一对应关系。将浏览器的 Cookie 设置为禁止后(在浏览器的菜单上选择"工具"→"Internet 选项"→"隐私"命令,将第三方 Cookie 设置成禁止),重新打开浏览器运行例 4-15 会得到不同的结果。也就是说,同一用户对应了多个 session 对象,这样服务器就无法知道在这些页面上访问的用户是否同一个用户。

如果用户的浏览器不支持 Cookie,可以通过 URL 重写来实现 session 对象的唯一性。

URL 重写，就是当用户从一个页面重新链接到另一个页面时，通过向这个新的 URL 添加参数，把 session 对象的 ID 传带过去，这样就可以保障用户在该网站各个页面中的 session 对象是完全相同的。可以使用 response 对象调用 encodeURL()或 encodeRedirectURL()方法实现 URL 重写。例如，如果从 tom.jsp 页面链接到 jerry.jsp 页面，并准备把 session 对象的 ID 传到 jerry.jsp 页面，那么 tom.jsp 页面首先实现对目标页面的 URL 重写。

```
String str = response.encodeRedirectURL("jerry.jsp");
```

然后将链接目标写成<%=str%>。

在例 4-16 中，jiafei.jsp、tom.jsp 和 jerry.jsp 之间实行 URL 重写，jiafei.jsp 存放在 ch4 中，tom.jsp 存放在 ch4 的子目录 two 中，jerry.jsp 存放在 ch4 的子目录 three 中。

例 4-16

jiafei.jsp

```
<%@ page contentType = "text/html; charset = GB2312" %>
<HTML><BODY bgcolor = cyan>
<%  String str = response.encodeURL("two/tom.jsp");
%>
     我是 jiafei.jsp 页面，输入你的姓名链接到 tom.jsp
     <%  String id = session.getId();
          out.println("<br>你的 session 对象的 ID 是：<br>" + id);
     %>
   <FORM action = "<% = str %>" method = post name = form>
        <INPUT type = "text" name = "boy">
        <INPUT TYPE = "submit" value = "送出" name = submit>
   </FORM>
</BODY></HTML>
```

tom.jsp

```
<%@ page contentType = "text/html; charset = GB2312" %>
<%  String str = response.encodeURL("/ch4/three/jerry.jsp");
%>
<HTML><BODY bgcolor = cyan>
   我是 tom.jsp 页面
   <%  String id = session.getId();
         out.println("你的 session 对象的 ID 是：<br>" + id);
   %>
<BR>单击超链接，链接到 jerry.jsp 的页面。
<BR><A HREF = "<% = str %>">欢迎去 jerry.jsp 页面！</A>
</BODY></HTML>
```

jerry.jsp

```
<%@ page contentType = "text/html; charset = GB2312" %>
<%  String str = response.encodeURL("/ch4/jiafei.jsp");
%>
<HTML><BODY bgcolor = cyan>
   我是 jerry.jsp 页面
   <%  String id = session.getId();
```

```
        out.println("你的 session 对象的 ID 是：<br>" + id);
    %>
<BR>单击超链接,链接到 jiafei.jsp 的页面。
<BR><A HREF = "<% = str %>">欢迎去 jiafei.jsp! </A>
</BODY></HTML>
```


4.3.3 session 对象存储数据

session 对象驻留在服务器端,该对象调用某些方法保存用户在访问某个 web 服务目录期间的有关数据。session 对象使用下列方法处理数据：

```
public void setAttribute(String key,Object obj)
```

session 对象类似于散列表,可以调用该方法将参数 Object 指定的对象 obj 添加到 session 对象中,并为添加的对象指定了一个索引关键字。如果添加的两个对象的关键字相同,则先前添加的对象被清除。

```
public Object getAttibute(String key)
```

获取 session 对象中含有的关键字是 key 的对象。由于任何对象都可以添加到 session 对象中,因此用该方法取回对象时,应强制转化为原来的类型。

```
public Enumeration getAttributeName()
```

session 对象调用该方法产生一个枚举对象,该枚举对象使用 nextElemets()遍历 session 中的各个对象所对应的关键字。

```
public long getCreationTime()
```

session 对象调用该方法可以获取该对象创建的时间,单位是毫秒(从 1970 年 1 月 1 日午夜(格林尼治时间)起至该对象创建时刻所走过的毫秒数)。

```
public void removeAttribute(String key)
```

session 对象调用该方法从当前 session 对象中删除关键字是 key 的对象。

例 4-17 中涉及 3 个页面：example4_17.jsp、shop.jsp 和 account.jsp,这里使用 session 对象存储顾客的姓名和购买的商品。example4_17.jsp、shop.jsp 和 account.jsp 的效果如图 4-12 所示。

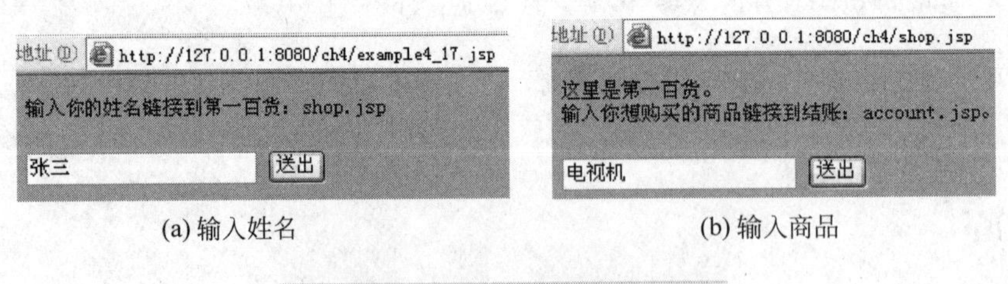

(a) 输入姓名　　(b) 输入商品

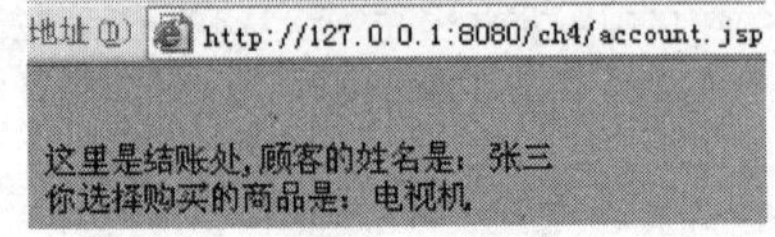

(c) 得到session中的数据

图 4-12　3 个页面的效果图

例 4-17

example4_17.jsp

```
<%@ page contentType="text/html;charset=GB2312" %>
<HTML><BODY bgcolor=cyan><FONT Size=2>
  <% session.setAttribute("customer","顾客");
  %>
 <P>输入你的姓名链接到第一百货:shop.jsp
   <FORM action="shop.jsp" method=post name=form>
       <INPUT type="text" name="boy">
       <INPUT TYPE="submit" value="送出" name=submit>
  </FORM>
</FONT></BODY></HTML>
```

shop.jsp

```
<%@ page contentType="text/html;charset=GB2312" %>
<HTML><BODY bgcolor=cyan><FONT Size=2>
   <%   String s=request.getParameter("boy");
         session.setAttribute("name",s);
   %>
  这里是第一百货。<br>输入你想购买的商品链接到结账:account.jsp。
  <FORM action="account.jsp" method=post name=form>
    <INPUT type="text" name="buy">
    <INPUT TYPE="submit" value="送出" name=submit>
  </FORM>
</FONT></BODY></HTML>
```

account.jsp

```
<%@ page contentType="text/html;charset=GB2312" %>
<%! //处理字符串的方法
   public String getString(String s){
     if(s==null)
        s="";
     try{ byte b[]=s.getBytes("ISO-8859-1");
          s=new String(b);
        }
     catch(Exception e){}
     return s;
   }
%>
<HTML><BODY bgcolor=cyan><FONT Size=2>
  <%   String s=request.getParameter("buy");
        session.setAttribute("goods",s);
  %>
<BR>
<% String 顾客=(String)session.getAttribute("customer");
   String 姓名=(String)session.getAttribute("name");
   String 商品=(String)session.getAttribute("goods");
   姓名=getString(姓名);
   商品=getString(商品);
```

```
%>
      这里是结账处,<% = 顾客 %>的姓名是:<% = 姓名 %>
<br>你选择购买的商品是:<% = 商品 %>
</FONT></BODY></HTML>
```


4.3.4 在Tag文件中使用session对象

JSP页面通过调用Tag文件可以实现代码的复用,那么在调用Tag文件时,Tag文件就可以对session对象中存储的数据进行处理,并将必要的结果返回给JSP页面。

例4-18是个猜数字的小游戏。当用户访问服务器上的example4_18.jsp页面时,服务器随机分配给用户一个1至100之间的整数,然后将这个整数存在用户的session对象中。用户在example4_18.jsp中单击超链接请求guess.jsp页面。guess.jsp通过调用一个Tag文件GuessTag.tag判断用户的猜测是否正确。Tag文件GuessTag.tag负责判断用户的猜测是否和session对象中存放的那个整数相同,如果猜测大于session对象中存放的那个整数,Tag文件就返回"你猜大了";如果猜测小于session对象中存放的那个整数,Tag文件就返回"你猜小了";如果猜测等于session对象中存放的那个整数,Tag文件就返回"你猜对了"。用户刷新或重新返回example4_18.jsp页面可获得一个新的随机数。example4_18.jsp页面和guess.jsp页面的效果如图4-13所示。

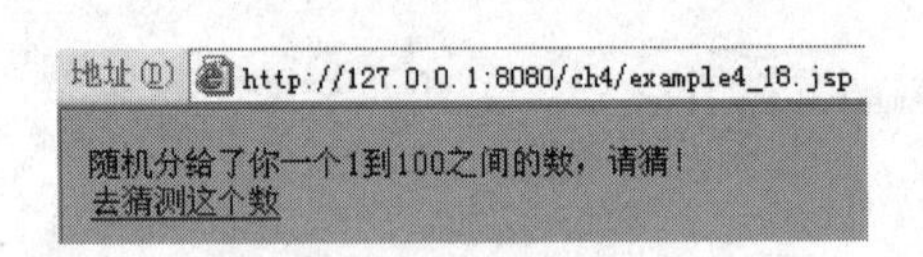

(a) 得到随机数

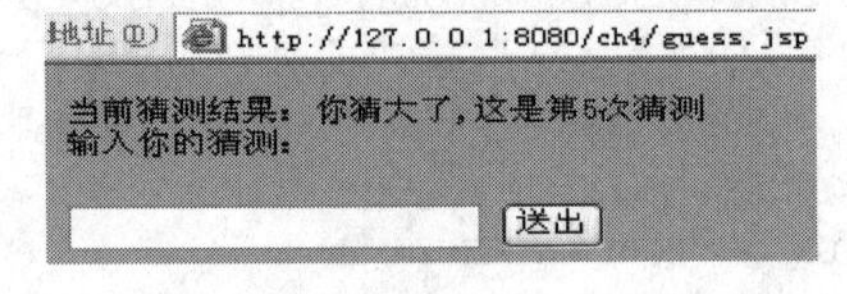

(b) 猜数

图4-13 两个页面的效果图

例4-18

example4_18.jsp

```
<%@ page contentType = "text/html; charset = GB2312" %>
<HTML><BODY bgcolor = cyan><FONT Size = 2>
  随机分给了你一个1到100之间的数,请猜!
  <%   int num = (int)(Math.random() * 100) + 1;
        session.setAttribute("count",new Integer(0));
        session.setAttribute("save",new Integer(num));
  %>
 <BR><A HREF = "guess.jsp">去猜测这个数</A>
</FONT></BODY></HTML>
```

guess.jsp

```
<%@ page contentType = "text/html; charset = GB2312" %>
<%@ taglib tagdir = "/WEB-INF/tags" prefix = "guessNumber" %>
<HTML><BODY bgcolor = cyan><FONT Size = 2>
 <%   String str = request.getParameter("guessNumber");
        if(str == null)
          str = " * ";
```

```
        if(str.length() == 0)
          str = " * ";
  %>
 < guessNumber：GuessTag number = "<% = str %>" />
 当前猜测结果：<% = message %>          <%-- message 是 Tag 文件返回的对象 --%>
 <% if(message.startsWith("你猜对了")){
  %>   < br >< A HREF = "example4_18.jsp">重新获得随机数</A >
 <% }
     else{
  %> < BR >输入你的猜测：
       < FORM action = "" method = "post" name = form >
          < INPUT type = "text" name = "guessNumber" >
          < INPUT TYPE = "submit" value = "送出" name = "submit">
      </FORM >
<% }
 %>
</FONT ></BODY ></HTML >
```

GuessTag. tag

```
<%@ tag pageEncoding = "GB2312" %>
<%@ attribute name = "number" required = "true" %>
<%@ variable name - given = "message" scope = "AT_END" %>
  <%   String mess = "";
        Integer integer = (Integer)session.getAttribute("save");
        int realnumber = integer.intValue();
        int guessNumber = 0;
        boolean boo = true;
        try{  guessNumber = Integer.parseInt(number);
        }
        catch(Exception exp){
              boo = false;
        }
        if(boo){
          if(guessNumber == realnumber){
             int n = ((Integer)session.getAttribute("count")).intValue();
             n = n + 1;
             session.setAttribute("count",new Integer(n));
             mess = "你猜对了,这是第" + n + "次猜测";
          }
          else if(guessNumber > realnumber){
             int n = ((Integer)session.getAttribute("count")).intValue();
             n = n + 1;
             session.setAttribute("count",new Integer(n));
             mess = "你猜大了,这是第" + n + "次猜测";
          }
          else if(guessNumber < realnumber){
             int n = ((Integer)session.getAttribute("count")).intValue();
             n = n + 1;
             session.setAttribute("count",new Integer(n));
             mess = "你猜小了,这是第" + n + "次猜测";
```

```
            }
            else if(number.equals("你还没开始猜测")) {
               mess = "你还没开始猜测";
            }
            jspContext.setAttribute("message",mess); //将 message 对象返回给 JSP 页面
         }
         else{
            jspContext.setAttribute("message","请输入你的猜测");
         }
%>
```

4.3.5 session 对象的生存期限

一个用户在某个 Web 服务目录的 session 对象的生存期限依赖于 session 对象是否调用 invalidate()方法使得 session 无效,依赖于 session 对象达到了设置的最长的"发呆"状态时间或服务器是否被关闭。Tomcat 服务器允许用户最长的"发呆"状态时间为 30 分钟,所谓"发呆"状态时间是指用户对某个 Web 服务目录发出的两次请求之间的间隔时间。比如,用户对某个 Web 服务目录下的 JSP 页面发出请求,并得到响应,如果用户不再对该 web 服务目录发出请求(可能去请求其他的 Web 服务目录),那么用户对该 Web 服务目录进入"发呆"状态,直到用户再次请求该 Web 服务目录时,"发呆"状态结束。

可以修改 Tomcat 服务器安装目录 conf 目录下的 web.xml,重新设置各个 Web 服务目录下的 session 对象的最长"发呆"时间。打开 Tomcat 安装目录 conf 目录下的配置文件 web.xml,找到

```
<session-config>
    <session-timeout>30</session-timeout>
</session-config>
```

将其中的 30 修改成所要求的值即可(单位为分钟),如果时间修改为负数,发呆时间不受限制。

session 对象可以使用下列方法获取或设置和生存时间有关的信息:

- public long getCreationTime() 获取 session 创建的时间,单位是毫秒(GMT 时间,1970 年 1 月 1 日午夜起至 session 创建时刻所走过的毫秒数)。
- public long getLastAccessedTime() 获取 session 最后一次被操作的时间,单位是毫秒。
- public int getMaxInactiveInterval() 获取 session 最长的"发呆"时间(单位是秒)。
- public void setMaxInactiveInterval(int interval) 设置 session 最长的"发呆"时间(单位是秒)。
- public boolean isNew() 判断 session 是否是一个新建的对象。
- invalidate() 使 session 无效。

在下面的例 4-19 中,session 对象使用 setMaxInactiveInterval(int interval)方法设置最长的"发呆"状态时间为 10 秒。用户可以通过刷新页面检查是否达到了最长的"发呆"时间,如果两次刷新之间的间隔超过 10 秒,用户先前的 session 将被取消,用户将获得一个新的 session 对象。

例 4-19

example4_19.jsp

```
<%@ page contentType="text/html;charset=GB2312" %>
<%@ page import="java.util.*" %>
<HTML><BODY bgcolor=yellow><FONT Size=3>
<% session.setMaxInactiveInterval(10);
     boolean boo=session.isNew();
     out.println("<br>如果你第一次访问当前 Web 服务目录,你的会话是新的");
     out.println("<br>如果你不是首次访问当前 Web 服务目录,你的会话不是新的");
     out.println("<br>会话是新的吗?:"+boo);
     out.println("<br>欢迎来到本页面,你的 session 允许的最长发呆时间为"+
                         session.getMaxInactiveInterval()+"秒");
     out.println("<br>你的 session 的创建时间是"+new Date(session.getCreationTime()));
     out.println("<br>你的 session 的 Id 是"+session.getId());
     Long lastTime=(Long)session.getAttribute("lastTime");
     if(lastTime==null){
         long n=session.getLastAccessedTime();
         session.setAttribute("lastTime",new Long(n));
     }
     else{
         long m=session.getLastAccessedTime();
         long n=((Long)session.getAttribute("lastTime")).longValue();
         out.println("<br>你的发呆时间大约是"+(m-n)+"毫秒,大约"+(m-n)/1000+"
秒");
         session.setAttribute("lastTime",new Long(m));
     }
%>
<FONT></BODY></HTML>
```

4.3.6 使用 session 设置时间间隔

当用户访问一个 JSP 页面时,该页面可能接收用户的数据,对数据进行处理后再对用户的请求做出响应,但在某些情况下,需要限制用户不断地使用该页面所提供的功能服务,因为在某些 Web 应用中,某个 JSP 页面在某个时刻可能有大量的用户要使用该页面提供的功能,如果一个用户在间隔很短的时间内不断地使用该页面所提供的某种功能,将会影响其他用户使用该页面所提供的功能。例如,用户可能提交一个考号,服务器根据该考号查询数据库中的有关成绩返回给该用户,如果用户不断地提交自己的考号,已经没有任何意义,因此应当限制该用户等待若干"时间"后,再次使用该页面所提供的功能。

example4_20.jsp 提供计算字符串长度的功能,用户可以通过表单提交一个字符串给当前页面。example4_20.jsp 利用用户的 session 来设置用户使用当前页面所提供功能的间隔时间(间隔时间是 5 秒)。

例 4-20

example4_20.jsp

```
<%@ page contentType="text/html; charset=GB2312" %>
<%@ page import="java.util.*" %>
```

```
<HTML><BODY>
<br>反复使用该页面提供的计算字符串长度功能的间隔时间必须大于5秒。<br>
<br>输入一个字符串,页面将计算它的长度:<br>
   <FORM action="" method="get" name=form>
     <INPUT TYPE="text" name="boy">
     <INPUT TYPE="submit" value="Enter">
   </FORM>
   <% int time=5;
        String str=null;
        str=request.getParameter("boy");
        if(str==null) str="";
        Date date=(Date)session.getAttribute("date");
        if(date==null){
            date=new Date();
            session.setAttribute("date",date);
        }
        date=(Date)session.getAttribute("date");
        Calendar calendar=Calendar.getInstance();
        calendar.setTime(date);
        long timeInSession=calendar.getTimeInMillis();
        long currentTime=0;
        if(session.isNew()==false)
          currentTime=System.currentTimeMillis();
        session.setAttribute("date",new Date(currentTime));
        long intervalTime=(currentTime-timeInSession)/1000;
        if(intervalTime<=time&&session.isNew()==false){
           out.println("请"+time+"秒后再访问本页");
        }
        else{
           out.println("你输入的字符串的长度:"+str.length());
        }
     %>
</BODY></HTML>
```

4.3.7 计数器

在第 2 章讲述过一个计数器的例子(例 2-2),但是那个例子并不能限制用户通过不断地刷新页面来增加计数器的计数。计数器就是记录某个 Web 服务目录(通常所说的网站)被访问的次数,但需要限制用户通过不断地刷新页面或再次访问该 Web 服务目录的其他的页面来增加计数器的计数。

使用 session 实现计数器的步骤是:当一个用户请求 Web 服务目录下的任何一个 JSP 页面时,首先检查该用户的 session 对象中是否已经有计数,如果没有计数,立刻将当前的计数增 1,并将计数存到用户的 session 中,否则不改变当前的计数。

在例 4-21 中,Web 服务目录 ch4 有两个 JSP 页面 one.jsp、two.jsp 和一个 tag 文件 Count.tag。Count.tag 文件负责计数。one.jsp、two.jsp 使用 Count.tag 实现计数。用户首次请求 one.jsp 和 two.jsp 的任何一个,都会使得网站的计数增 1。

例 4-21

one.jsp

```
<%@ page contentType="text/html; charset=GB2312" %>
<%@ taglib prefix="person" tagdir="/WEB-INF/tags" %>
<HTML><BODY size=3>
<P>欢迎访问本站
   <person:Count/>
   <A href="two.jsp">欢迎去 two.jsp 参观</A>
</BODY></HTML>
```

two.jsp

```
<%@ page contentType="text/html; charset=GB2312" %>
<%@ taglib prefix="person" tagdir="/WEB-INF/tags" %>
<HTML><BODY bgcolor=yellow size=3>
<P>欢迎访问本站
   <person:Count/>
   <A href="one.jsp">欢迎去 one.jsp 参观</A>
</BODY></HTML>
```

Count.tag

```
<%@ tag import="java.io.*" %>
<FONT Size=4>
<%! int number=0;
       File file=new File("count.txt");
       synchronized void countPeople(){ //计算访问次数的同步方法
           if(!file.exists()){
              number++;
              try { file.createNewFile();
                    FileOutputStream out=new FileOutputStream("count.txt");
                    DataOutputStream dataOut=new DataOutputStream(out);
                    dataOut.writeInt(number);
                    out.close();
                    dataOut.close();
              }
              catch(IOException ee){}
              }
           else{
              try{  FileInputStream in=new FileInputStream("count.txt");
                    DataInputStream dataIn=new DataInputStream(in);
                    number=dataIn.readInt();
                    number++;
                    in.close();
                    dataIn.close();
                    FileOutputStream out=new FileOutputStream("count.txt");
                    DataOutputStream dataOut=new DataOutputStream(out);
                    dataOut.writeInt(number);
                    out.close();
                    dataOut.close();
              }
```

```
                catch(IOException ee){}
            }
        }
  %>
<%   String str = (String)session.getAttribute("count");
       if(str == null){
          countPeople();
          String personCount = String.valueOf(number);
          session.setAttribute("count",personCount);
       }
  %>
<P><P>你是第<% = (String)session.getAttribute("count") %>
          个访问本网站的用户。
</FONT>
```

4.4 out 对象

out 对象是一个输出流,用来向用户端输出数据。在前面的许多例子里曾多次使用 out 对象进行数据的输出。out 对象可调用如下的方法用于各种数据的输出。例如:

- out.print(Boolean),out.println(Boolean):用于输出一个布尔值。
- out.print(char),out.println(char):输出一个字符。
- out.print(double),out.println(double):输出一个双精度的浮点数。
- out.print(float),out.println(float):用于输出一个单精度的浮点数。
- out.print(long),out.println(long):输出一个长整型数据。
- out.print(String),out.println(String):输出一个字符串对象的内容。
- out.newLine():输出一个换行符。
- out.flush():输出缓冲区里的内容。
- out.close():关闭流。

例 4-22 使用 out 对象向用户输出包括表格等内容的信息,example4_22.jsp 页面的效果如图 4-14 所示。

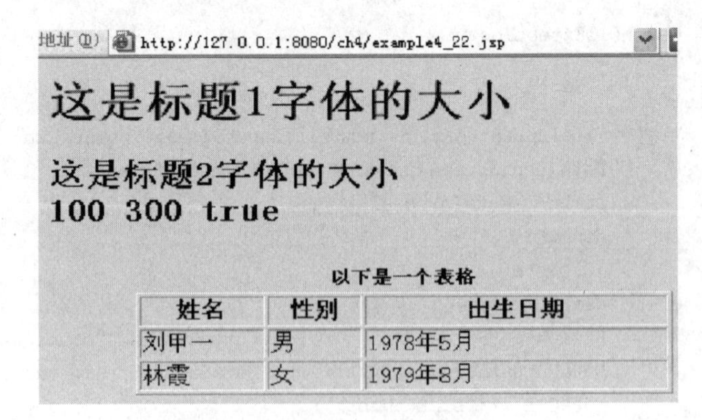

图 4-14 使用 out 流输出数据

例 4-22

example4_22.jsp

```
<%@ page contentType = "text/html; charset = GB2312" %>
<%@ page import = "java.util.*" %>
<HTML><BODY bgcolor = yellow>
  <% int a = 100; long b = 300; boolean c = true;
       out.println("<H1>这是标题 1 字体的大小</HT1>");
       out.println("<H2>这是标题 2 字体的大小</HT2>");
       out.print("<BR>");
       out.println(a); out.println(b); out.println(c);
  %>
  <Center>
  <p><FONT Size = 2>以下是一个表格</FONT>
  <% out.print("<Font face = 隶书 size = 2>");
      out.println("<Table Border>");
      out.println("<tr>");
            out.println("<th width = 80>" + "姓名" + "</th>");
            out.println("<th width = 60>" + "性别" + "</th>");
            out.println("<th width = 200>" + "出生日期" + "</th>");
      out.println("</tr>");
      out.println("<tr>");
            out.println("<td>" + "刘甲一" + "</td>");
            out.println("<td>" + "男" + "</td>");
            out.println("<td>" + "1978 年 5 月" + "</td>");
      out.println("</tr>");
      out.println("<tr>");
            out.println("<td>" + "林霞" + "</td>");
            out.println("<td>" + "女" + "</td>");
            out.println("<td>" + "1979 年 8 月" + "</td>");
     out.println("</tr>");
     out.println("</Table>");
     out.print("</FONT>") ;
  %>
  </Center>
</BODY></HTML>
```

4.5 application 对象

当一个用户第一次访问 Web 服务目录上的一个 JSP 页面时，JSP 引擎创建一个和该用户相对应的 session 对象，而且不同用户的 session 对象是互不相同的。当用户在所访问的 Web 服务目录的各个页面之间浏览时，这个 session 对象都是同一个，直到用户关闭浏览器或 session 对象达到了最长的生命期限，这个 session 对象才被取消。

与 session 对象不同的是 application 对象，Tomcat 服务器启动后，就产生了这个 application 对象。当一个用户访问 Web 服务目录上的一个 JSP 页面时，JSP 引擎为该用户分配这个 application 对象，当用户在所访问的 Web 服务目录的各个页面之间浏览时，这个 application 对象都是同一个，直到 Tomcat 服务器关闭，这个 application 对象才被取消。与

session 对象不同的是，所有用户在同一 Web 服务目录中的 application 对象是相同的一个，即每个 Web 服务目录下的 application 对象被访问该服务目录的所有的用户共享。需要注意的是，不同 Web 服务目录下的 application 是互不相同的。

4.5.1 application 对象的常用方法

1. public void setAttribute(String key,Object obj)

application 对象可以调用该方法将参数 Object 指定的对象 obj 添加到 application 对象中，并为添加的对象指定一个索引关键字。如果添加的两个对象的关键字相同，则先前添加的对象被清除。

2. public Object getAttribute(String key)

application 对象可以调用该方法获取 application 对象中含有的关键字是 key 的对象。由于任何对象都可以添加到 application 对象中，因此用该方法取回对象时，应强制转化为原来的类型。

3. public Enumeration getAttributeNames()

application 对象调用该方法产生一个枚举对象。该枚举对象使用 nextElemets()遍历 application 中的各个对象所对应的关键字。

4. public void removeAttribute(String key)

application 对象可以调用该方法从当前 application 对象中删除关键字是 key 的对象。

5. public String getServletInfo()

application 对象可以调用该方法获取 Servlet 编译器的当前版本的信息。

由于一个 Web 服务目录下的 application 对象被所有访问该 Web 服务目录的用户共享，所以任何用户对 application 对象中存储的数据的改变都会影响到其他用户。因此，在某些情况下，对 application 对象的操作需要实现同步处理。

有些服务器不直接支持使用 application 对象，必须先用 ServletContext 类声明这个对象，再使用 getServletContext()方法对这个 application 对象进行初始化。

4.5.2 用 application 制作留言板

在例 4-23 中，用户通过 submit. jsp 页面向 messagePane. jsp 页面提交姓名、留言标题和留言内容。messagePane. jsp 页面获取这些内容后，用同步方法将这些内容添加到一个向量中，然后将这个向量再添加到 application 对象中。当用户单击查看留言板时，showMessage. jsp 页面负责显示所有用户的留言内容，即从 application 对象中取出向量，然后遍历向量中存储的信息。

在这里使用了向量这种数据结构，Java 的 java. util 包中的 Vector 类负责创建一个向量对象。如果已经学会使用数组，那么很容易就会使用向量。当创建一个向量时不用像数组那样必须要给出数组的大小。向量创建后，例如，Vector＜String＞ a＝new Vector＜String＞()；a 可以使用方法 add(String str)把 String 对象 str 添加到向量的末尾，向量的大小会自动地增加；可以使用方法 add(int index,String str)把一个 String 对象 str 追加到该向量的指定位置；可以使用方法 elementAt(int index)获取指定索引处的向量的元素(索引初始位置是 0)；可以使用方法 size()获取向量所含有的元素的个数。

submit. jsp、messagePane. jsp、showMessage. jsp 的页面效果如图 4-15 所示。

地址(D) http://127.0.0.1:8080/ch4/submit.jsp

输入你的名字： 李四

输入你的留言标题： 求Java购书

输入你的留言：

本人因学习JSP需要Java的知识，哪位有Java书，本人想买，联系E-mail: we@qq.com

提交信息

查看留言板

(a) 提交留言的页面

地址(D) http://127.0.0.1:8080/ch4/messagePane.jsp

你的信息已经提交！ 返回留言板 查看留言板

(b) 保存留言的页面

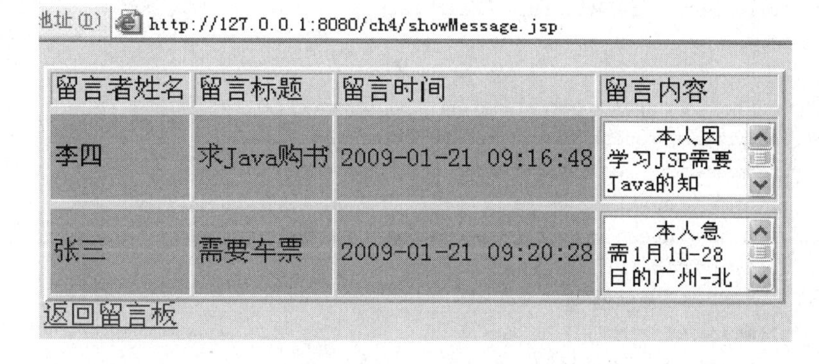

(c) 查看留言的页面

图 4-15　3 个页面的效果图

例 4-23

submit. jsp

```
<%@ page contentType = "text/html; charset = GB2312" %>
<HTML><BODY>
 <FORM action = "messagePane.jsp" method = "post" name = "form">
     <P>输入你的名字：
     <INPUT type = "text" name = "peopleName">
       <BR>
       <P>输入你的留言标题：
     <INPUT type = "text" name = "Title">
       <BR>
       <P>输入你的留言：
     <BR>
     <TEXTAREA name = "messages" ROWs = "10" COLS = 36 WRAP = "physical" >
     </TEXTAREA>
     <BR>
     <INPUT type = "submit" value = "提交信息" name = "submit">
```

```
  </FORM>
  <FORM action = "showMessage.jsp" method = "post" name = "form1">
      <INPUT type = "submit" value = "查看留言板" name = "look">
  </FORM>
</BODY></HTML>
```

messagePane.jsp

```
<%@ page contentType = "text/html; charset = GB2312" %>
<%@ page import = "java.util.*" %>
<%@ page import = "java.text.SimpleDateFormat" %>
<HTML><BODY>
   <%! Vector<String> v = new Vector<String>();
          int i = 0;
          ServletContext application;
          synchronized void sendMessage(String s){
              application = getServletContext(); ;
              v.add(s);
              application.setAttribute("Mess",v);
          }
    %>
    <% String name = request.getParameter("peopleName");
        String title = request.getParameter("Title");
        String messages = request.getParameter("messages");
            if(name == null)
                name = "guest" + (int)(Math.random() * 10000);
            if(title == null)
                title = "无标题";
            if(messages == null)
                messages = "无信息";
           SimpleDateFormat matter = new SimpleDateFormat("yyyy - MM - dd HH:mm:ss");
           String time = matter.format(new Date());
           String s = name + "#" + title + "#" + time + "#" + messages;
           sendMessage(s);
           out.print("你的信息已经提交!");
    %>
  <A HREF = "submit.jsp">返回留言板</A>
  <A HREF = "showMessage.jsp">查看留言板</A>
</BODY></HTML>
```

showMessage.jsp

```
<%@ page contentType = "text/html; charset = GB2312" %>
<%@ page import = "java.util.*" %>
<HTML><BODY bgcolor = yellow>
 <% Vector<String> v = (Vector)application.getAttribute("Mess");
        out.print("<table border = 2>");
          out.print("<tr>");
              out.print("<td bagcolor = cyan>" + "留言者姓名" + "</td>");
              out.print("<td bagcolor = cyan>" + "留言标题" + "</td>");
              out.print("<td bagcolor = cyan>" + "留言时间" + "</td>");
              out.print("<td bagcolor = cyan>" + "留言内容" + "</td>");
```

```
            for(int i = 0; i < v.size(); i++){
                out.print("<tr>");
                String message = v.elementAt(i);
                byte bb[] = message.getBytes("iso-8859-1");
                message = new String(bb);
                String a[] = message.split("#");
                out.print("<tr>");
                int number = a.length - 1;
                for(int k = 0; k <= number; k++) {
                    if(k < number)
                        out.print("<td bgcolor = cyan>" + a[k] + "</td>");
                    else
                      out.print("<td><TextArea rows = 3 cols = 12>" + a[k] + "</TextArea></td
>");
                }
                out.print("</tr>");
            }
        out.print("</table>");
      %>
  <A HREF = "submit.jsp">返回留言板</A>
</BODY></HTML>
```

4.6　实验 1：request 对象

1. 相关知识点

HTTP 通信协议是用户与服务器之间一种请求与响应(request/response)的通信协议。在 JSP 中，内置对象 request 封装了用户请求信息时所提交的信息，那么该对象调用相应的方法可以获取封装的信息，即使用该对象可以获取用户提交的信息。

2. 实验目的

本实验的目的是让学生掌握怎样在 JSP 中使用内置对象 request。

3. 实验要求

编写一个 JSP 页面 inputNumber.jsp，该页面提供一个表单，用户可以通过表单输入两个数和四则运算符号提交给该页面。用户提交表单后，JSP 页面 inputNumber.jsp 将计算任务交给一个 Tag 文件 Computer.tag 去完成。

1) inputNumber.jsp 的具体要求

inputNumber.jsp 页面提供一个表单，要求表单中提供两个 text 输入框，供用户输入数字；提供一个 select 下拉列表，该下拉列表有加、减、乘、除四个选项，供用户选择运算符号(如图 4-16 所示)。用户在表单中输入数字、选择运算符号提交给 inputNumber.jsp 页面。inputNumber.jsp 使用 Tag 标记调用 Tag 文件 Computer.tag，并将表单提交的数字和运算符号传递给 Computer.tag。

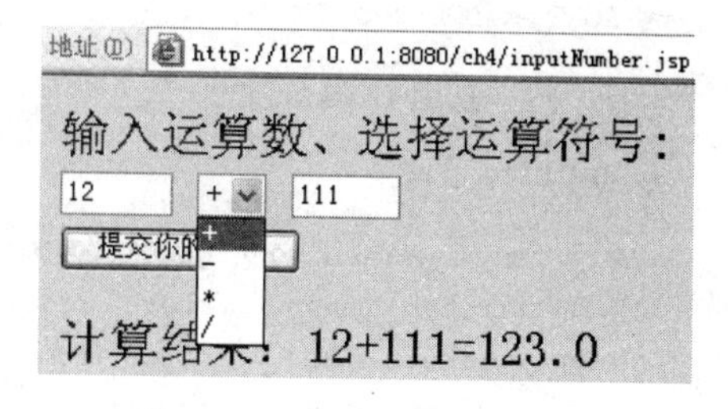

图 4-16　inputNumber.jsp 页面

2) Computer.tag 的具体要求

要求 Computer.tag 使用 attribute 指令得到 JSP 页面传递过来的值,使用 variable 指令将计算结果返回给 JSP 页面 inputNumber.jsp。

4. 参考代码

代码仅供参考,学生可按着实验要求,参考本代码编写代码。

JSP 页面参考代码如下:

inputNumber.jsp

```
<%@ page contentType = "text/html; charset = GB2312" %>
<%@ taglib tagdir = "/WEB-INF/tags" prefix = "computer" %>
<HTML>
<BODY bgcolor = yellow>
  <FONT Size = 5>
  <FORM action = "" method = post name = form>
    输入运算数、选择运算符号:<br>
    <Input type = text name = "numberOne" size = 6>
      <Select name = "operator">
        <Option value = " + "> +
        <Option value = " - "> -
        <Option value = " * "> *
        <Option value = "/">/
      </Select>
    <Input type = text name = "numberTwo" size = 6>
    <BR><INPUT TYPE = "submit" value = "提交你的选择" name = "submit">
  </FORM>
<% String a = request.getParameter("numberOne");
    String b = request.getParameter("numberTwo");
    String operator = request.getParameter("operator");
    if(a == null||b == null){
       a = "";
       b = "";
    }
    if(a.length()> 0&&b.length()> 0){
%> <computer:Computer numberA = "<% = a%>" numberB = "<% = b%>" operator = "<% = operator%
>"/>
       计算结果:<% = a%><% = operator%><% = b%>=<% = result%>
<% }
%>
</FONT></BODY></HTML>
```

Tag 文件参考代码如下:

Computer.Tag

```
<%@ tag pageEncoding = "gb2312"%>
<%@ attribute name = "numberA" required = "true" %>
<%@ attribute name = "numberB" required = "true" %>
<%@ attribute name = "operator" required = "true" %>
<%@ variable name-given = "result" scope = "AT_END" %>
<% try{
```

```
            double a = Double.parseDouble(numberA);
            double b = Double.parseDouble(numberB);
            double r = 0;
            if(operator.equals(" + "))
               r = a + b;
            else if(operator.equals(" - "))
               r = a - b;
            else if(operator.equals(" * "))
               r = a * b;
            else if(operator.equals("/"))
               r = a/b;
            jspContext.setAttribute("result",String.valueOf(r));
        }
        catch(Exception e){
           jspContext.setAttribute("result","发生异常：" + e);
        }
    %>
```

4.7 实验 2：response 对象

1. 相关知识点

response 对象对用户的请求作出动态响应，向用户端发送数据。response 对象调用 setContentType(String s)方法可以动态改变响应的 contentType 属性的值。response 对象调用 addHeader(String head,String value)方法可以动态改变响应头和响应头的值。response 对象调用 setStatus(int n)方法可以动态改变响应的状态行的内容。response 对象调用 sendRedirect(URL url)方法可以实现用户的重定向。

2. 实验目的

本实验的目的是让学生掌握怎样使用 response 对象动态响应用户的请求。

3. 实验要求

编写两个 JSP 页面 input.jsp 和 result.jsp。input.jsp 页面提交一个数字给 result.jsp 页面，result.jsp 页面使用 response 对象作出动态响应。

1）input.jsp 的具体要求

input.jsp 提供表单，用户在表单中输入一个数字，提交给 result.jsp 页面。

2）result.jsp 的具体要求

result.jsp 页面首先使用 request 对象获得 input.jsp 页面提交的数字，然后根据数字的大小作出不同的响应。如果数字小于 0，response 对象调用 setContentType(String s)方法将 contentType 属性的值设置为 text/plain，同时输出数字的平方；如果数字大于等于 0 并且小于 100，response 对象调用 setContentType(String s)方法将 contentType 属性的值设置为 application/msword，同时输出数字的立方；如果数字大于等于 100，response 对象调用 setStatus(int n)方法将状态行的内容设置为 404；如果用户在 input.jsp 页面输入了非数字，response 对象调用 sendRedirect(URL url)方法将用户重定向到 input.jsp 页面。

4. 参考代码

代码仅供参考，学生可按着实验要求，参考本代码编写代码。

JSP 页面参考代码如下：

input.jsp

```
<%@ page contentType="text/html; charset=GB2312" %>
<HTML>
<BODY bgcolor=yellow>
  <FONT Size=2>
  <FORM action="result.jsp" method=post name=form>
    输入数字：<Input type=text name="number" size=6>
   <INPUT TYPE="submit" value="提交" name="submit">
  </FORM>
</FONT>
</BODY>
</HTML>
```

result.jsp

```
<%@ page contentType="text/html; charset=GB2312" %>
<HTML>
<BODY bgcolor=cyan><Font size=3>
<% String str=request.getParameter("number");
       try{
          double number=Double.parseDouble(str);
          if(number<0){
            response.setContentType("text/plain; charset=GB2312");
            out.println(number+"的平方："+(number*number));
          }
          else if(number>=0&&number<100){
            response.setContentType("application/msword; charset=GB2312");
            out.println(number+"的立方："+(number*number*number));
          }
          else{
            response.setStatus(404);
          }
       }
       catch(Exception e){
          response.sendRedirect("input.jsp");
       }
 %>
</FONT></BODY></HTML>
```

4.8 实验 3：session 对象

1. 相关知识点

HTTP 是一种无状态协议。一个用户向服务器发出请求(request)，然后服务器返回响应(response)，链接就被关闭。所以，Tomcat 服务器必须使用内置 session 对象(会话)记录有关链接的信息。同一个用户在某个 Web 服务目录中的 session 是相同的；同一个用户在不同的 Web 服务目录中的 session 是互不相同的；不同用户的 session 是互不相同的。一个用户在某个 Web 服务目录的 session 对象的生存期限依赖于用户是否关闭浏览器,依赖于 session 对象是否调用 invalidate()方法使得 session 无效或 session 对象达到了设置的最长的"发呆"时间。

2. 实验目的

本实验的目的是让学生掌握怎样使用 session 对象存储和用户有关的数据。

3. 实验要求

编写 5 个 JSP 页面 inputGuess.jsp、result.jsp、small.jsp、large.jsp 和 success.jsp,实现猜数字游戏。具体要求如下。

1) inputGuess.jsp 的具体要求

用户请求 inputGuess.jsp 时,随机分配给该用户一个 1 到 100 之间的数。该页面同时负责将这个数字存在用户的 session 对象中。该页面提供表单,用户可以使用该表单输入自己的猜测,并提交给 result.jsp 页面。

2) result.jsp 的具体要求

result.jsp 页面负责判断 inputGuess.jsp 提交的猜测是否和用户的 session 对象中存放的那个数字相同,如果相同就将用户重定向到 success.jsp;如果不相同就将用户重定向到 large.jsp 或 small.jsp。

3) small.jsp 和 large.jsp 的具体要求

small.jsp 和 large.jsp 页面提供表单,用户可以使用该表单继续输入自己的猜测,并提交给 result.jsp 页面。

4) success.jsp 的具体要求

success.jsp 页面负责显示用户成功的消息,并负责输出用户 session 对象中的数据。

4. 参考代码

代码仅供参考,学生可按着实验要求,参考本代码编写代码。

JSP 页面参考代码如下:

inputGuess.jsp

```
<%@ page contentType = "text/html; charset = GB2312" %>
<HTML>
<BODY bgcolor = cyan><FONT Size = 2>
<P>随机分给了你一个 1 到 100 之间的数,请猜!
  <%
      int number = (int)(Math.random() * 100) + 1;
      session.setAttribute("count",new Integer(0));
      session.setAttribute("save",new Integer(number));
  %>
   <FORM action = "result.jsp" method = "post" name = form>
    输入你的猜测:<INPUT type = "text" name = "boy" >
    <INPUT TYPE = "submit" value = "送出" name = "submit">
   </FORM>
</FONT></BODY></HTML>
```

result.jsp

```
<% String str = request.getParameter("boy");
```

```
    if(str.length() == 0){
        response.sendRedirect("inputGuess.jsp");
    }
    int guessNumber = -1;
    try{
        guessNumber = Integer.parseInt(str);
        Integer integer = (Integer)session.getAttribute("save");
        int realnumber = integer.intValue();
        if(guessNumber == realnumber){
            int n = ((Integer)session.getAttribute("count")).intValue();
            n = n + 1;
            session.setAttribute("count",new Integer(n));
            response.sendRedirect("success.jsp");
        }
        else if(guessNumber > realnumber){
            int n = ((Integer)session.getAttribute("count")).intValue();
            n = n + 1;
            session.setAttribute("count",new Integer(n));
            response.sendRedirect("large.jsp");
        }
        else if(guessNumber < realnumber){
            int n = ((Integer)session.getAttribute("count")).intValue();
            n = n + 1;
            session.setAttribute("count",new Integer(n));
            response.sendRedirect("small.jsp");
        }
    }
    catch(Exception e){
        response.sendRedirect("inputGuess.jsp");
    }
%>
```

small.jsp

```
<%@ page contentType = "text/html; charset = GB2312" %>
<HTML>
<BODY bgcolor = cyan>
 <FONT Size = 2>
  <FORM action = "result.jsp" method = "get" name = form>
  猜小了,请再猜:<INPUT type = "text" name = "boy">
  <INPUT TYPE = "submit" value = "送出" name = "submit">
  </FORM>
</FONT></BODY></HTML>
```

large.jsp

```
<%@ page contentType = "text/html; charset = GB2312" %>
<HTML>
<BODY bgcolor = cyan>
 <FONT Size = 2>
  <FORM action = "result.jsp" method = "get" name = form>
  猜大了,请再猜:<INPUT type = "text" name = "boy">
```

```
  <INPUT TYPE = "submit" value = "送出" name = "submit">
  </FORM>
</FONT></BODY></HTML>
```

success.jsp

```
<%@ page contentType = "text/html; charset = GB2312" %>
<HTML>
<BODY bgcolor = cyan><FONT Size = 2>
<%
   int count = ((Integer)session.getAttribute("count")).intValue();
   int num = ((Integer)session.getAttribute("save")).intValue();
%>
<P>恭喜你,猜对了
<BR>你共猜了<% = count %>次,这个数字就是<% = num %>。
<BR>单击超链接返回到 inputGuess.jsp 页面:
<BR><A href = "inputGuess.jsp"> inputGuess.jsp </A>
</FONT></BODY></HTML>
```

习 题 4

1. 假设 JSP 使用的表单中有如下的 GUI(多选择框):

```
<input type = "checkbox" name = "item" value = "dog" >狗
<input type = "checkbox" name = "item" value = "stone" >石头
<input type = "checkbox" name = "item" value = "cat" >猫
<input type = "checkbox" name = "item" value = "water" >水
```

该表单所请求的 JSP 可以使用内置对象 request 获取该表单提交的数据,那么,下列哪些是 request 获取该表单提交的值的正确语句(　　)。

(A) String a = request.getParameter("item");
(B) String b = request.getParameter("checkbox");
(C) String c[] = request.getParameterValues("item");
(D) String d[] = request.getParameterValues("checkbox");

2. 如果表单提交的信息中有汉字,接受该信息的页面应做怎样的处理?

3. 编写两个 JSP 页面 inputString.jsp 和 computer.jsp,用户可以使用 inputString.jsp 提供的表单输入一个字符串,并提交给 computer.jsp 页面,该页面通过内置对象获取 inputString.jsp 页面提交的字符串,并显示该字符串的长度。

4. response 调用 sendRedirect(URL url)方法的作用是什么?

5. 回答下列问题:

(1) 一个用户在不同 Web 服务目录中的 session 对象相同吗?
(2) 一个用户在同一 Web 服务目录的不同子目录中的 session 对象相同吗?
(3) 如果用户长时间不关闭浏览器,用户的 session 对象可能消失吗?
(4) 用户关闭浏览器后,用户的 session 对象一定消失吗?

6. 参照例 4-18 编写一个猜英文 26 个小写字母的 Web 游戏。

第5章 JSP中的文件操作

本章导读

主要内容

- File 类
- 字节流
- 字符流
- RandomAccess 流
- 文件上传
- 文件下载

难点

- RandomAccess 流
- 文件上传

关键实践

- 读/写文件
- 加密文件

有时服务器需要将用户提交的信息保存到文件或根据用户的要求将服务器上的文件的内容显示到用户端。JSP 通过 Java 的输入输出流来实现文件的读/写操作。

本章中，为了体现代码复用的重要性，某些例子将采用 JSP 调用 Tag 文件的模式来学习文件的操作，即将有关文件的读/写指派给 Tag 文件（如图 5-1 所示）。

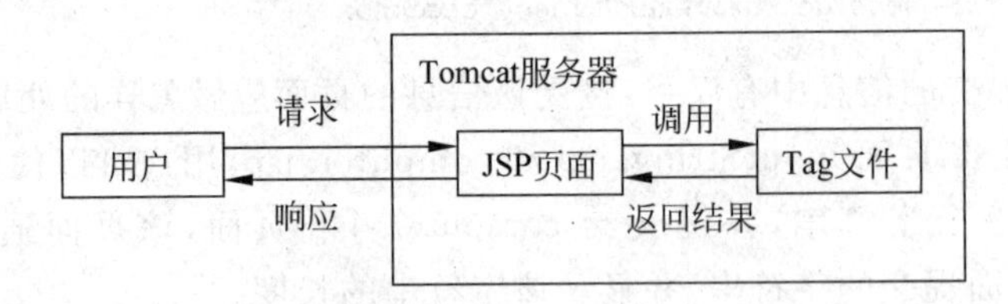

图 5-1　调用 Tag 文件

本章使用的 Web 服务目录是 ch5，除非特别声明，本章的 JSP 页面保存在 ch5 中。为了使用 Tag 文件，在当前 Web 服务目录 ch5 下建立如下的目录结构：

```
ch5\WEB-INF\tags
```

本章的 Tag 文件保存在上述目录中。

5.1 File 类

File 对象用来获取文件本身的一些信息。例如文件所在的目录、文件的长度、文件读/写权限等，不涉及对文件的读/写操作。创建一个 File 对象的构造方法有 3 个：

- File(String filename)；
- File(String directoryPath，String filename)；
- File(File f，String filename)。

对于第一个构造方法，filename 是文件名字或文件的绝对路径，如 filename="Hello.txt"或 filename="c：/mybook/A. txt"；对于第二个构造方法，directoryPath 是文件的路径，filename 是文件名字，如 directoryPath="c：/mybook/"，filename="A. txt"；对于第三个构造方法，参数 f 是指定成一个目录的文件，filename 是文件名字，如 f=new File("c：/mybook")，filename="A. txt"。

需要注意的是，当使用第一个构造方法 File(String filename)创建文件时，filename 是文件名字，那么该文件被认为是与当前应用程序在同一目录中，由于 JSP 引擎是在 bin 下启动执行的，所以该文件被认为在下列目录中：

```
D：\apache - tomcat - 6.0.13\bin
```

5.1.1 获取文件的属性

经常使用 File 类的下列方法获取文件本身的一些信息：

- public String getName()：获取文件的名字。
- public boolean canRead()：判断文件是否是可读的。
- public boolean canWrite()：判断文件是否可被写入。
- public boolean exists()：判断文件是否存在。
- public long length()：获取文件的长度(单位是字节)。
- public String getAbsolutePath()：获取文件的绝对路径。
- public String getParent()：获取文件的父目录。
- public boolean isFile()：判断文件是否是一个正常文件，而不是目录。
- public boolean isDirectroy()：判断文件是否是一个目录。
- public boolean isHidden()：判断文件是否是隐藏文件。
- public long lastModified()：获取文件最后修改的时间(时间是从 1970 年午夜(格林尼治时间)至文件最后修改时刻的毫秒数)。

例 5-1 中，使用上述的一些方法，获取某些文件的信息。

例 5-1

example5_1. jsp

```
<%@ page contentType = "text/html; charset = GB2312" %>
<%@ page import = "java.io. * " %>
<HTML><BODY bgcolor = cyan><FONT Size = 2>
  <% File f1 = new
```

```
        File("D:/apache-tomcat-6.0.13/webapps/ch5","example5_1.jsp");
        File f2 = new File("jasper.sh");
    %>
  <br> 文件<% = f1.getName() %>是可读的吗? <% = f1.canRead() %>
  <br>文件<% = f1.getName() %>的长度:<% = f1.length() %>字节
  <BR> jasper.sh 是目录吗? <% = f2.isDirectory() %>
  <BR><% = f1.getName() %>的父目录是:<% = f1.getParent() %>
  <BR><% = f2.getName() %>的绝对路径是:<% = f2.getAbsolutePath() %>
</FONT></BODY></HTML>
```

5.1.2 创建目录

1. 创建目录

File 对象调用方法 public boolean mkdir()创建一个目录,如果创建成功就返回 true,否则返回 false(如果该目录已经存在将返回 false)。

例 5-2 中,我们在 ch5 目录下创建一个名字是 Students 的目录。

例 5-2

example5_2.jsp

```
<%@ page contentType = "text/html; charset = GB2312" %>
<%@ page import = "java.io.*" %>
<HTML><BODY><FONT Size = 2>
  <% File dir = new
        File("D:/apache-tomcat-6.0.13/webapps/ch5","Students");
    %>
 <br> 在 ch5 下创建一个新的目录:Students,<br>成功创建了吗?
    <% = dir.mkdir() %>
 <br> Students 是目录吗? <% = dir.isDirectory() %>
</FONT></BODY></HTML>
```

2. 列出目录中的文件

如果 File 对象是一个目录,那么该对象可以调用下述方法列出该目录下的文件和子目录:

- public String[] list():用字符串形式返回目录下的全部文件。
- public File [] listFiles():用 File 对象形式返回目录下的全部文件。

例 5-3 中,输出了 ch5 目录下的全部文件和全部子目录。

例 5-3

example5_3.jsp

```
<%@ page contentType = "text/html; charset = GB2312" %>
<%@ page import = "java.io.*" %>
<HTML><BODY bgcolor = cyan><FONT Size = 2>
  <% File dir = new File("D:/apache-tomcat-6.0.13/webapps/ch5");
      File file[] = dir.listFiles();
    %>
  <br>目录有:
      <% for(int i = 0; i<file.length; i++){
            if(file[i].isDirectory())
```

```
            out.print("<br>" + file[i].toString());
        }
    %>
<br>文件名字：
    <% for(int i = 0; i < file.length; i++){
            if(file[i].isFile())
            out.print("<br>" + file[i].toString());
        }
    %>
</FONT></BODY></HTML>
```

3. 列出指定类型的文件

有时需要列出目录下指定类型的文件，例如包含.jsp、.txt等扩展名的文件。可以使用File类的下述两个方法，列出指定类型的文件。

- public String[] list(FilenameFilter obj)：该方法用字符串形式返回目录下的指定类型的所有文件。
- public File [] listFiles(FilenameFilter obj)：该方法用File对象返回目录下的指定类型的所有文件。

例5-4中，列出ch5目录下的部分JSP文件的名字。

例 5-4

example5_4.jsp

```
<%@ page contentType = "text/html; charset = GB2312" %>
<%@ page import = "java.io.*" %>
<HTML><BODY bgcolor = cyan><FONT Size = 2>
<%! class FileJSP implements FilenameFilter{
        String str = null;
        FileJSP(String s){
            str = "." + s;
        }
        public boolean accept(File dir,String name){
            return name.endsWith(str);
        }
    }
%>
<br>ch5 目录中的 jsp 文件：
 <% File dir = new File("D:/apache-tomcat-6.0.13/webapps/ch5");
    FileJSP file_jsp = new FileJSP("jsp");
    String file_name[] = dir.list(file_jsp);
    for(int i = 0; i < file_name.length; i++)
        out.print("<BR>" + file_name[i]);
 %>
</FONT></BODY></HTML>
```

5.1.3 删除文件和目录

File对象调用方法public boolean delete()可以删除当前对象代表的文件或目录。如果File对象表示的是一个目录，则该目录必须是一个空目录，删除成功将返回true。

例 5-5 删除 ch5 目录下的 example5_1.jsp 文件和 Students 目录。

例 5-5

example5_5.jsp

```
<%@ page contentType="text/html; charset=GB2312" %>
<%@ page import ="java.io.*" %>
<HTML><BODY>
  <% File f=new File("D:/apache-tomcat-6.0.13/webapps/ch5","example5_1.jsp");
     File dir=new File("D:/apache-tomcat-6.0.13/webapps/ch5","Students");
     boolean b1=f.delete();
     boolean b2=dir.delete();
  %>
<P>文件<%=f.getName()%>成功删除了吗?<%=b1%>
<P>目录<%=dir.getName()%>成功删除了吗?<%=b2%>
</BODY></HTML>
```

5.2 使用字节流读/写文件

Java 的 I/O 流提供一条通道程序,可以使用这条通道把源中的数据送给目的地。输入流的指向称作源,程序从指向源的输入流中读取源中的数据。输出流的指向是数据要去的一个目的地,程序通过向输出流中写入数据把信息传递到目的地,如图 5-2 和图 5-3 所示。

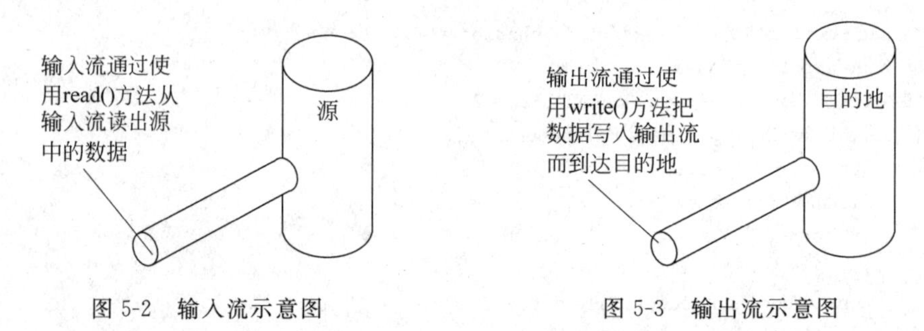

图 5-2 输入流示意图　　图 5-3 输出流示意图

java.io 包提供大量的流类,其中 InputStream、OutputStream、Reader 和 Writer 类是最重要的四个抽象类。称 InputStream 类及其子类对象为字节输入流类,称 OutputStream 类及其子类对象为字节输出流类,称 Reader 类及其子类对象为字符输入流类,称 Writer 类及其子类对象为字符输出流类。本节介绍字节流,下一节介绍字符流。

InputStream 类的常用方法:

- int read():输入流调用该方法从源中读取单个字节的数据,该方法返回字节值(0～255 之间的一个整数)。如果未读出字节就返回－1。
- int read(byte b[]):输入流调用该方法从源中试图读取 b.length 个字节到字节数组 b 中,返回实际读取的字节数目。如果到达文件的末尾,则返回－1。
- int read(byte b[],int off,int len):输入流调用该方法从源中试图读取 len 个字节到字节数组 b 中,并返回实际读取的字节数目。如果到达文件的末尾,则返回－1。参数 off 指定从字节数组的某个位置开始存放读取的数据。
- void close():输入流调用该方法关闭输入流。

- long skip(long numBytes)：输入流调用该方法跳过 numBytes 个字节，并返回实际跳过的字节数目。

OutputStream 类的常用方法：

- void write(int n)：输出流调用该方法向输出流写入单个字节。
- void write(byte b[])：输出流调用该方法向输出流写入一个字节数组。
- void write(byte b[],int off,int len)：从给定字节数组中起始于偏移量 off 处取 len 个字节写入到输出流。
- void close()：关闭输出流。

5.2.1 FileInputStream 类和 FileOutputStream 类

FileInputStream 类是从 InputStream 类中派生出来的简单输入流类。该类的所有方法都是从 InputStream 类继承来的。为了创建 FileInputStream 类的对象，用户可以调用它的构造方法。下面显示了两个构造方法：

- FileInputStream(String name)；
- FileInputStream(File file)。

第一个构造方法使用给定的文件名 name 创建一个 FileInputStream 对象。第二个构造方法使用 File 对象创建 FileInputStream 对象。参数 name 和 file 指定的文件称作输入流的源，输入流通过调用 read 方法读出源中的数据。

FileInputStream 文件输入流打开一个到达文件的输入流(源就是这个文件，输入流指向这个文件)。例如，为了读取一个名为 myfile.dat 的文件，建立一个文件输入流对象，如下所示：

```
try{ FileInputStream istream = new FileInputStream("myfile.dat");
}
catch (IOException e){
     System.out.println("File read error: " + e);
}
```

文件输入流构造方法的另一种格式是允许使用文件对象来指定要打开哪个文件。例如，下面这段代码使用文件输入流构造方法来建立一个文件输入流：

```
try{   File f = new File("myfile.dat");
       FileInputStream istream = new FileInputStream(f);
}
catch (IOException e){
     System.out.println("File read error: " + e);
}
```

当使用文件输入流构造方法建立通往文件的输入流时，可能会出现错误(也被称为异常)。例如，试图要打开的文件可能不存在。当出现 I/O 错误，Java 生成一个出错信号，它使用一个 IOException 对象来表示这个出错信号。

与 FileInputStream 类相对应的类是 FileOutputStream 类。FileOutputStream 类提供了基本的文件写入能力。除了从 OutputStream 类继承来的方法以外，FileOutputStream 类还有两个常用的构造方法。这两个构造方法如下所示：

- FileOutputStream(String name)；
- FileOutputStream(File file)。

第一个构造方法使用给定的文件名 name 创建一个 FileOutputStream 对象。第二个构造方法使用 File 对象创建 FileOutputStream 对象。参数 name 和 file 指定的文件称作输出流的目的地，通过向输出流中写入数据把信息传递到目的地。创建输出流对象也能发生 IOException 异常，必须在 try、catch 块语句中创建输出流对象。

使用 FileInputStream 的构造方法 FileInputStream(String name)创建一个输入流时，以及使用 FileOutputStream 的构造方法 FileOutputStream(String name)创建一个输出流时，如果参数仅仅是文件的名字(不带路径)，就要保证参数表示的文件和当前应用程序在同一目录下，由于 JSP 引擎是在 bin 下启动执行的，所以文件必须在 bin 目录中。

5.2.2 BufferedInputStream 类和 BufferedOutputStream 类

为了提高读/写的效率，FileInputStream 流经常和 BufferedInputStream 流配合使用，FileOutputStream 流经常和 BufferedOutputStream 流配合使用。BufferedInputStream 类的一个常用的构造方法是：BufferedInputStream(InputStream in)，该构造方法创建缓存输入流，该输入流的指向是一个输入流。当要读取一个文件，例如 A.txt 时，可以先建立一个指向该文件的文件输入流：

```
FileInputStream in = new FileInputStream("A.txt");
```

然后再创建一个指向文件输入流 in 的输入缓存流(就好像把两个输入水管接在一起)：

```
BufferedInputStream bufferRead = new BufferedInputStream(in);
```

这时，就可以让 bufferRead 调用 read 方法读取文件的内容。bufferRead 在读取文件的过程中，会进行缓存处理，提高读取的效率。同样，当要向一个文件，例如 B.txt，写入字节时，可以先建立一个指向该文件的文件输出流：

```
FileOutputStream out = new FileOutputStream("B.txt");
```

然后再创建一个指向输出流 out 的输出缓存流：

```
BufferedOutputStream bufferWriter = new BufferedOutputStream(out);
```

这时，bufferWriter 调用 write 方法向文件写入内容时会进行缓存处理，提高写入的效率。需要注意的是，写入完毕后，须调用 flush 方法将缓存中的数据存入文件。

例 5-6 中，example5_6.jsp 将若干内容写入一个文件，然后读取这个文件，并将文件的内容显示给用户，example5_6.jsp 的效果如图 5-4 所示。

地址(D) http://127.0.0.1:8080/ch5/example5_6.jsp
你们好，很高兴认识你们呀！
nice to meet you

图 5-4　使用字节流读/写文件

例 5-6

example5_6.jsp

```
<%@ page contentType = "text/html; charset = GB2312" %>
<%@ page import = "java.io.*" %>
<HTML><BODY bgcolor = cyan><FONT size = 2>
<% File dir = new File("D:/","Students");
```

```
        dir.mkdir();
        File f = new File(dir,"hello.txt");
        try{
            FileOutputStream outfile = new FileOutputStream(f);
            BufferedOutputStream bufferout = new BufferedOutputStream(outfile);
            byte b[] = "你们好,很高兴认识你们呀!<BR>nice to meet you".getBytes();
            bufferout.write(b);
            bufferout.flush();
            bufferout.close();
            outfile.close();
            FileInputStream in = new FileInputStream(f);
            BufferedInputStream bufferin = new BufferedInputStream(in);
            byte c[] = new byte[90];
            int n = 0;
            while((n = bufferin.read(c))!= -1){
               String temp = new String(c,0,n);
               out.print(temp);
            }
            bufferin.close();
            in.close();
        }
        catch(IOException e){}
    %>
</FONT></BODY></HTML>
```

5.3 使用字符流读/写文件

前面学习了使用字节流读/写文件,但是字节流不能直接操作 Unicode 字符,所以 Java 提供了字符流。由于汉字在文件中占用两个字节,如果使用字节流,读取不当会出现乱码现象,采用字符流就可以避免这个现象。在 Unicode 字符中,一个汉字被看作一个字符。

所有字符输入流类都是 Reader(输入流)抽象类的子类,而所有字符输出流类都是 Writer(输出流)抽象类的子类。

Reader 类中常用方法:

- int read():输入流调用该方法从源中读取一个字符,该方法返回一个整数(0~65535 之间的一个整数,Unicode 字符值)。如果未读出字符就返回-1。
- int read(char b[]):输入流调用该方法从源中读取 b.length 个字符到字符数组 b 中,返回实际读取的字符数目。如果到达文件的末尾,则返回-1。
- int read(char b[],int off,int len):输入流调用该方法从源中读取 len 个字符并存放到字符数组 b 中,返回实际读取的字符数目。如果到达文件的末尾,则返回-1。其中,off 参数指定 read 方法从字符数组 b 中的某个地方存放数据。
- void close():输入流调用该方法关闭输入流。
- long skip(long numBytes):输入流调用该方法跳过 numBytes 个字符,并返回实际跳过的字符数目。

Writer 类中常用方法:

- void write(int n):向输出流写入一个字符。

- void write(char b[])：向输出流写入一个字符数组。
- void write(char b[],int off,int length)：从给定字符数组中起始于偏移量 off 处取 len 个字符写入到输出流。
- void close()：关闭输出流。

5.3.1 FileReader 类和 FileWriter 类

FileReader 类是从 Reader 类中派生出来的简单输入类，该类的所有方法都是从 Reader 类继承来的。为了创建 FileReader 类的对象，用户可以调用它的构造方法。下面显示了两个构造方法：

- FileReader(String name)；
- FileReader(File file)。

第一个构造方法使用给定的文件名 name 创建一个 FileReader 对象。第二个构造方法使用 File 对象创建 FileReader 对象。参数 name 和 file 指定的文件称作输入流的源，输入流通过调用 read 方法读出源中的数据。

与 FileReader 类相对应的类是 FileWriter 类。FileWriter 类提供了基本的文件写入能力。除了从 Writer 类继承来的方法以外，FileWriter 类还有两个常用的构造方法。这两个构造方法如下所示：

- FileWriter(String name)；
- FileWriter (File file)。

第一个构造方法使用给定的文件名 name 创建一个 FileWriter 对象。第二个构造方法使用 File 对象创建 FileWriter 对象。参数 name 和 file 指定的文件称作输出流的目的地，通过向输出流中写入数据把信息传递到目的地。创建输入、输出流对象会发生 IOException 异常，必须在 try、catch 块语句中创建输入、输出流对象。

5.3.2 BufferedReader 类和 BufferedWriter 类

为了提高读/写的效率，FileReader 流经常和 BufferedReader 流配合使用，FileWriter 流经常和 BufferedWriter 流配合使用。BufferedReader 流还可以使用方法 String readLine() 读取一行，BufferedWriter 流还可以使用方法 void write(String s,int off,int length)将字符串 s 的一部分写入文件，使用方法 newLine()向文件写入一个行分隔符。

例 5-7 中，用户可以在 JSP 页面 exampleWrite.jsp 页面提供的文本区中输入文本内容提交给当前 JSP 页面，exampleWrite.jsp 调用 Tag 文件 WriteTag.tag 将用户提交的内容写入一个文件。用户可以在 JSP 页面 exampleRead.jsp 页面调用 Tag 文件 ReadTag.tag 读取曾写入的文件，并将这个文件的内容显示在一个文本区中。

exampleWrite.jsp 和 exampleRead.jsp 效果如图 5-5 所示。

例 5-7

exampleWrite.jsp

```
<%@ page contentType = "text/html; charset = GB2312" %>
<%@ taglib tagdir = "/WEB-INF/tags" prefix = "file" %>
<HTML><BODY bgcolor = yellow><Font size = 3>
```

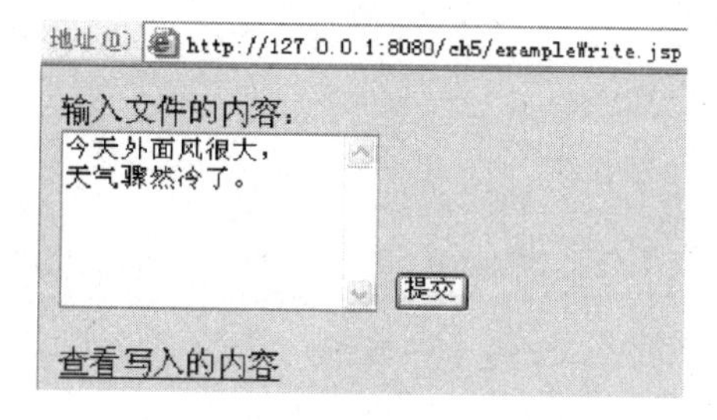

(a) 写文件

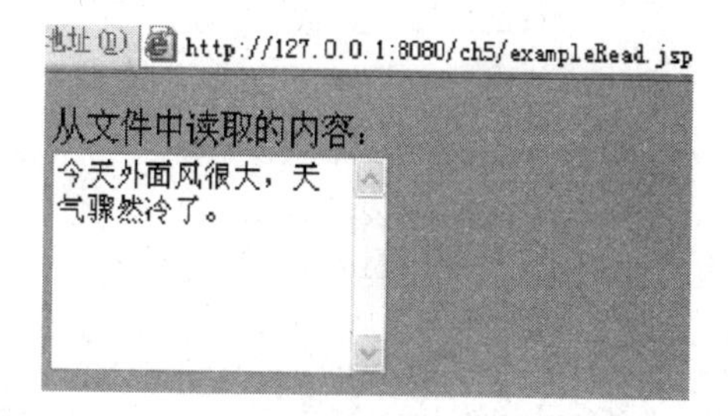

(b) 读文件

图 5-5　两个页面的效果图

```
<FORM action = "" Method = "post" >
   输入文件的内容：
   <br>
    <TextArea name = "write" Rows = "6" Cols = "20" ></TextArea>
    <Input type = submit value = "提交">
</FORM>
  <% String str = request.getParameter("write");
       if(str == null)
          str = "";
       byte bb[] = str.getBytes("iso - 8859 - 1");
       str = new String(bb);
  %>
  <file:WriteTag dir = "D:/2000" fileName = "hello.txt" content = "<% = str %>" />
  <a href = "exampleRead.jsp">查看写入的内容</a>
</FONT></BODY></HTML>
```

exampleRead.jsp

```
<%@ page contentType = "text/html; charset = GB2312" %>
<%@ taglib tagdir = "/WEB - INF/tags" prefix = "file" %>
<HTML><BODY bgcolor = cyan><Font size = 3>
  <file:ReadTag dir = "D:/2000" fileName = "hello.txt" />
  从文件中读取的内容：
  <br><TextArea name = "read" Rows = "6" Cols = "20"><% = result %></TextArea>
  <% -- result 是 Tag 文件返回的对象 --%>
</FONT></BODY></HTML>
```

WriteTag.tag

```
<%@ tag pageEncoding = "GB2312" %>
<%@ tag import = "java.io.* " %>
<%@ attribute name = "dir" required = "true" %>
<%@ attribute name = "fileName" required = "true" %>
<%@ attribute name = "content" type = "java.lang.String" required = "true" %>
<%!
   public void writeContent(String str,File f){
      try{ FileWriter outfile = new FileWriter(f);
           BufferedWriter bufferout = new BufferedWriter(outfile);
           bufferout.write(str);
           bufferout.close();
```

```
                outfile.close();
            }
            catch(IOException e){}
        }
    %>
    <%  File mulu = new File(dir);
        mulu.mkdir();
        File f = new File(mulu,fileName);
        if(content.length()> 0){
            writeContent(content,f);
            out.println("成功写入");
        }
    %>
```

ReadTag.tag

```
<%@ tag import = "java.io.*" %>
<%@ attribute name = "dir" required = "true" %>
<%@ attribute name = "fileName" required = "true" %>
<%@ variable name-given = "result" scope = "AT_END" %>
<%!
    public String readContent(File f)
    { StringBuffer str = new StringBuffer();
        try{  FileReader in = new FileReader(f);
              BufferedReader bufferin = new BufferedReader(in);
              String temp;
              while((temp = bufferin.readLine())!= null)
                  str.append(temp);
              bufferin.close();
              in.close();
        }
        catch(IOException e){}
        return new String(str);
    }
%>
<%  File f = new File(dir,fileName);
    String fileContent = readContent(f);
    jspContext.setAttribute("result",fileContent); //返回对象 result
%>
```

5.4 RandomAccessFile 类

前面学习了用来处理文件的几个文件输入、输出流,而且通过一些例子,已经了解了一些流的功能。

RandomAccessFile 类创建的流与前面的输入、输出流不同。RandomAccessFile 类既不是输入流类 InputStream 类的子类,也不是输出流类 OutputStream 类的子类。习惯上,仍然称 RandomAccessFile 类创建的对象为一个流。RandomAccessFile 流的指向既可以作为源也可以作为目的地。换句话说,当想对一个文件进行读/写操作时,可以创建一个指向该文件的 RandomAccessFile 流，这样既可以从这个流读取文件的数据，也可以通过这个流向文件写入数据。

RandomAccessFile 类的两个构造方法：

- RandomAccessFile(String name,String mode)：参数 name 用来确定一个文件名，给出创建的流的源(也是流的目的地)。参数 mode 取"r"(只读)或"rw"(可读/写)，决定创建的流对文件的访问权利。
- RandomAccessFile(File file,String mode)：参数 file 是一个 File 对象，给出创建的流的源(也是流的目的地)。参数 mode 取"r"(只读)或"rw"(可读/写)，决定创建的流对文件的访问权利。创建对象时应捕获 IOException 异常。

RandomAccessFile 流对文件的读/写方式比前面学习过的采用顺序读/写方式的文件输入、输出流更为灵活。例如，RandomAccessFile 类中有一个方法 seek(long a)，该方法可以用来移动 RandomAccessFile 流在文件中的读/写位置，其中参数 a 确定读/写位置，即距离文件开头的字节数目。另外，RandomAccessFile 流还可以调用 getFilePointer()方法获取当前流在文件中的读/写位置。

RandomAccessFile 类的常用方法：

- getFilePointer()：获取当前流在文件中的读/写的位置。
- length()：获取文件的长度。
- readByte()：从文件中读取一个字节。
- readDouble()：从文件中读取一个双精度浮点值(8 个字节)。
- readInt()：从文件中读取一个 int 值(4 个字节)。
- readLine()：从文件中读取一个文本行。
- readUTF()：从文件中读取一个 UTF 字符串。
- seek(long a)：定位当前流在文件中的读/写的位置。
- write(byte b[])：写 b.length 个字节到文件。
- writeDouble(double v)：向文件写入一个双精度浮点值。
- writeInt(int v)：向文件写入一个 int 值。
- writeUTF(String s)：写入一个 UTF 字符串。

例 5-8 实现网上小说创作，每个用户都可以参与一部小说的写作，也就是说一个用户须接着前一个用户的曾写内容继续写作。在服务器的某个目录下有 4 部小说，小说的内容完全由用户来决定。用户首先在 example5_8.jsp 页面选择一部小说的名字，然后链接到 continueWrite.jsp 页面。continueWrite.jsp 页面显示了小说的已有内容，用户可以在该页面输入续写的内容，再提交给 continue.jsp 页面。continue.jsp 页面负责将续写的内容存入文件。continue.jsp 页面的 isThreadSafe 属性值设置为 false，使得该页面同一时刻只能处理响应一个用户的请求，其他用户须排队等待。

example5_8.jsp、continueWrite.jsp 和 continue.jsp 页面的效果如图 5-6 所示。

例 5-8

example5_8.jsp

```
<%@ page contentType = "text/html; charset = GB2312" %>
<%@ page import = "java.io.*" %>
```

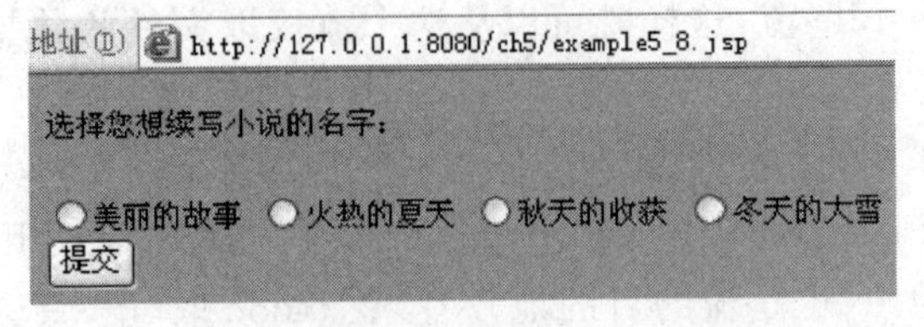

(a) 选择一部小说

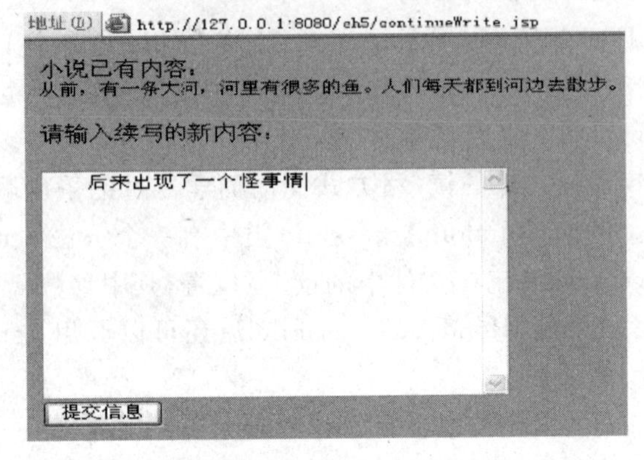

(b) 输入内容

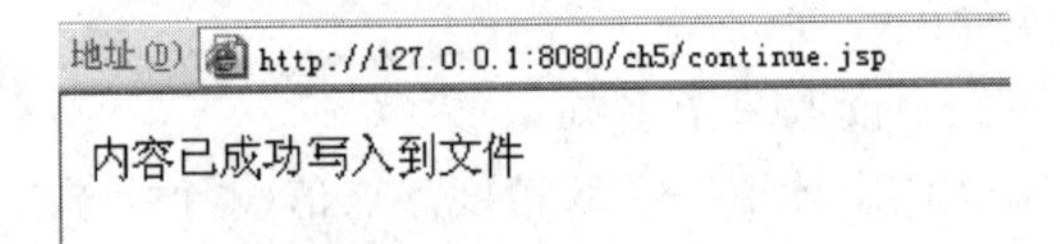

(c) 写入内容

图 5-6　3 个页面的效果图

```
<HTML><BODY bgcolor = cyan><FONT size = 2>
<% String str = response.encodeURL("continueWrite.jsp");
%>
   <P>选择你想续写小说的名字：
   <FORM action = "<% = str %>" method = post name = form>
     <INPUT type = "radio" name = "R" value = "spring.doc" >美丽的故事
     <INPUT type = "radio" name = "R" value = "summer.doc" >火热的夏天
     <INPUT type = "radio" name = "R" value = "autumn.doc" >秋天的收获
     <INPUT type = "radio" name = "R" value = "winter.doc" >冬天的大雪
   <BR><INPUT type= submit name = "g" value = "提交">
</FORM>
</FONT></BODY></HTML>
```

continueWrite.jsp

```
<%@ page contentType = "text/html; charset = GB2312" %>
<%@ page import = "java.io.*" %>
<%@ page info = "story" %>
<HTML><BODY bgcolor = cyan><P>小说已有内容：<Font size = 2 Color = Navy>
<% String str = response.encodeURL("continue.jsp");
%>
<%-- 获取用户提交的小说的名字 --%>
```

```
<%
    String name = (String)request.getParameter("R");
    if(name == null)
        name = "";
    byte c[] = name.getBytes("ISO - 8859 - 1");
    name = new String(c);
    session.setAttribute("name",name);
    String dir = getServletInfo();
    File storyFileDir = new File(dir);
    storyFileDir.mkdir();
    File f = new File(storyFileDir,name);
    //列出小说的内容
    try{   RandomAccessFile file =
              new RandomAccessFile(f,"r");
              String temp = null;
              while((temp = file.readUTF()) != null){
                  byte d[] = temp.getBytes("ISO - 8859 - 1");
                  temp = new String(d);
                  out.print("<BR>" + temp);
              }
              file.close();
    }
    catch(IOException e){}
 %>
 </FONT>
 <P>请输入续写的新内容:
  <Form action = "<% = str %>" method = post name = form>
    <TEXTAREA name = "messages" ROWs = "12" COLS = 40 WRAP = "physical">
    </TEXTAREA>
    <BR>
    <INPUT type = "submit" value = "提交信息" name = "submit">
  </FORM>
</BODY></HTML>
```

continue.jsp

```
<%@ page contentType = "text/html; charset = GB2312" %>
<%@ page isThreadSafe = "false" %>
<%@ page import = "java.io.*" %>
<%@ page info = "story" %>
<HTML>
<BODY>
  <%!   String writeContent(File f,String s){
              try{
                    RandomAccessFile out = new RandomAccessFile(f,"rw");
                    out.seek(out.length()); //定位到文件的末尾
                    out.writeUTF(s);
                    out.close();
                    return "内容已成功写入到文件";
              }
```

```
            catch(IOException e){
                return "不能写入到文件";
            }
        }
    %>
    <%-- 获取用户提交的小说的名字 --%>
    <%   String name = (String)session.getAttribute("name");
         byte c[] = name.getBytes("ISO - 8859 - 1");
         name = new String(c);
         //获取用户续写的内容
         String content = (String)request.getParameter("messages");
         if(content == null)
           content = "";
         String dir = getServletInfo();
         File storyFileDir = new File(dir);
         storyFileDir.mkdir();
         File f = new File(storyFileDir,name);
         String message = writeContent(f,content);
         out.print(message);
    %>
</BODY></HTML>
```


5.5 文 件 上 传

用户通过一个 JSP 页面上传文件给服务器时,该 JSP 页面必须含有 File 类型的表单,并且表单必须将 ENCTYPE 的属性值设成 multipart/form-data。File 类型表单如下所示:

```
<FORM action = "接受上传文件的页面" method = "post" ENCTYPE = "multipart/form - data"
    <Input type = "File" name = "picture" >
</FORM>
```

JSP 引擎可以让内置对象 request 调用方法 getInputStream()获得一个输入流,通过这个输入流读入用户上传的全部信息,包括文件的内容以及表单域的信息。

下面的例 5-9 中,用户通过 example5_9.jsp 页面上传如下的文本文件 a.txt。

```
a.txt:
```

request 获得一个输入流读取用户上传的全部信息,包括表单的头信息以及上传文件的内容。以后将讨论如何去掉表单的信息,获取文件的内容。

在 accept.jsp 页面,内置对象 request 调用方法 getInputStream()获得一个输入流 in,用 FileOutputStream 类再创建一个输出流 o。输入流 in 读取用户上传的信息,输出流 o 将读取的信息写入文件 B.txt。用户上传的全部信息,包括文件 a.txt 的内容以及表单域的信息存放于服务器的 C:/1000 目录中在 B.txt 文件中。文件 B.txt 的前 4 行(包括一个空行)以及倒数 5 行(包括一个空行)是表单域的内容,中间部分是上传文件 a.txt 的内容,B.txt 的内容如图 5-7(a)所示。example5_9.jsp,accept.jsp 的效果如图 5-7(b)、图 5-7(c)所示。

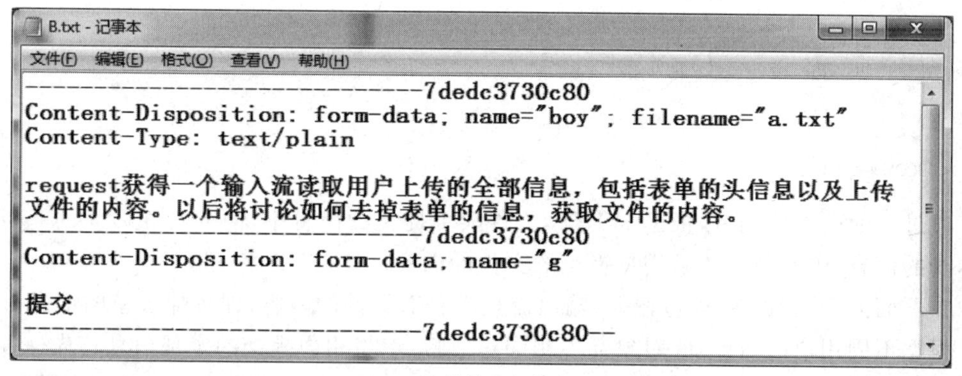

(a) 上传的全部信息

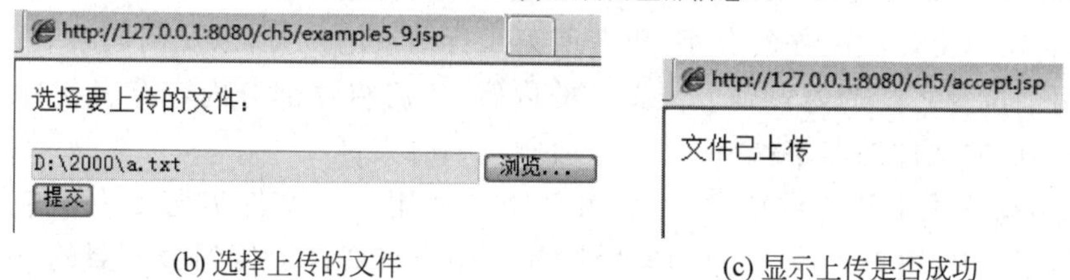

(b) 选择上传的文件　　　　(c) 显示上传是否成功

图 5-7　3 个页面的效果图

例 5-9

example5_9.jsp

```
<%@ page contentType="text/html;charset=GB2312" %>
<HTML><BODY>
<P>选择要上传的文件：<BR>
  <FORM action="accept.jsp" method="post" ENCTYPE="multipart/form-data">
    <INPUT type=FILE name="boy" size="38">
    <BR><INPUT type="submit" name ="g" value="提交">
</FORM>
</BODY></HTML>
```

accept.jsp

```
<%@ page contentType="text/html;charset=GB2312" %>
<%@ page import ="java.io.*" %>
<HTML><BODY>
 <% try{  InputStream in=request.getInputStream();
          File dir=new File("C:/1000");
          dir.mkdir();
          File f=new File(dir,"B.txt");
          FileOutputStream o=new FileOutputStream(f);
          byte b[]=new byte[1000];
          int n;
          while((n=in.read(b))!=-1)
             o.write(b,0,n);
          o.close();
          in.close();
          out.print("文件已上传");
```

```
    }
    catch(IOException ee){
          out.print("上传失败" + ee);
    }
  %>
</BODY></HTML>
```

通过上面的讨论知道,根据 HTTP,文件表单提交的信息中,前 4 行和后面的 5 行是表单本身的信息,中间部分才是用户提交的文件的内容。

在下面的例 5-10 中,通过输入、输出流技术获取文件的内容,即去掉表单的信息。

根据不同用户的 session 对象互不相同这一特点,将用户提交的全部信息首先保存成一个临时文件,该临时文件的名字是用户的 session 对象的 Id,然后读取该临时文件的第 2 行,因为这一行中含有用户上传的文件的名字,再获取第 4 行结束的位置,以及倒数第 6 行结束的位置,因为这两个位置之间的内容是上传文件的内容,然后将这部分内容存入文件,该文件的名字和用户上传的文件的名字保持一致,最后删除临时文件。

Web 应用经常要提供上传文件功能,因此例 5-10 将使用 Tag 文件实现文件上传。

在下面的例 5-10 中,用户上传一个图像文件,允许用户将文件上传到服务器的 Web 服务目录 ch5 下的某个子目录中。example5_10. jsp 负责提供上传文件的表单,用户通过 example5_10. jsp 将要上传的文件提交给 acceptFile. jsp,acceptFile. jsp 使用 Tag 标记调用 Tag 文件 UPFile. tag,该 Tag 文件负责上传文件。

example5_10. jsp、acceptFile. jsp 的效果如图 5-8(a)、图 5-8(b)所示。

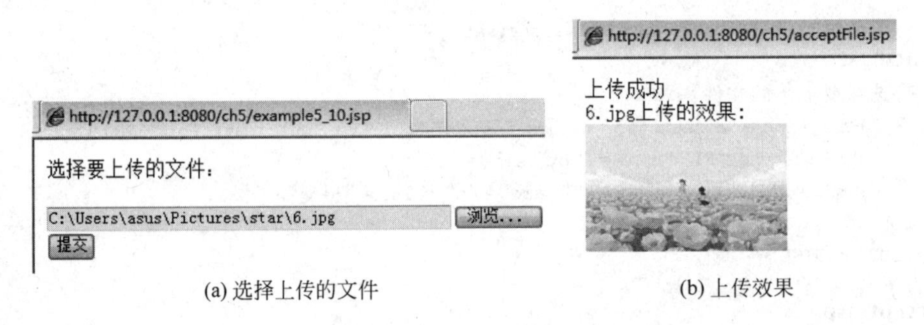

(a) 选择上传的文件　　(b) 上传效果

图 5-8　两个页面的效果图

例 5-10

example5_10. jsp

```
<%@ page contentType = "text/html;charset = GB2312" %>
<%@ page import = "java.io.*" %>
<%@ taglib tagdir = "/WEB-INF/tags" prefix = "ok" %>
<HTML><BODY>
<P>选择要上传的文件:<BR>
<FORM action = "acceptFile.jsp" method = "post" ENCTYPE = "multipart/form-data">
      <INPUT type = FILE name = "boy" size = "45">
      <BR><INPUT type = "submit" name = "boy" value = "提交">
</FORM>
</BODY></HTML>
```

acceptFile.jsp

```
<%@ page contentType="text/html;charset=GB2312" %>
<%@ page import ="java.io.*" %>
<%@ taglib tagdir="/WEB-INF/tags" prefix="upload" %>
<HTML><BODY color = "pink">
<upload:UpFile subdir="ch5/image" />    <%--ch5是当前的Web服务目录--%>
   <%=message%>                          <%--message是Tag文件返回的对象--%>
   <br><%=fileName%>上传的效果：    <%--fileName是Tag文件返回的对象--%>
   <br><image src="image/<%=fileName%>" width=160 height=100>
      </image>
</BODY></HTML>
```

UpFile.tag

```
<%@ tag pageEncoding="GB2312" %>
<%@ tag import="java.io.*" %>
<%@ attribute name="subdir" required="true" %>
<%@ variable name-given="message" scope="AT_END" %>
<%@ variable name-given="fileName" scope="AT_END" %>
<% jspContext.setAttribute("message","");
    String fileName=null;
    try{  //用客户的session对象的Id建立一个临时文件
          String tempFileName=(String)session.getId();
          File file = new File(""); //该文件被认为在Tomcat服务器的/bin中,见5.1节
          String parentDir=file.getAbsolutePath();
          parentDir=parentDir.substring(0,parentDir.lastIndexOf("bin")-1);
          String saveDir = parentDir+"/webapps/"+subdir;
          File dir = new File(saveDir);
          dir.mkdir();
          //建立临时文件f1
          File f1=new File(dir,tempFileName);
          FileOutputStream o=new FileOutputStream(f1);
          //将客户上传的全部信息存入f1
          InputStream in=request.getInputStream();
          byte b[]=new byte[10000];
          int n;
          while( (n=in.read(b))!=-1){
             o.write(b,0,n);
          }
          o.close();
          in.close();
          //读取临时文件f1,从中获取上传文件的名字和上传文件的内容
          RandomAccessFile randomRead=new RandomAccessFile(f1,"r");
          //读出f1的第2行,析取出上传文件的名字
          int second=1;
          String secondLine=null;
          while(second<=2) {
               secondLine=randomRead.readLine();
               second++;
          }
          //获取f1中第2行中"filename"之后"="出现的位置
```

```
        int position = secondLine.lastIndexOf("=");
         //客户上传的文件的名字是
        fileName = secondLine.substring(position + 2,secondLine.length() - 1);
        randomRead.seek(0); //再定位到文件 f1 的开头
        //获取第 4 行 Enter 符号的位置
        long forthEndPosition = 0;
        int forth = 1;
        while((n = randomRead.readByte())! = - 1&&(forth <= 4)){
           if(n == '\n'){
              forthEndPosition = randomRead.getFilePointer();
              forth++;
           }
        }
        //根据客户上传文件的名字,将该文件存入磁盘
        byte cc[] = fileName.getBytes("ISO - 8859 - 1");
        fileName = new String(cc);
        File f2 =  new File(dir,fileName);
        RandomAccessFile randomWrite = new RandomAccessFile(f2,"rw");
        //确定出文件 f1 中包含客户上传的文件的内容的最后位置,即倒数第 6 行
        randomRead.seek(randomRead.length());
        long endPosition = randomRead.getFilePointer();
        long mark = endPosition;
        int j = 1;
        while((mark >= 0)&&(j <= 6)) {
             mark --;
             randomRead.seek(mark);
             n = randomRead.readByte();
             if(n == '\n'){
                endPosition = randomRead.getFilePointer();
                j++;
             }
        }
        //将 randomRead 流指向文件 f1 的第 4 行结束的位置
        randomRead.seek(forthEndPosition);
        long startPoint = randomRead.getFilePointer();
      //从 f1 读出客户上传的文件存入 f2(读取第 4 行结束位置和倒数第 6 行之间的内容)
        while(startPoint < endPosition - 1){
             n = randomRead.readByte();
             randomWrite.write(n);
             startPoint = randomRead.getFilePointer();
        }
        randomWrite.close();
        randomRead.close();
        jspContext.setAttribute("message","上传成功"); //将 message 返回 JSP 页面
        jspContext.setAttribute("fileName",fileName);  //将 fileName 返回 JSP 页面
        f1.delete(); //删除临时文件
    }
    catch(Exception ee) {
        jspContext.setAttribute("message","没有选择文件或上传失败");
    }
%>
```

5.6 文件下载

JSP 内置对象 response 调用方法 getOutputStream()可以获取一个指向用户的输出流，服务器将文件写入这个流，用户就可以下载这个文件。当提供下载功能时，应当使用 response 对象向用户发送 HTTP 头信息，这样用户的浏览器就会调用相应的外部程序打开下载的文件，response 调用 setHeader 方法添加下载头的格式如下：

```
response.setHeader("Content-disposition","attachment; filename="下载文件名");
```

在例 5-11 中，用户在 example5_11.jsp 页面选择一个要下载的文件，将该文件的名字提交给 load.jsp 页面，load.jsp 页面调用 Tag 文件：LoadFile.tag 下载文件。example5_11.jsp、load.jsp 的效果如图 5-9 所示。

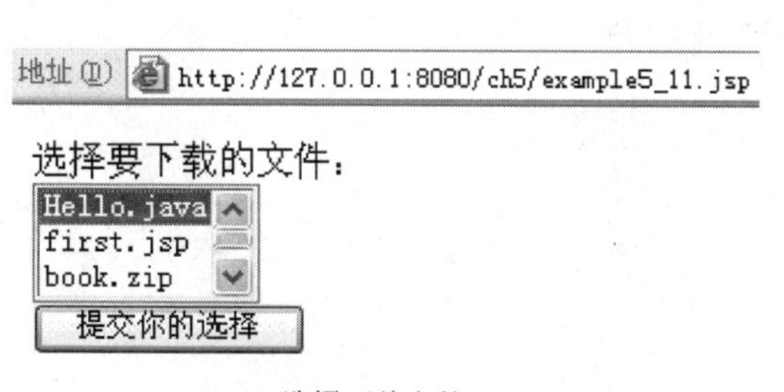

(a) 选择下载文件

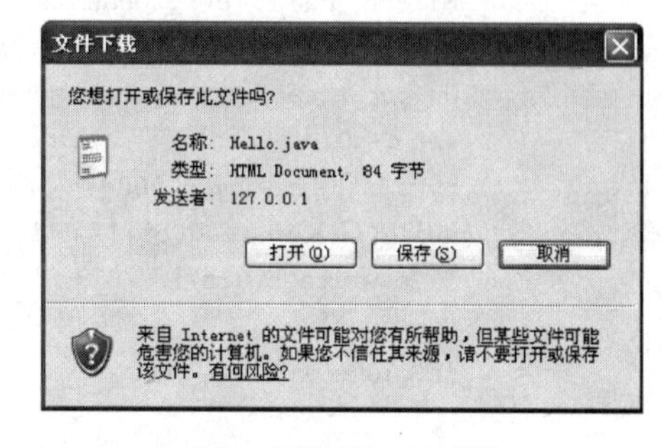

(b) "文件下载"对话框

图 5-9　例 5-11 中两个页面的效果图

例 5-11

example5_11.jsp

```
<%@ page contentType="text/html; charset=GB2312" %>
<HTML><BODY>
<P>
 <FORM action="load.jsp" method=post name=form>
        选择要下载的文件：<br>
        <Select name="filePath" size=3>
            <Option Selected value="d:/2000/Hello.java">Hello.java
            <Option value="d:/2000/first.jsp">first.jsp
            <Option value="d:/2000/book.zip">book.zip
             <Option value="d:/2000/A.txt">A.txt
        </Select>
        <br><INPUT TYPE="submit" value="提交你的选择">
   </FORM>
</BODY></HTML>
```

load.jsp

```
<%@ page contentType="text/html; charset=GB2312" %>
<%@ taglib tagdir="/WEB-INF/tags" prefix="download" %>
<HTML><BODY bgcolor=cyan><FONT size=2>
```

```
<% String path = request.getParameter("filePath");
 %>
 <download: LoadFile filePath = "<% = path %>" />
</FONT></BODY></HTML>
```

LoadFile.tag

```
<%@ tag import = "java.io.*" %>
<%@ attribute name = "filePath" required = "true" %>
<%
    String fileName = filePath.substring(filePath.indexOf("/") + 1);
    response.setHeader("Content - disposition","attachment; filename = " + fileName);
   //下载的文件
    try{
        //读取文件,并发送给用户下载
        File f = new File(filePath);
        FileInputStream in = new FileInputStream(f);
        OutputStream o = response.getOutputStream();
        int n = 0;
        byte b[] = new byte[500];
        while((n = in.read(b))! = -1)
           o.write(b,0,n);
        o.close();
        in.close();
     }
     catch(Exception exp){
     }
 %>
```

5.7 实验 1:使用文件字节流读/写文件

1. 相关知识点

FileInputStream 流以字节(byte)为单位顺序地读取文件,只要不关闭流,每次调用 read 方法就顺序地读取源中其余的内容,直到源的末尾或流被关闭。

FileOutStream 流以字节(byte)为单位顺序地写文件,只要不关闭流,每次调用 writer 方法就顺序地向输出流写入内容。

2. 实验目的

本实验的目的是让学生掌握使用文件输入、输出字节流读/写文件。

3. 实验要求

编写四个 JSP 页面 giveContent.jsp、writeContent.jsp、lookContent.jsp、readContent.jsp 以及两个 Tag 文件 Write.tag 和 Read.tag。

1) giveContent.jsp 的具体要求

giveContent.jsp 页面提供一个表单,要求该表单提供一个 text 文本输入框、select 下拉列表和一个 TextArea 文本区,用户可以在 text 输入框输入文件的名字,在 select 下拉列表选择一个目录(下拉列表的选项必须是 Tomcat 服务器所驻留计算机上的目录),通过

TextArea 输入多行文本。单击表单的提交按钮将 text 中输入的文件名字，select 下拉列表中选中的目录以及 TextArea 文本区中的内容提交给 writeContent.jsp 页面。giveContent.jsp 页面的效果如图 5-10 所示。

2）writeContent.jsp 的具体要求

writeContent.jsp 页面首先获得 giveContent.jsp 页面提交的文件所在目录、名字以及 TextArea 文本区中的内容，然后使用 Tag 标记调用 Tag 文件 Write.tag，并将文件所在目录、名字以及 TextArea 文本区中的内容传递给 Write.tag。效果如图 5-11 所示。

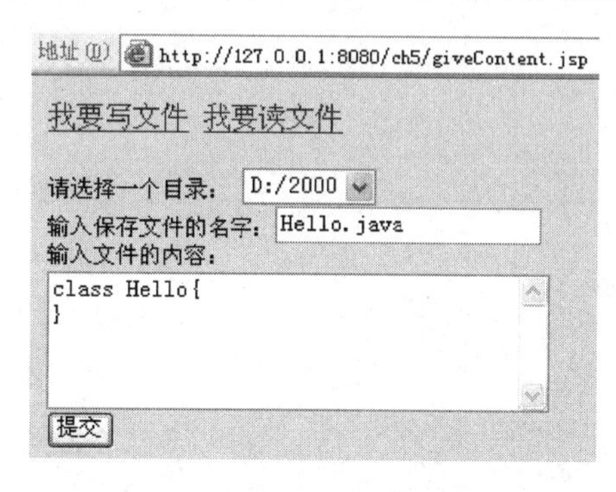

图 5-10　giveContent.jsp 页面的效果

地址(D) http://127.0.0.1:8080/ch5/writeContent.jsp
文件写入成功！
文件所在目录:D:/2000
文件的名字:Hello.java

图 5-11　writeContent.jsp 页面的效果

3）lookContent.jsp 的具体要求

lookContent.jsp 页面提供一个表单，该表单提供两个 text 文本输入框，用户可以向这两个 text 文本输入框输入目录和文件名字。单击表单的提交按钮将 text 中输入的文件目录以及文件名字提交给 readContent 页面。效果如图 5-12 所示。

4）readContent.jsp 的具体要求

readContent.jsp 页面首先获得 lookContent.jsp 页面提交的文件目录、名字，然后使用 Tag 标记调用 Tag 文件 Read.tag。并将文件所在目录、名字传递给 Read.tag。效果如图 5-13 所示。

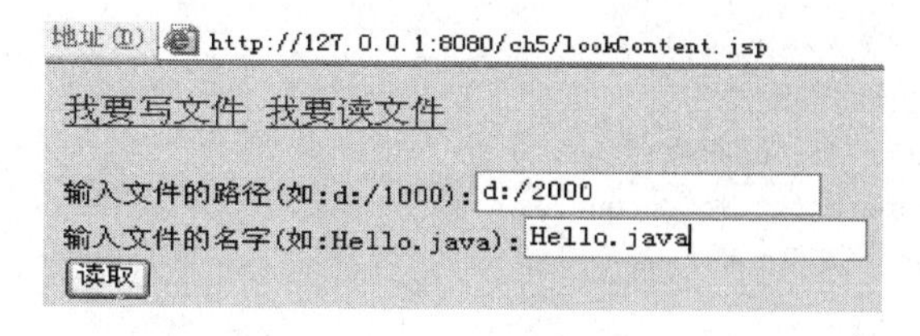

图 5-12　lookContent.jsp 页面的效果

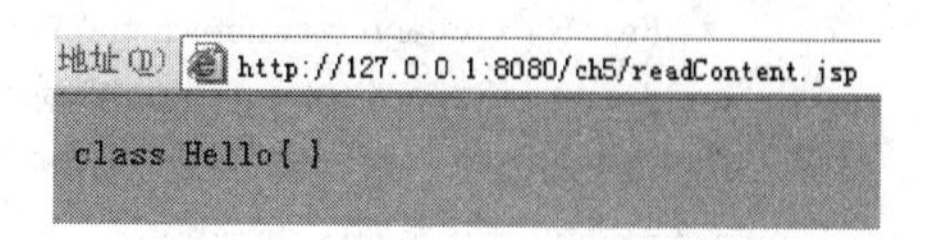

图 5-13　readContent.jsp 页面的效果

5）Write.tag 的具体要求

Write.tag 文件使用 attribute 指令获得 writeContent.jsp 页面传递过来的文件目录、文件名字和文件内容，然后使用文件字节输出流将文件内容写入到文件中，该文件所在目录就是 writeContent.jsp 页面传递过来的文件目录，名字就是 writeContent.jsp 页面传递过来的文件名字。

6）Read.tag 的具体要求

Read.tag 文件使用 attribute 指令获得 readContent.jsp 页面传递过来的文件目录和文

件名字,然后使用文件字节输入流读取文件,并负责显示所读取的内容。

4. 参考代码

代码仅供参考,学生可按着实验要求,参考本代码编写代码。

JSP 页面参考代码如下:

giveContent.jsp

```
<%@ page contentType="text/html; charset=GB2312" %>
<%@ taglib tagdir="/WEB-INF/tags" prefix="file" %>
<head>
  <A href="giveContent.jsp">我要写文件</A>
  <A href="lookContent.jsp">我要读文件</A>
</head>
<HTML><BODY bgcolor=yellow>
<Font size=2>
  <FORM action="writeContent.jsp" method=post>
     请选择一个目录:
     <Select name="fileDir">
        <Option value="C:/1000">C:/1000
        <Option value="D:/2000">D:/2000
        <Option value="D:/1000">D:/1000
     </Select>
  <BR>输入保存文件的名字:<Input type=text name="fileName">
  <BR>输入文件的内容:<BR>
   <TextArea name="fileContent" Rows="5" Cols="38"></TextArea>
  <BR><Input type=submit value="提交">
  </FORM></FONT>
</BODY></HTML>
```

writeContent.jsp

```
<%@ page contentType="text/html; charset=GB2312" %>
<%@ taglib tagdir="/WEB-INF/tags" prefix="file" %>
<HTML><BODY bgcolor=cyan>
<Font size=2>
<%   String fileDir=request.getParameter("fileDir");
      String fileName=request.getParameter("fileName");
      String fileContent=request.getParameter("fileContent");
      byte c[]=fileContent.getBytes("iso-8859-1");
      fileContent=new String(c);
%>
<file:Write fileDir="<%=fileDir%>" fileName="<%=fileName%>"
            fileContent="<%=fileContent%>" />
</FONT></BODY></HTML>
```

lookContent.jsp

```
<%@ page contentType="text/html; charset=GB2312" %>
<head>
  <A href="giveContent.jsp">我要写文件</A>
  <A href="lookContent.jsp">我要读文件</A>
</head>
```

```
<HTML><BODY bgcolor=yellow>
  <Font size=2>
  <FORM action="readContent.jsp" method="post" name="form">
     输入文件的路径(如：d:/1000)：<INPUT type="text" name="fileDir">
    <BR>输入文件的名字(如：Hello.java)：<INPUT type="text" name="fileName">
    <BR><INPUT type="submit" value="读取" name="submit">
   </FORM>
  </Font>
</BODY></HTML>
```

readContent.jsp

```
<%@ page contentType="text/html; charset=GB2312" %>
<%@ taglib tagdir="/WEB-INF/tags" prefix="file" %>
<HTML><BODY bgcolor=cyan>
  <Font size=2>
  <%   String fileDir=request.getParameter("fileDir");
        String fileName=request.getParameter("fileName");
   %>
  <file:Read fileDir="<%=fileDir%>" fileName="<%=fileName%>"/>
</FONT>
</BODY></HTML>
```

Tag 文件参考代码如下：

Write.tag

```
<%@ tag pageEncoding="GB2312" %>
<%@ tag import="java.io.*" %>
<%@ attribute name="fileContent" required="true" %>
<%@ attribute name="fileDir" required="true" %>
<%@ attribute name="fileName" required="true" %>
<%
    File f=new File(fileDir,fileName);
    try{
          FileOutputStream output=new FileOutputStream(f);
          byte bb[]=fileContent.getBytes();
          output.write(bb,0,bb.length);
          output.close();
          out.println("文件写入成功!");
          out.println("<br>文件所在目录："+fileDir);
          out.println("<br>文件的名字："+fileName);
    }
    catch(IOException e){
          out.println("文件写入失败"+e);
    }
%>
```

Read.tag

```
<%@ tag pageEncoding="GB2312" %>
<%@ tag import="java.io.*" %>
<%@ attribute name="fileDir" required="true" %>
```

```
<%@ attribute name = "fileName" required = "true" %>
<%
    File dir = new File(fileDir);
    File f = new File(dir,fileName);
    try{
        FileInputStream in = new FileInputStream(f);
        int m = - 1;
        byte bb[] = new byte[1024];
        String content = null;
        while((m = in.read(bb))!= - 1){
            content = new String(bb,0,m);
            out.println(content);
        }
        in.close();
    }
    catch(IOException e){
          out.println("文件读取失败" + e);
    }
%>
```

5.8 实验 2：使用文件字符流加密文件

1. 相关知识点

FileInputStream 和 FileReader 流都顺序地读取文件，只要不关闭流，每次调用 read 方法就顺序地读取源中其余的内容，直到源的末尾或流被关闭；二者的区别是，FileInputStream 流以字节(byte)为单位读取文件；FileReader 流以字符(char)为单位读取文件。

FileOutStream 流和 FileWriter 流顺序地写文件，只要不关闭流，每次调用 writer 方法就顺序地向输出流写入内容，直到流被关闭。二者的区别是，FileOutStream 流以字节(byte)为单位写文件；FileWriter 流以字符(char)为单位写文件。

2. 实验目的

本实验的目的是让学生掌握使用文件字符输入、输出流读/写文件。

3. 实验要求

编写 3 个 JSP 页面 inputContent.jsp，write.jsp、read.jsp 以及两个 Tag 文件 SecretWrite.tag 和 SecretRead.tag。具体要求如下。

1) inputContent.jsp 的具体要求

inputContent.jsp 页面提供一个表单，要求该表单提供 TextArea 的输入界面，用户可以在通过 TextArea 的输入界面输入多行文本提交给 write.jsp 页面。inputContent.jsp 页面效果如图 5-14 所示。

2) write.jsp 的具体要求

write.jsp 页面调用一个 Tag 文件 SecretWrite.tag，将 inputContent.jsp 页面提交的文本信息加密后写入到文件 save.txt 中。页面效果如图 5-15 所示。

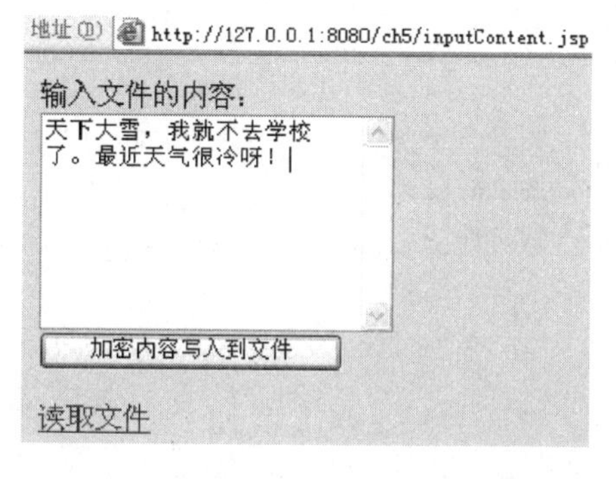

图 5-14　inputContent.jsp 页面的效果

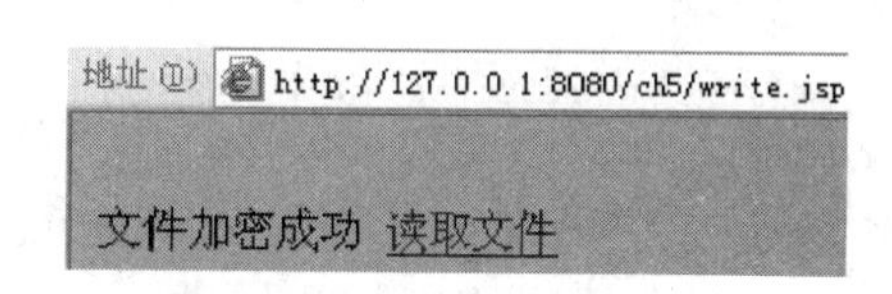

图 5-15　write.jsp 页面的效果

3）read.jsp 的具体要求

read.jsp 页面提供一个表单，该表单提供两个单选按钮，名字分别是“读取加密的文件”和“读取解密的文件”，该页面选中的单选按钮的值提交给本页面。如果该页面提交的值是单选按钮“读取加密的文件”的值，该页面就调用 Tag 文件 SecretRead.tag 读取文件 save.txt；如果该页面提交的值是单选按钮“读取解密的文件”的值，该页面就调用 Tag 文件 SecretRead.tag 读取文件 save.txt，并解密该文件。read.jsp 页面负责显示 SecretRead.tag 文件返回的有关信息。页面效果如图 5-16 所示。

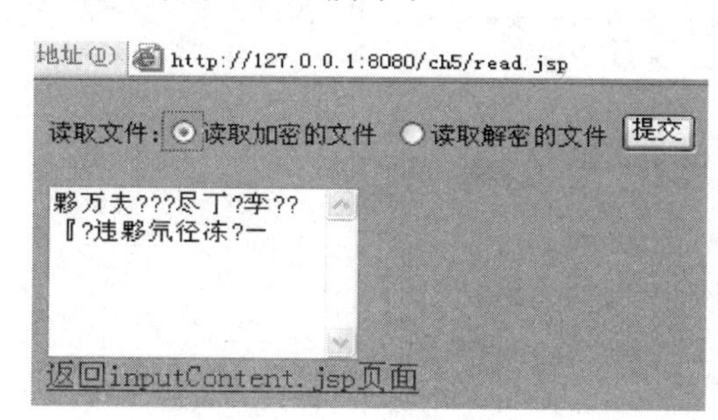

图 5-16　read.jsp 页面的效果

4）SecretWrite.tag 的具体要求

SecretWrite.tag 文件使用 attribute 指令获得 write.jsp 页面传递过来的文本信息，并使用文件输出流将其写入到文件 save.txt。

5）SecretRead.tag 的具体要求

SecretRead.tag 文件使用文件输入流读取文件 save.txt，并根据 read.jsp 的要求决定是否进行解密处理，然后使用 variable 指令将有关信息返回给 read.jsp 页面。

4. 参考代码

代码仅供参考，学生可按着实验要求，参考本代码编写代码。

JSP 页面参考代码如下：

inputContent.jsp

```
<%@ page contentType = "text/html; charset = GB2312" %>
<%@ taglib tagdir = "/WEB-INF/tags" prefix = "file" %>
<HTML><BODY bgcolor = yellow>
```

```
<Font size = 3>
   <FORM action = "write.jsp" Method = "post">
   输入文件的内容:
   <br>
    <TextArea name = "ok" Rows = "8" Cols = "26"></TextArea>
   <br><Input type = submit value = "加密内容写入到文件">
    </FORM>
   <A href = "read.jsp">读取文件</A>
</FONT>
</BODY></HTML>
```

write.jsp

```
<%@ page contentType = "text/html; charset = GB2312" %>
<%@ taglib tagdir = "/WEB-INF/tags" prefix = "file" %>
<HTML><BODY bgcolor = cyan>
<Font size = 3>
<% String str = request.getParameter("ok");
    if(str.length()> 0){
      byte bb[] = str.getBytes("iso-8859-1");
      str = new String(bb);
 %>  <file:SecretWrite content = "<% = str %>" />
<%      out.println("<br>" + message);
    }
 %>
 <A href = "read.jsp">读取文件</A>
</FONT>
</BODY></HTML>
```

read.jsp

```
<%@ page contentType = "text/html; charset = GB2312" %>
<%@ taglib tagdir = "/WEB-INF/tags" prefix = "file" %>
<HTML><BODY bgcolor = cyan>
<Font size = 2>
   <FORM action = "" method = post name = form>
        读取文件:<INPUT type = "radio" name = "R" value = "secret">读取加密的文件
                 <INPUT type = "radio" name = "R" value = "unsecret">读取解密的文件
       <INPUT TYPE = "submit" value = "提交" name = "submit">
   </FORM>
</FONT>
<% String condition = request.getParameter("R");
    if(condition!= null){
 %>  <file:SecretRead method = "<% = condition %>"/>
       <TextArea rows = 6 cols = 20><% = content %><%-- content 是 Tag 文件返回的对象 --%>
       </TextArea>
<% }
 %>
<br><A href = "inputContent.jsp">返回 inputContent.jsp 页面</A>
</BODY></HTML>
```

Tag 文件参考代码如下：

SecretWrite. tag

```
<%@ variable name-given="message" scope="AT_END" %>
<%@ tag pageEncoding="GB2312" %>
<%@ tag import="java.io.*" %>
<%@ attribute name="content" required="true" %>
<%  File dir=new File("C:/","Students");
     dir.mkdir();
     File f=new File(dir,"save.txt");
     try{ FileWriter outfile=new FileWriter(f);
          BufferedWriter bufferout=new BufferedWriter(outfile);
          char a[]=content.toCharArray();
          for(int i=0; i<a.length; i++)
               a[i]=(char)(a[i]^12);
          content=new String(a);
          bufferout.write(content);
          bufferout.close();
          outfile.close();
          jspContext.setAttribute("message","文件加密成功");
        }
     catch(IOException e){
          jspContext.setAttribute("message","文件加密失败");
     }
%>
```

SecretRead. tag

```
<%@ tag pageEncoding="GB2312" %>
<%@ tag import="java.io.*" %>
<%@ attribute name="method" required="true" %>
<%@ variable name-given="content" scope="AT_END" %>
<%  File dir=new File("C:/","Students");
     File f=new File(dir,"save.txt");
     StringBuffer mess=new StringBuffer();
     String str;
     try{
          FileReader in=new FileReader(f);
          BufferedReader bufferin=new BufferedReader(in);
          String temp;
          while((temp=bufferin.readLine())!=null)
               mess.append(temp);
          bufferin.close();
          in.close();
          str=new String(mess);
          if(method.equals("secret"))
             jspContext.setAttribute("content",str);
          else if(method.equals("unsecret")){
             char a[]=str.toCharArray();
             for(int i=0; i<a.length; i++)
                a[i]=(char)(a[i]^12);
```

```
                str = new String(a);
                jspContext.setAttribute("content",str);
            }
            else
               jspContext.setAttribute("content","");

        }
        catch(IOException e){
            jspContext.setAttribute("content","");
        }
%>
```

习 题 5

1. File 对象能读/写文件吗?
2. File 对象怎样获取文件的长度?
3. 准备读取文件 A.java,下列哪些是正确创建输入流的代码?(　　)

```
(A)    try{ FileInputStream in = new FileInputStream("A.java");
        }
        catch(IOException e){}
(B)    try{ InputStream in = new InputStream("A.java");
        }
        catch(IOException e){}
(C)    try{ FileReader in = new FileInputStream(new File("A.java"));
        }
        catch(IOException e){}
(D)    try{ FileReader in = new FileReader (new File("A.java"));
        }
        catch(IOException e){}
```

4. RandomAccessFile 类创建的流在读/写文件时有什么特点?

5. 编写两个 JSP 页面 input.jsp 和 read.jsp,input.jsp 通过表单提交一个目录和该目录下的一个文件名给 read.jsp,read.jsp 根据 input.jsp 提交的目录和文件名调用 Tag 文件 Read.tag 读取文件的内容。

第6章 在JSP中使用数据库

本章导读

主要内容

- MySQL数据库管理系统
- JDBC
- 连接MySQL数据库
- 查询、更新、删除记录
- 用结果集更新数据库
- 预处理
- 事务
- 常见数据库连接

难点

- 用结果集更新数据库
- 预处理
- 事务

关键实践

- 查询记录
- 更新记录
- 删除记录

在许多Web应用中，服务器需要和用户进行必要的数据交互，比如，服务器需要将用户提供的数据永久、安全地保存在服务器端，需要为用户提供数据查询等。此时，Web应用就可能需要和数据库打交道，其原因是数据库在数据查询、修改、保存、安全等方面有着其他数据处理手段无法替代的地位。许多优秀的数据库管理系统在数据管理，特别是在基于Web的数据管理方面正在扮演着重要的角色。

本章并非讲解数据库原理，而是讲解如何在JSP中使用JDBC提供的API和数据库进行交互信息，特点是，只要掌握与某种数据库管理系统所管理的数据库交互信息，就会很容易地掌握和其他数据库管理系统所管理的数据库交互信息。所以，为了便于教学，本书使用的数据库管理系统是MySQL数据库管理系统（读者可以选择任何熟悉的数据库管理系统学习本章的内容，见6.11）。

本章为了更好地体现一个Web应用将数据的处理和显示相分离，除个别例子为了说明基本知识外，大部分例子采用JSP+Tag模式，即JSP页面调用Tag文件来完成对数据库的操作,如图 6-1 所示。

本章使用的 Web 服务目录是 ch6,为了使用 Tag 文件,在 web 服务目录 ch6 下建立如下的目录结构:

```
WEB - INF\tags
```

本章的 Tag 文件保存在上述目录中。

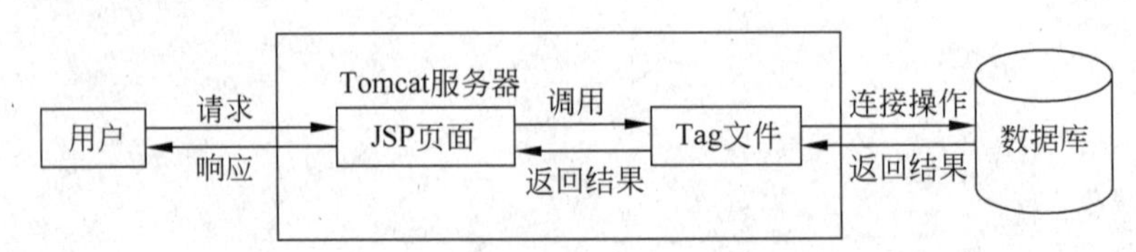

图 6-1 使用 Tag 文件操作数据库

6.1 MySQL 数据库管理系统

MySQL 数据库管理系统,简称 MySQL,是世界上最流行的开源数据库管理系统,其社区版(MySQL Community Edition)是世界上最流行的免费下载的开源数据库管理系统。MySQL 最初由瑞典 MySQL AB 公司开发,目前由 Oracle 公司负责源代码的维护和升级,Oracle 将 MySQL 分为社区版和商业版,并保留 MySQL 开放源码这一特点。目前许多 Web 开发项目都选用 MySQL,其主要原因是 MySQL 的社区版(MySQL Community Edition)性能卓越,满足许多 Web 应用已经绰绰有余,而且 MySQL 的社区版是开源数据库管理系统,可以降低软件的开发和使用成本。

6.1.1 下载、安装与启动 MySQL

1. 下载

MySQL 是开源项目,很多网站都提供免费下载。可以使用任何搜索引擎搜索关键字:"MySQL 下载"获得有关的下载地址。这里选择的地址是 MySQL 的官方网站:www.mysql.com,该网站免费提供 MySQL 最新版本的下载以及相关技术文章。下载的版本是:MySQL Community Server 5.6.16.zip,该版本可以安装在 Window 操作系统平台上(如果是其他的操作系统平台,请按网站的提示下载相应的版本)。

登录 www.mysql.com 后选择导航条上的 products,在出现的页面的左侧选择"MySQL Community Edition"或在出现的页面的右侧选择"下载 MySQL 社区版"。然后在出现的下载页面中选择下载适合相应平台的 MySQL 即可,这里下载的是 mysql-5.6.16-win32.zip,如图 6-2 所示。

2. 安装

将下载的 mysql-5.6.16-win32.zip 解压缩到本地计算机即可,比如解压缩到 D:\。本教材将下载的 mysql-5.6.16-win32.zip 解压缩到 D:\,形成的安装目录结构如图 6-3 所示。

3. 启动

MySQL 是一个网络数据库管理系统,可以使远程的计算机访问它所管理的数据库。

图 6-2　下载 MySQL 社区版 MySQL Community Server

安装好 MySQL 后，需启动 MySQL 提供的数据库服务器，以便使远程的计算机访问它所管理的数据库。

为了启动 MySQL 数据库服务器，需要执行 MySQL 安装目录的 bin 子目录中的 mysqld.exe 文件。需要打开 MS-DOS 命令行窗口，并使用 MS-DOS 命令进入到 bin 目录中。例如：

```
cd D:\mysql-5.6.16-win32\bin
```

然后在命令行输入：

```
mysqld
```

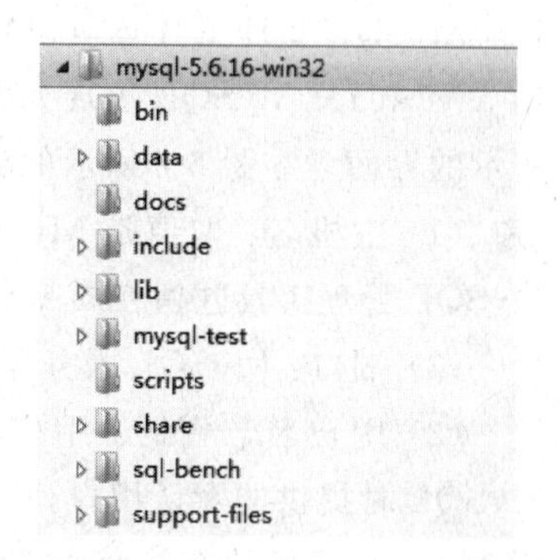

图 6-3　MySQL 的安装目录结构

或

```
mysqld -nt
```

启动 MySQL 数据库服务器。如果启动成功，MySQL 数据库服务器将占用当前 MS-DOS 窗口，如图 6-4 所示。

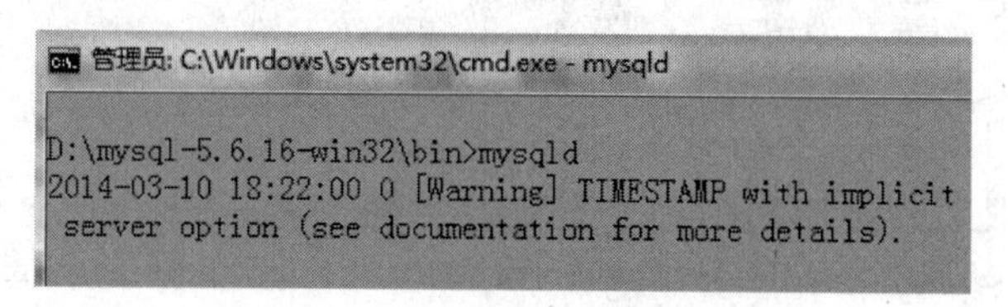

图 6-4　启动 MySQL 服务器

需要注意的是，直接关闭启动 MySQL 数据库服务器所占用的命令行窗口不能关闭 MySQL 数据库服务器，可以使用操作系统提供的“任务管理器”(按 Ctrl+Shift+Esc 键打开任务管理器)来关闭 MySQL 数据库服务器。如果当前计算机已经启动 MySQL 数据库服务器，那么必须关闭 MySQL 数据库服务器后，才能再次在命令行窗口重新启动 MySQL 数据库服务器。

6.1.2 建立数据库

启动 MySQL 数据库服务器后，就可以建立数据库，并在数据库中创建表。

1. 使用 MySQL 管理工具

可以下载图形界面的 MySQL 管理工具，并使用该工具进行创建数据库、在数据库中创建表等操作，MySQL 管理工具有免费的也有需要购买的。

可以在搜索引擎中搜索 MySQL 管理工具，选择一款 MySQL 管理工具。本教材使用的是 Navicat MySQL 管理工具，可以登录：http://www.navicat.com.cn/download 下载试用版或购买商业版。

下载 navicat9_mysql_cs.exe 后，安装即可。

1) 建立连接

MySQL 管理工具必须和数据库服务器建立连接后才可以建立数据库及相关操作。启动 navicat for MySQL 出现如图 6-5 所示界面，单击图 6-5 所示界面上的连接，出现如图 6-6 所示的界面。在图 6-6 所示的界面输入如下信息：

(1) 主机名：取值是 MySQL 服务器所在计算机的域名或 IP，如果 MySQL 服务器和 MySQL 管理工具驻留在同一台计算机上，主机名取值可以是 localhost 或 127.0.0.1。

(2) 端口：MySQL 服务器占用的端口是 3306。

(3) 用户名与密码：用户名必须是 MySQL 授权的用户名，可取值 root，root 是安装 MySQL 时提供的默认用户，其密码是空(没有密码)。和数据库服务器建立连接后，可以使用 MySQL 提供的有关命令或使用 MySQL 管理工具(单击 MySQL 管理工具界面上的“管理用户”)修改 root 的密码或增加新的用户，本教材使用 root 用户和默认的密码已经能满足学习的要求，因此不再介绍修改密码和增加用户这些内容(不属于本书范畴)。

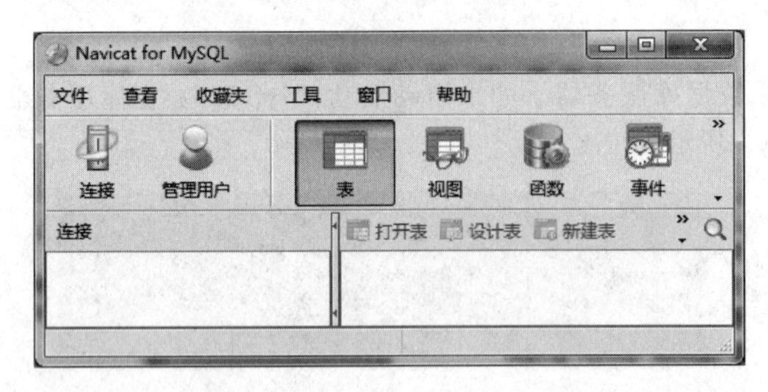

图 6-5 启动 navicat for MySQL 客户端管理工具

建立的连接名称是 gengxiangyi，用户名取 root，密码是空(如图 6-6 所示)。

2) 建立数据库

在新建的连接(gengxiangyi)上单击鼠标右键，选择“打开连接”，以便通过该连接在 MySQL 数据库服务器中建立数据库。

打开连接后，在新建的连接(gengxiangyi)上单击鼠标右键，然后选择“新建数据库”，在弹出的新建数据库对话框中输入有关信息，比如输入数据库的名称，选择使用的字符编码。

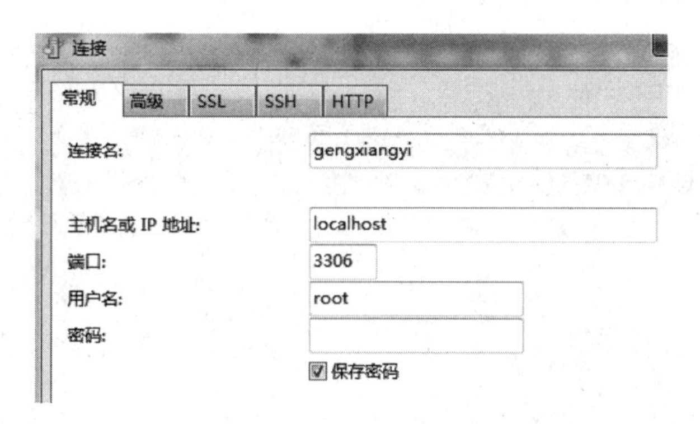

图 6-6 建立一个新连接 gengxiangyi

这里建立的数据库的名字是 warehouse,使用的字符编码是 gb2312(GB2312 Simplified Chinese),创建新数据库后,在连接(gengxiangyi)上可以看到新建立的数据库名称(warehouse),如图 6-7 所示。

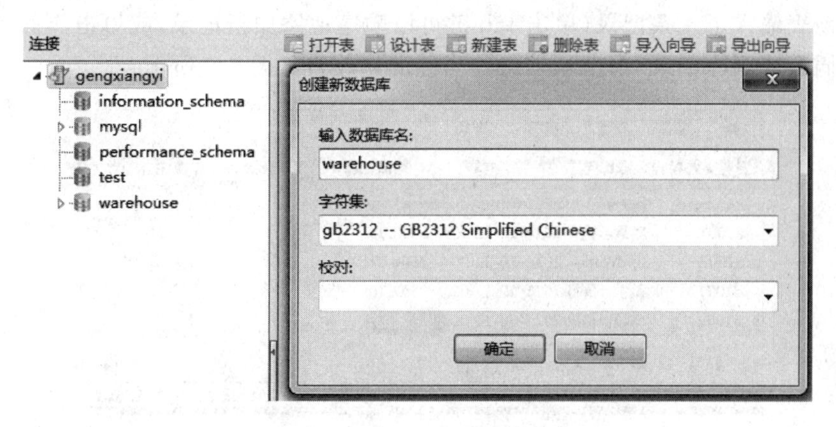

图 6-7 新建 warehouse 数据库

3) 创建表

创建好数据库后,就可以在该数据库中建立若干个表。启动 navicat for MySQL 管理器(如果已经关闭,需要重新打开),用鼠标单击曾建立的 gengxiangyi 连接打开该连接。创建表的步骤如下:

(1) 打开数据库。

如果曾使用某个连接建立了数据库,那么打开该连接后就可以在管理界面上看到曾建立的数据库的名字,在数据库名称上单击鼠标右键,选择“打开数据库”就可以打开该数据库,然后就可以在该数据库中建立表。

(2) 建立表。

打开数据库后,在其“表”选择项上单击鼠标右键,选择“新建表”,将出现创建表的对话框。这里建立的表是 product,该表的字段(列)为:

```
number(char) name(char) madeTime(datetime) price(float)
```

其中 number 字段为主键,如图 6-8 所示。

图 6-8　在数据库中创建表

(3) 管理表。

在“表”选择项单击鼠标左键,可以展开“表”,以便管理曾建立的表,比如管理曾建立的 product 表。这里在曾建立的 product 表上单击鼠标右键,选择“打开表”,然后在弹出的对话框中向该表输入了 4 条记录(单击 tab 键可以顺序地添加新记录,或单击界面下面的“+”或“-”号插入或删除记录,单击“√”保存当前的修改),如图 6-9 所示。

文件 编辑 查看 窗口 帮助

导入向导 导出向导 筛选向导 网格查看 表单查看 备注 十六进位

number	name	madeTime	price
a1001	苹果手机	2014-12-12 00:	8763.98
a1002	android手	2014-10-31 00:	7608.87
b1001	戴尔计算机	2015-12-12 00:	3587.76
b1002	华硕计算机	2015-08-12 00:	2986

INSERT INTO `product` (`number`　　第 4 条记录 (共 4 条) 于 1 页

图 6-9　管理表

2. 使用 MySQL 监视器

启动 MySQL 数据库服务器后,也可以用命令行方式创建数据库(要求有比较好的 SQL 语句基础)。MySQL 提供的监视器(MySQL monitor),允许用户使用命令行方式管理数据库。如果有比较好的数据库知识,特别是 SQL 语句的知识,那么使用命令行方式管理 MySQL 数据库也是很方便的,本节只介绍几个简单的命令,能满足本教材的需求即可。

注意:可以在网络上搜索到 MySQL 命令详解,详细讲解 MySQL 本身的知识内容不属于本书的范畴。

为了启动 MySQL 监视器(即和 MySQL 数据库服务器建立连接),需要执行 MySQL 安装目录的 bin 子目录中的 mysql.exe 文件,执行格式为:

```
mysql -u 用户名 密码
```

或

```
mysql -h localhost -u root -p
```

然后按要求输入密码即可(如果密码是空,可以不输入密码)。

需要再打开一个 MS-DOS 命令行窗口,并使用 MS-DOS 命令进入到 bin 目录中,然后使用默认的 root 用户启动 MySQL 监视器(在安装 MySQL 时 root 用户是默认的一个用户,没有密码)。命令如下:

```
mysql  - u root
```

成功启动 MySQL 监视器后,MS-DOS 窗口出现“mysql＞”字样效果,如图 6-10 所示。如果想关闭 MySQL 监视器,输入 exit 即可。

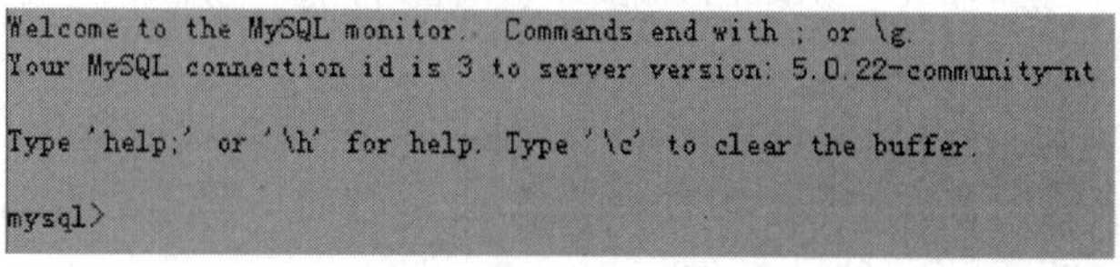

```
Welcome to the MySQL monitor.  Commands end with ; or \g.
Your MySQL connection id is 3 to server version: 5.0.22-community-nt

Type 'help;' or '\h' for help. Type '\c' to clear the buffer.

mysql>
```

图 6-10　启动 MySql 监视器

1) 创建数据库

启动 MySQL 监视器后就可以使用 SQL 语句进行创建数据库、表等操作。在 MS-DOS 命令行窗口输入 SQL 语句需要用“;”号结束,在编辑 SQL 语句的过程中可以使用\c 终止当前 SQL 语句的编辑。需要提醒的是,可以把一个完整的 SQL 语句命令分成几行来输入,最后用分号作结束标志即可。

下面使用 MySQL 监视器创建一个名字为 Book 的数据库,在当前 MySQL 监视器占用的命令行窗口输入创建数据库的 SQL 语句(如图 6-11 所示):

```
create database Book;
```

2) 为数据库建表

创建数据库后就可以使用 SQL 语句在该库中创建表。为了在数据库中创建表,必须首先进入该数据库(即使用数据库),命令格式是:“user 数据库名;”或“user 数据库名”。在当前 MySQL 监视器占用的命令行窗口输入(按 Enter 键确认,进入数据库也可以没有分号):

```
use Book
```

进入数据库 Book,操作如图 6-12 所示。

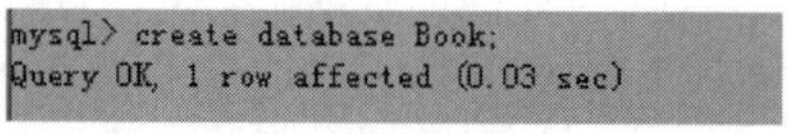

```
mysql> create database Book;
Query OK, 1 row affected (0.03 sec)
```

图 6-11　创建数据库

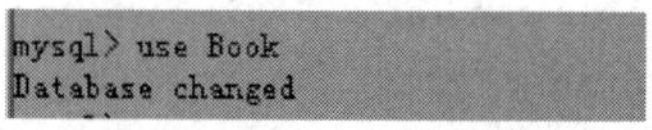

```
mysql> use Book
Database changed
```

图 6-12　进入 Book 数据库

下面在数据库 Book 建立一个名字为 bookForm 表,该表的字段为:

```
ISBN(varchar) name(varchar) price(float)
```

输入创建 bookForm 表的 SQL 语句(创建表的 SQL 语句操作效果如图 6-13 所示):

```
CREATE TABLE bookList (
ISBN varchar(100) not null ,
name varchar(100) CHARACTER SET gb2312,
```

```
price float ,
PRIMARY KEY (ISBN)
);
```



```
mysql> CREATE TABLE bookList (
    -> ISBN varchar(100) not null ,
    -> name varchar(100) CHARACTER SET gb2312,
    -> price float ,
    -> PRIMARY KEY (ISBN)
    -> );
Query OK, 0 rows affected (0.01 sec)
```

图 6-13　创建 bookForm 表

创建 bookForm 表之后(如果已经退出数据库,需要再次进入数据库)就可以使用 SQL 语句对 bookForm 表进行添加、更新和查询操作。在当前 MySQL 监视器占用的命令行窗口键入插入记录的 SQL 语句(每次只能插入一条记录):

```
insert into bookList values('7 - 302 - 01465 - 5','高等数学',28.67);
insert into booklist values('7 - 352 - 01465 - 8','大学英语',58.5);
```

操作过程如图 6-14(a)所示。

在当前 MySQL 监视器占用的命令行窗口输入查询记录的 SQL 语句:

```
select * from bookList;
```

操作过程如图 6-14(b)所示。

3) 导入. sql 文件中的 SQL 语句

在使用 MySQL 监视器时,如果觉得在命令行输入 SQL 语句不方便,那么可以事先将需要的 SQL 语句保存在一个扩展名是. sql 的文本文件中,然后在 MySQL 监视器占用的命令行窗口使用 source 命令导入. sql 的文本文件中的 SQL 语句。

建议再建立数据库之后使用这样的方式操作数据库。例如,已经建立了名字是 Car 的数据库,并开始使用这个数据库,那么就可以将操作数据库的有关 SQL 语句,比如在数据库建立表,插入记录等 SQL 语句,存放在一个. sql 文本文件中,比如 group. sql 文本文件中(group. sql 保存在 D:\1000 中),group. sql 文本文件的内容如下:

group. sql

```
drop table carList ;
create table carList(
number char(60) CHARACTER SET gb2312 not null,
name char(50) CHARACTER SET gb2312 ,
price float,
year date,
PRIMARY KEY(number)
);
insert into carList values('加 A89CQ8','奔驰','820000','2015 - 12 - 26');
insert into carList values('洲 C12456','宝马','620000','2015 - 10 - 10');
select * from carList;
```

然后在当前 MySQL 监视器占用的命令行窗口输入如下命令:

```
source d:/1000/group.sql
```

导入 SQL 语句。如果.sql 文件中存在错误的 SQL 语句,将提示错误信息,否则将成功执行这些 SQL 语句(效果如图 6-14(c)所示)。

```
mysql> insert into bookList values('7-302-01465-5','高等数学',28.67);
Query OK, 1 row affected (0.45 sec)

mysql> insert into booklist values('7-352-01465-8','大学英语',58.5);
Query OK, 1 row affected (0.00 sec)
```

(a) 向bookList表添加记录

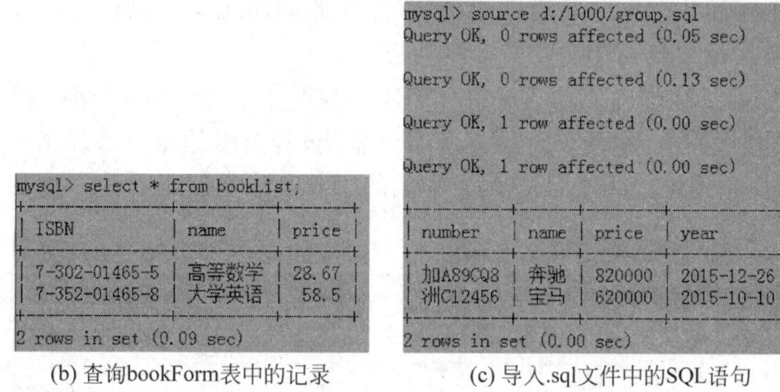

(b) 查询bookForm表中的记录　　(c) 导入.sql文件中的SQL语句

图 6-14　三个 MySQL 监视器页面效果图

4）删除数据库或表

删除数据库的命令：drop database <数据库名>。例如：删除名为 tiger 的数据库：

```
drop database tiger;
```

删除表的命令：drop table <表名>。例如,使用 book 数据库后,执行

```
drop table bookList;
```

将删除 book 数据库中的 bookList 表。

6.2　JDBC

JDBC(Java DataBase Connectivity)提供了访问数据库的 API,即由一些 Java 类和接口组成,是 Java 运行平台的核心类库中的一部分。在 JSP 中可以使用 JDBC 实现对数据库中表的记录的查询、修改和删除等操作。JDBC 技术在数据库开发中占有很重要的地位,JDBC 操作不同的数据库仅仅是连接方式上的差异而已,使用 JDBC 的应用程序一旦和数据库建立连接,就可以使用 JDBC 提供的 API 操作数据库,如图 6-15 所示。

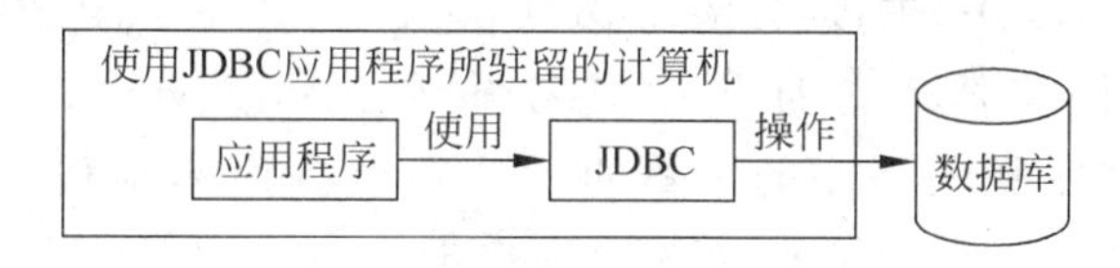

图 6-15　使用 JDBC 操作数据库

经常使用 JDBC 进行如下的操作：

(1) 与一个数据库建立连接。

(2) 向已连接的数据库发送 SQL 语句。

(3) 处理 SQL 语句返回的结果。

6.3 连接 MySQL 数据库

应用程序为了能和数据库交互信息，必须首先和数据库建立连接。目前在开发中常用的连接数据库的方式是加载 JDBC-数据库驱动程序。

用 Java 语言编写的数据库驱动程序称作 JDBC-数据库驱动程序。JDBC 提供的 API 将 JDBC-数据库驱动程序转换为 DBMS(数据库管理系统)所使用的专用协议来实现和特定的 DBMS 交互信息，简单地说，JDBC 可以调用本地的 JDBC-数据库驱动程序和相应的数据库建立连接，如图 6-16 所示。

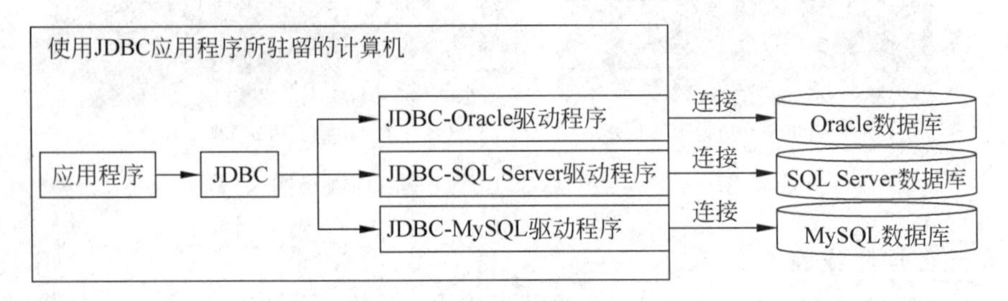

图 6-16 使用 JDBC-数据库驱动程序

使用 JDBC-数据库驱动程序方式和数据库建立连接需要经过两个步骤：

(1) 加载 JDBC-数据库驱动程序。

(2) 和指定的数据库建立连接。

6.3.1 加载 JDBC-数据库驱动程序

目前，许多数据库厂商都提供了相应的 JDBC-数据库驱动程序。当使用 JDBC-数据库驱动程序访问数据库时，必须要保证连接数据库的应用程序所驻留的计算机上安装有相应 JDBC-数据库驱动程序。比如，Tomcat 服务器上的某个 Web 应用程序想访问 MySQL 数据库管理系统所管理的数据库，那么 Tomcat 服务器所驻留的计算机上必须要安装 MySQL 提供的 JDBC-数据库驱动程序。

可以登录 MySQL 的官方网站：www. mysql. com，下载 JDBC-数据库驱动程序(JDBC Driver for MySQL)。登录 www. mysql. com 后，在页面的导航条上选择 products，然后在页面的右侧区中选择 MySQL 连接器(MySQL Connectors)，进入到数据库驱动程序的下载页面，也可以直接在浏览器的地址栏中直接输入 http://www. mysql. com/products/connector/进入到数据库驱动程序的下载页面。在数据库驱动程序的下载页面上选择 “JDBC Driver for MySQL (Connector/J)”然后单击 download 下载按钮即可。

本教材下载的是 mysql-connector-java-5. 1. 28. zip，将该 zip 文件解压至硬盘，在解压后的目录下的 mysql-connector-java-5. 1. 28-bin. jar 文件就是连接 MySQL 数据库的 JDBC-数

据库驱动程序。将该驱动程序复制到 Tomcat 服务器所使用的 JDK 的扩展目录中(即 java_home 环境变量指定的 JDK，见第 1 章的 1.2.1)。比如：D:\jdk1.7\jre\lib\ext，或复制到 Tomcat 服务器安装目录的\common\lib 文件夹中。比如：D:\apache-tomcat-8.0.3\common\lib。

应用程序加载 MySQL 的 JDBC-数据库驱动程序代码如下：

```
try{ Class.forName("com.mysql.jdbc.Driver");
}
catch(Exception e){}
```

6.3.2 建立连接

Java.sql 包中的 DriverManager 类有两个用于建立连接的类方法(static 方法)：

```
Connection getConnection(java.lang.String, java.lang.String, java.lang.String)
Connection getConnection(java.lang.String)
```

上述方法都可能抛出 SQLException 异常，DriverManager 类调用上述方法可以和数据库建立连接，即可以返回一个 Connection 对象。

为了能和 MySQL 数据库服务器管理的数据库建立连接，必须保证该 MySQL 数据库服务器已经启动，如果没有更改过 MySQL 数据库服务器的配置，那么该数据库服务器占用的端口是 3306。假设 MySQL 数据库服务器所驻留的计算机的 IP 地址是 192.168.100.1。应用程序要和 MySQL 数据库服务器管理的数据库 warehouse 建立连接，而有权访问数据库 warehouse 的用户的 id 和密码分别是 root、99，那么使用

```
Connection getConnection(java.lang.String, java.lang.String, java.lang.String)
```

方法建立连接的代码如下：

```
try{   String uri = "jdbc:mysql:// 192.168.100.1:3306/warehouse";
       String user = "root";
       String password = "99";
       con = DriverManager.getConnection(uri,user,password);
   }
catch(SQLException e){
       System.out.println(e);
}
```

如果 root 用户没有设置密码，那么将上述

```
String password = "99";
```

更改为：

```
String password = "";
```

使用

```
Connection getConnection(java.lang.String)
```

方法建立连接的代码如下：

```
try{  String uri = " jdbc:mysql:// 192.168.100.1:3306/warehouse? user = root&password = 99";
      con = DriverManager.getConnection(uri);
   }
catch(SQLException e){
      System.out.println(e);
}
```

如果 root 用户没有设置密码,那么将上述 uri 中的

```
&password = 99
```

更改为:

```
&password =
```

在某些 Web 程序中需要避免操作数据库出现中文乱码(细节见 6.3.3 节),那么需要使用

```
Connection getConnection(java.lang.String)
```

方法建立连接,连接中的代码是(假设用户是 root,其密码是 99):

```
String uri = "jdbc:mysql://127.0.0.1/warehouse?" +
             "user = root&password = 99&characterEncoding = gb2312";
con = DriverManager.getConnection(uri);
```

应用程序一旦和某个数据库建立连接,就可以通过 SQL 语句和该数据库中的表交互信息,比如查询、修改、更新表中的记录。

注意:如果用户要和连接 MySQL 驻留在同一计算机上,使用的 IP 地址可以是 127.0.0.1 或 localhost。另外,由于 3306 是 MySQL 数据库服务器的默认端口号,连接数据库时允许应用程序省略默认的 3360 端口号。

6.3.3 MySQL 乱码解决方案

用 JSP 页面访问 MySQL 数据库时(更新和插入记录)可能出现"中文乱码"问题。本节介绍的办法可以防止 JSP 页面访问数据库时出现"中文乱码"问题。

1. 数据库和表使用支持中文的字符编码

如果使用 MySQL 管理工具(界面直观方便),那么在创建数据库时会提示客户设置数据库使用的字符编码,在创建表时也会提示用户指定字段(列)使用的字符编码(见 6.1.2),用户可以把数据库和表的字段(如果需要支持中文字符)使用的字符编码设置成 gb2312 或 gbk。

如果使用 MySQL 监视器创建数据库,可以在创建数据库时指定数据库使用的字符编码:

```
create 数据库名 CHARACTER SET 字符编码
```

例如,创建名字是 people 的数据库(CHARACTER SET 也可以小写,这里是为了强调使用了大写):

```
create people CHARACTER SET gb2312
```

创建表时,可以指定某个字段使用的字符编码(否则是默认的字符编码不支持中文):

```
字段名 类型 CHARACTER SET 字符编码
```

例如，建立名字是 myList 的表，并让其中的 name 字段使用 gb2312 编码：

```
create table myList (
id' int NOT NULL auto_increment,
name varchar(100) CHARACTER SET gb2312,
PRIMARY KEY (id)
);
```

2. 连接数据库支持中文编码

JSP 中连接 MySQL 数据库时，需要使用

```
Connection getConnection(java.lang.String)
```

方法建立连接，而且向该方法参数传递的字符串是：

```
"jdbc:mysql://地址/数据库? user = 用户 &password = 密码 &characterEncoding = gb2312";
```

例如，假设数据名是 book，用户是 root，密码为 99，那么连接方式是：

```
Connection con;
String uri = " jdbc: mysql://127. 0. 0. 1/book? user = root&password = 99&characterEncoding =
gb2312";
con = DriverManager.getConnection(uri);
```

如果 root 用户没有设置密码，那么将上述代码中的

```
&password = 99
```

更改为：

```
&password =
```

编辑代码时，如果 uri 含有较多的字符，建议使用字符串并置连接符号"＋"连接字符串，让代码更清晰些。例如：

```
String uri = "jdbc:mysql://127.0.0.1/book?" +
             "user = root&password = 99&characterEncoding = gb2312";
con = DriverManager.getConnection(uri);
```

例 6-1 是一个简单的 JSP 页面，该页面中的 Java 程序片代码负责加载 JDBC-驱动程序，并连接到数据库 warehouse，查询 product 表中的全部记录(见 6.1 节曾建立的 warehouse 数据库)。页面运行效果如图 6-17 所示。

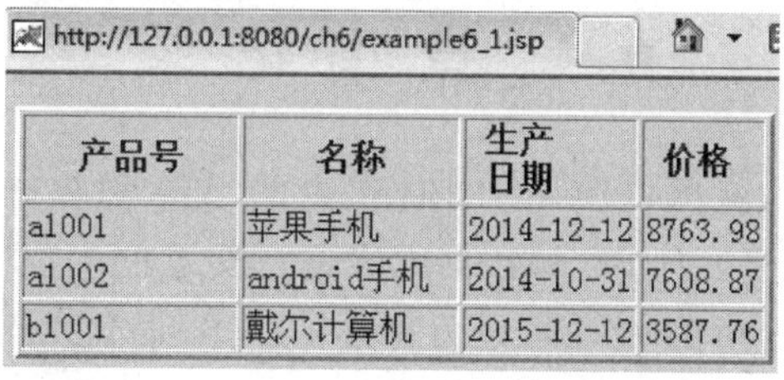

http://127.0.0.1:8080/ch6/example6_1.jsp

产品号	名称	生产日期	价格
a1001	苹果手机	2014-12-12	8763.98
a1002	android手机	2014-10-31	7608.87
b1001	戴尔计算机	2015-12-12	3587.76

图 6-17 访问 warehouse 数据库

例 6-1

example6_1.jsp

```
<%@ page contentType = "text/html;charset = GB2312" %>
<%@ page import = "java.sql.*" %>
<HTML><BODY bgcolor = cyan>
<% Connection con;
   Statement sql;
   ResultSet rs;
   try{ Class.forName("com.mysql.jdbc.Driver");
   }
   catch(Exception e){}
   try { String uri = "jdbc:mysql://127.0.0.1/warehouse";
         String user = "root";
         String password = "";
         con = DriverManager.getConnection(uri,user,password);
         //也可以写成 con = DriverManager.getConnection(uri + "? user = root&password = ");
         sql = con.createStatement();
         rs = sql.executeQuery("SELECT * FROM product ");
         out.print("<table border = 2>");
         out.print("<tr>");
           out.print("<th width = 100>" + "产品号");
           out.print("<th width = 100>" + "名称");
           out.print("<th width = 50>" + "生产日期");
           out.print("<th width = 50>" + "价格");
         out.print("</TR>");
         while(rs.next()){
           out.print("<tr>");
             out.print("<td>" + rs.getString(1) + "</td>");
             out.print("<td>" + rs.getString(2) + "</td>");
             out.print("<td>" + rs.getDate("madeTime") + "</td>");
             out.print("<td>" + rs.getFloat("price") + "</td>");
           out.print("</tr>") ;
         }
         out.print("</table>");
         con.close();
   }
   catch(SQLException e){
         out.print(e);
   }
 %>
</BODY></HTML>
```

6.4 查询记录

和数据库建立连接后,就可以使用 JDBC 提供的 API 和数据库交互信息。比如查询、修改和更新数据库中的表等。JDBC 和数据库表进行交互的主要方式是使用 SQL 语句(其他方式见 6.8 节),JDBC 提供的 API 可以将标准的 SQL 语句发送给数据库,实现和数据库的交互。

对一个数据库中表进行查询操作的具体步骤如下:

1. 向数据库发送 SQL 查询语句

首先使用 Statement 声明一个 SQL 语句对象,然后让已创建的连接对象 con 调用方法 createStatement()创建这个 SQL 语句对象,代码如下:

```
try{   Statement sql = con.createStatement();
   }
catch(SQLException e ){
      System.out.println(e);
}
```

2. 处理查询结果

有了 SQL 语句对象后,这个对象就可以调用相应的方法查询数据库中的表,并将查询结果存放在一个 ResultSet 对象中。也就是说 SQL 查询语句对数据库的查询操作将返回一个 ResultSet 对象,ResultSet 对象是以统一形式的列组织的数据行组成。例如,对于

```
ResultSet rs = sql.executeQuery("SELECT * FROM product");
```

内存的结果集对象 rs 的列数是 4 列,刚好和 product 的列数相同,第 1 列至第 4 列分别是 number、name、madeTime 和 price 列; 而对于

```
ResultSet rs = sql.executeQuery("SELECT name,price FROM product");
```

内存的结果集对象 rs 只有两列,第 1 列是 name 列、第 2 列是 price 列。

ResultSet 对象一次只能看到一个数据行,使用 next()方法走到下一数据行,获得一行数据后,ResultSet 对象可以使用 getXxx 方法获得字段值(列值),将位置索引(第 1 列使用 1,第 2 列使用 2 等)或列名传递给 getXxx 方法的参数即可。表 6-1 给了出了 ResultSet 对象的若干方法。

表 6-1 ResultSet 类的若干方法

返 回 类 型	方 法 名 称
boolean	next()
byte	getByte(int columnIndex)
Date	getDate(int columnIndex)
double	getDouble(int columnIndex)
float	getFloat(int columnIndex)
int	getInt(int columnIndex)
long	getLong(int columnIndex)
String	getString(int columnIndex)
byte	getByte(String columnName)
Date	getDate(String columnName)
double	getDouble(String columnName)
float	getFloat(String columnName)
int	getInt(String columnName)
long	getLong(String columnName)
String	getString(String columnName)

注意：无论字段是何种属性，总可以使用 getString(int columnIndex) 或 getString(String columnName)方法返回字段值的串表示。

当使用 ResultSet 的 getXxx 方法查看一行记录时，不可以颠倒字段的顺序。例如，不可以：

```
rs.getFloat(4);
rs.getDate(3)
```

6.4.1 顺序查询

查询数据库中的一个表的记录时，希望知道表中字段的个数以及各个字段的名字。由于无论字段是何种属性，总可以使用 getSring 方法返回字段值的串表示，因此，只要知道了表中字段的个数或字段的名字，就可以方便地查询表中的记录。

怎样知道一个表中有哪些字段呢？通过使用 JDBC 提供的 API，可以在查询之前知道表中的字段的个数和名字，这有助于编写可复用的查询代码。

当和数据库建立连接对象 con 之后，该连接对象调用 getMetaData()方法可以返回一个 DatabaseMetaData 对象。例如：

```
DatabaseMetaData metadata = con.getMetaData();
```

Metadata 对象再调用 getColumns()方法可以将表的字段信息以行列的形式存储在一个 ResultSet 对象中。例如：

```
ResultSet tableMessage = metadata.getColumns(null,null, "product",null);
```

如果 product 表有 *n* 个字段，tableMessage 就刚好有 *n* 行、每行 4 列。每行分别含有和相应字段有关的信息，信息的次序为："数据库名"、"数据库扩展名"、"表名"，"字段名"。例如，product 表有 4 个字段，那么上述 tableMessage 刚好有 4 行，每行有 4 列，如图 6-18 所示。

Warehouse	dbo	product	number
Warehouse	dbo	product	name
Warehouse	dbo	product	madeTime
Warehouse	dbo	product	price

图 6-18 tableMessage 对象

tableMessage 对象调用 next 方法使游标向下移动一行(游标的初始位置在第 1 行之前)，然后 tableMessage 调用 getXxx 方法可以查看该行中列的信息，其中最重要的信息是第 4 列，该列上的信息为字段的名字。

在下面的例 6-2 中，有一个负责查询数据库的 Tag 文件，该 Tag 文件负责连接数据库、查询表。用户通过 JSP 页面 example6_2.jsp 提供的表单输入数据库以及表的名字，并提交给 inquire.jsp 页面，然后 inquire.jsp 页面调用 Tag 文件查询数据库。Tag 文件将查询结果用表格的形式返回给 inquire.jsp 页面。example6_2.jsp 和 inquire.jsp 页面的效果如图 6-19(a)、(b)所示。

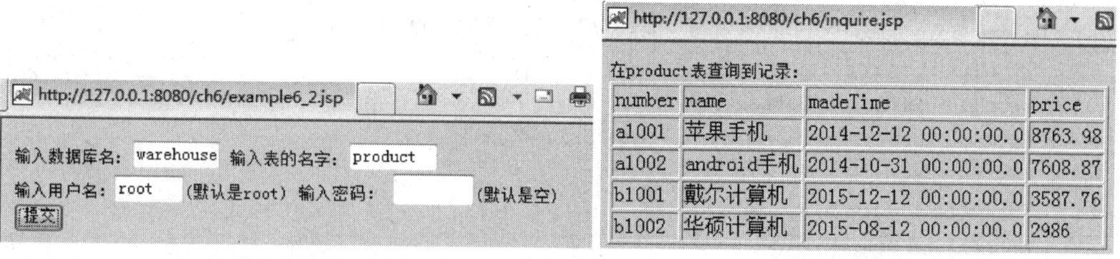

number	name	madeTime	price
a1001	苹果手机	2014-12-12 00:00:00.0	8763.98
a1002	android手机	2014-10-31 00:00:00.0	7608.87
b1001	戴尔计算机	2015-12-12 00:00:00.0	3587.76
b1002	华硕计算机	2015-08-12 00:00:00.0	2986

(a) 输入查询信息　　　　(b) 显示查询结果

图 6-19　两个顺序查询效果图

例 6-2

example6_2. jsp

```
<%@ page contentType = "text/html;charset = GB2312" %>
<%@ taglib tagdir = "/WEB - INF/tags" prefix = "inquire" %>
<HTML><Body bgcolor = cyan><Font size = 2>
<FORM action = "inquire.jsp" Method = "post" >
   输入数据库名：<Input type = text name = "dataBaseName" size = 8>
   输入表的名字：<Input type = text name = "tableName" size = 8>
  <br>输入用户名：<Input type = text name = "user" size = 6>(默认是 root)
   输入密码：<Input type = "password" name = "password" size = 6>(默认是空)
  <br><Input type = submit name = "g" value = "提交">
</Form>
</Font></Body></HTML>
```

inquire. jsp

```
<%@ page contentType = "text/html;charset = GB2312" %>
<%@ taglib tagdir = "/WEB - INF/tags" prefix = "inquire" %>
<HTML><Body bgcolor = cyan><Font size = 2>
  <% String database = request.getParameter("dataBaseName");
     String tName = request.getParameter("tableName");
     String id = request.getParameter("user");
     String secret = request.getParameter("password");
  %>
  <inquire:QueryTag dataBaseName = "<% = database %>"
                    tableName = "<% = tName %>"
                    user = "<% = id %>"/>
   在<% = biao %>表查询到记录：<% -- biao 是 Tag 文件返回的对象 -- %>
   <BR><% = queryResult %><% -- queryResult 是 Tag 文件返回的对象 -- %>
</Font></Body></HTML>
```

QueryTag. tag

```
<%@ tag pageEncoding = "GB2312" %>
<%@ tag import = "java.sql.*" %>
<%@ attribute name = "dataBaseName" required = "true" %>
<%@ attribute name = "tableName" required = "true" %>
<%@ attribute name = "user" required = "true" %>
<%@ attribute name = "password" required = "false" %>
```

```
<%@ variable name-given="biao" scope="AT_END" %>
<%@ variable name-given="queryResult" scope="AT_END" %>
<% StringBuffer result;
    result=new StringBuffer();
    try{ Class.forName("com.mysql.jdbc.Driver");
    }
    catch(Exception e){}
    Connection con;
    Statement sql;
    ResultSet rs;
    try{  result.append("<table border=1>");
          String uri="jdbc:mysql://127.0.0.1/"+dataBaseName;
          con=DriverManager.getConnection(uri,user,password);
          DatabaseMetaData metadata=con.getMetaData();
          ResultSet rs1=metadata.getColumns(null,null,tableName,null);
          int 字段个数=0;
          result.append("<tr>");
          while(rs1.next()){
             字段个数++;
             String clumnName=rs1.getString(4);
             result.append("<td>"+clumnName+"</td>");
          }
          result.append("</tr>");
          sql=con.createStatement();
          rs=sql.executeQuery("SELECT * FROM "+tableName);
          while(rs.next()){
                result.append("<tr>");
                for(int k=1;k<=字段个数;k++)
                   result.append("<td>"+rs.getString(k)+"</td>");
                result.append("</tr>");
          }
          result.append("</table>");
          con.close();
    }
    catch(SQLException e){
          result.append("请输入正确的用户名和密码");
}
//返回对象 queryResult
    jspContext.setAttribute("queryResult",new String(result));
    jspContext.setAttribute("biao",tableName);              //返回 biao 对象
%>
```

6.4.2 随机查询

前面学习了使用ResultSet类的next()方法顺序地查询数据，但有时候需要在结果集中前后移动、显示结果集指定的一条记录或随机显示若干条记录等。这时，必须要返回一个可滚动的结果集。为了得到一个可滚动的结果集，需使用下述方法先获得一个Statement对象：

```
Statement stmt=con.createStatement(int type,int concurrency);
```

然后，根据参数的 type、concurrency 的取值情况，stmt 返回相应类型的结果集：

```
ResultSet re = stmt.executeQuery(SQL 语句);
```

type 的取值决定滚动方式，取值可以是：

- ResultSet.TYPE_FORWORD_ONLY：结果集的游标只能向下滚动。
- ResultSet.TYPE_SCROLL_INSENSITIVE：结果集的游标可以上下移动，当数据库变化时，当前结果集不变。
- ResultSet.TYPE_SCROLL_SENSITIVE：返回可滚动的结果集，当数据库变化时，当前结果集同步改变。

Concurrency 取值决定是否可以用结果集更新数据库，Concurrency 取值：

- ResultSet.CONCUR_READ_ONLY：不能用结果集更新数据库中的表。
- ResultSet.CONCUR_UPDATABLE：能用结果集更新数据库中的表。

滚动查询经常用到 ResultSet 的下述方法：

- public boolean previous()：将游标向上移动，该方法返回 boolean 型数据，当移到结果集第一行之前时返回 false。
- public void beforeFirst：将游标移动到结果集的初始位置，即在第 1 行之前。
- public void afterLast()：将游标移到结果集最后一行之后。
- public void first()：将游标移到结果集的第 1 行。
- public void last()：将游标移到结果集的最后一行。
- public boolean isAfterLast()：判断游标是否在最后一行之后。
- public boolean isBeforeFirst()：判断游标是否在第 1 行之前
- public boolean ifFirst()：判断游标是否指向结果集的第 1 行。
- public boolean isLast()：判断游标是否指向结果集的最后一行。
- public int getRow()：得到当前游标所指行的行号，行号从 1 开始，如果结果集没有行，返回 0。
- public boolean absolute(int row)：将游标移到参数 row 指定的行号。

注意：如果 row 取负值，就是倒数的行数。absolute(-1)表示移到最后一行，absolute(—2)表示移到倒数第 2 行。当移动到第 1 行前面或最后一行的后面时，该方法返回 false。

在下面的例 6-3 中，负责查询数据库的 Tag 文件首先将游标移动到最后一行，然后再获取最后一行的行号，以便获得表中的记录数目，该 Tag 文件能随机输出若干条记录。用户可以在 JSP 页面 example6_3.jsp 页面输入数据库名、表名和随机查询的记录数目，并提交给 random.jsp，random.jsp 调用 Tag 文件完成查询操作，并将有关结果返回给 random.jsp 页面。在例 6-3 中，输入的数据库名是 warehouse，表是 product 表，随机查询的记录数目是 3。example6_3.jsp 和 random.jsp 的效果如图 6-20(a)、(b)所示。

例 6-3

example6_3.jsp

```
<%@ page contentType = "text/html;charset = GB2312" %>
<%@ taglib tagdir = "/WEB - INF/tags" prefix = "inquire" %>
<HTML><Body bgcolor = cyan><Font size = 2>
<P>随机查询记录
```

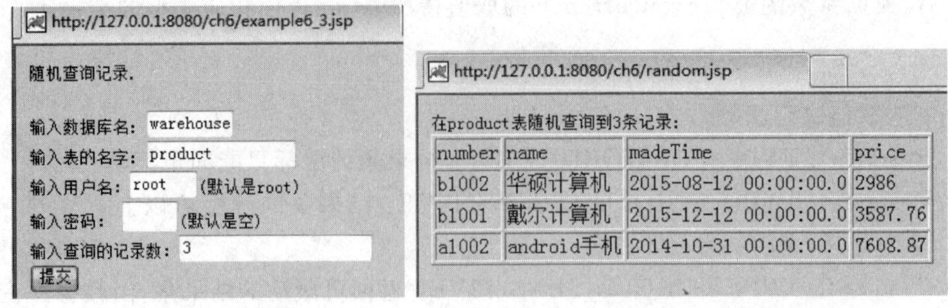

(a) 输入查询信息　　　　(b) 随机查询到3条记录

图 6-20　两个随机查询页面效果图

```
<FORM action = "random.jsp" Method = "post" >
   输入数据库名：<Input type = text name = "databaseName" size = 8 >
   <br>输入表的名字：<Input type = text name = "tableName" size = 15 >
   <br>输入用户名：<Input type = text name = "user" size = 6 >(默认是 root)
   <br>输入密码：<Input type = "password" name = "password" size = 3 >(默认是空)
   <br>输入查询的记录数：<Input type = text name = "count" value = 3 >
   <br><Input type = submit name = "g" value = "提交">
</Form>
</Font></Body></HTML>
```

random.jsp

```
<%@ page contentType = "text/html;charset = GB2312" %>
<%@ taglib tagdir = "/WEB-INF/tags" prefix = "inquire" %>
<HTML><Body bgcolor = cyan><Font size = 2>
  <% String dName = request.getParameter("databaseName");
     String tName = request.getParameter("tableName");
     String id = request.getParameter("user");
     String secret = request.getParameter("password");
     String n = request.getParameter("count");
  %>
  <inquire:RandomQuery databaseName = "<% = dName %>"
                       tableName = "<% = tName %>"
                       user = "<% = id %>"
                       password = "<% = secret %>"
                       count = "<% = n %>"/>
   在<% = biao %>表随机查询到<% = randomCount %>条记录：
   <% -- biao 和 randomCount 是 Tag 文件返回的对象 -- %>
   <BR><% = queryResult %>          <% -- queryResult 是 Tag 文件返回的对象 -- %>
</Font></Body></HTML>
```

RandomQuery.tag

```
<%@ tag pageEncoding = "GB2312" %>
<%@ tag import = "java.sql.*" %>
<%@ tag import = "java.util.*" %>
<%@ attribute name = "databaseName" required = "true" %>
<%@ attribute name = "tableName" required = "true" %>
<%@ attribute name = "user" required = "true" %>
```

```
<%@ attribute name = "password" required = "true" %>
<%@ attribute name = "count" required = "true" %>
<%@ variable name-given = "biao" scope = "AT_END" %>
<%@ variable name-given = "queryResult" scope = "AT_END" %>
<%@ variable name-given = "randomCount" scope = "AT_END" %>
<%
   Vector vector = new Vector();
   StringBuffer result;
   result = new StringBuffer();
   try{  Class.forName("com.mysql.jdbc.Driver");
   }
   catch(Exception e){}
   Connection con;
   Statement sql;
   ResultSet rs;
   int n = 0;
   try{  result.append("<table border = 1>");
         String uri = "jdbc:mysql://127.0.0.1/" + databaseName;
         con = DriverManager.getConnection(uri,user,password);
         DatabaseMetaData metadata = con.getMetaData();
         ResultSet rs1 = metadata.getColumns(null,null,tableName,null);
         int 字段个数 = 0;
         result.append("<tr>");
         while(rs1.next()){
            字段个数++;
            String clumnName = rs1.getString(4);
            result.append("<td>" + clumnName + "</td>");
         }
         result.append("</tr>");
         sql = con.createStatement(ResultSet.TYPE_SCROLL_SENSITIVE,
                                   ResultSet.CONCUR_READ_ONLY);
         rs = sql.executeQuery("SELECT * FROM " + tableName);
         rs.last();
         int rowNumber = rs.getRow();
         int number = rowNumber;                        //获取记录数
         for(int i = 1;i <= number;i++)
            vector.add(new Integer(i));
         int m = Math.min(Integer.parseInt(count),number);
         n = m;
         while(m > 0){
             int i = (int)(Math.random() * vector.size());
             //从 vector 中随机抽取一个元素:
             int index = ((Integer)vector.elementAt(i)).intValue();
             rs.absolute(index);                        //游标移到这一行
             result.append("<tr>");
             for(int k = 1;k <= 字段个数;k++)
                result.append("<td>" + rs.getString(k) + "</td>");
             result.append("</tr>");
             m--;
             vector.removeElementAt(i);                 //将抽取过的元素从 vector 中删除
          }
```

```
            result.append("</table>");
            con.close();
        }
        catch(SQLException e){
            result.append("请输入正确的用户名和密码");
    }
    //返回 queryResult 对象
        jspContext.setAttribute("queryResult",new String(result));
    jspContext.setAttribute("biao",tableName);              //返回 biao 对象
    //返回 randomCount 对象
        jspContext.setAttribute("randomCount",String.valueOf(n));

%>
```

6.4.3 条件查询

例 6-4 是查询数据库 warehouse 中的 product 表(见 6.1.2 创建的数据库)。用户通过 JSP 页面 example6_4.jsp 输入查询条件,如果按产品号查询,查询条件提交给 byNumber.jsp 页面;如果按价格查询,查询条件提交给 byPrice.jsp 页面。该例 6-4 中的 Tag 文件 NumberCondtion.tag 根据产品号查询记录,PriceConditon.tag 根据价格查询记录。example6_4.jsp、byNumber.jsp 和 byPrice.jsp 的效果如图 6-21(a)、(b)、(c)所示。

为了避免出现中文乱码,例子中数据库的连接方式采用(见 6.3.3):

```
"jdbc:mysql://地址/数据库?user=用户&password=密码&characterEncoding=gb2312";
```

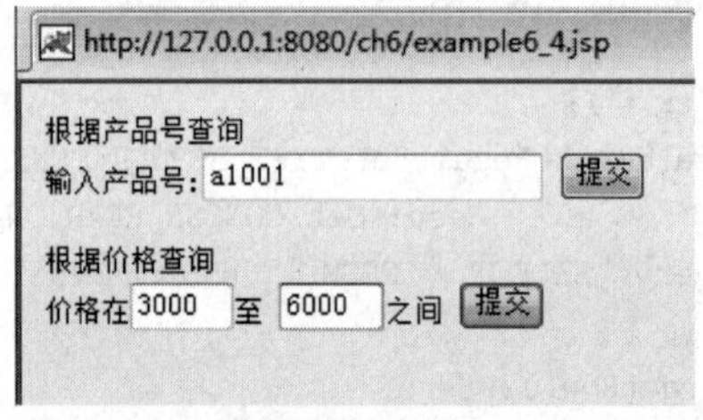

(a) 输入查询条件

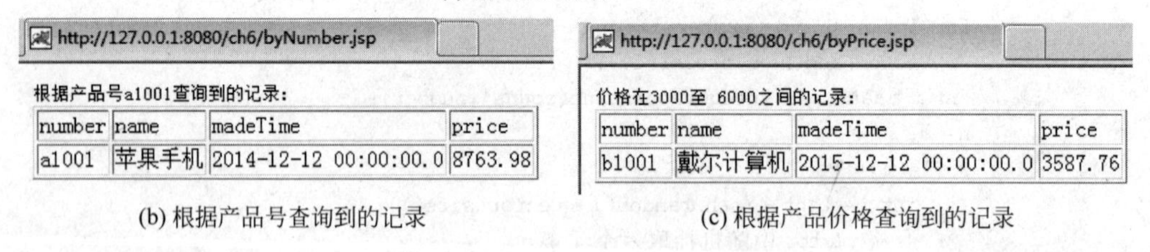

(b) 根据产品号查询到的记录　　　(c) 根据产品价格查询到的记录

图 6-21　3 个条件查询页面效果图

例 6-4

example6_4.jsp

```
<%@ page contentType="text/html;charset=GB2312" %>
<HTML><Body bgcolor=yellow><Font size=2>
<FORM action="byNumber.jsp" Method="post">
根据产品号查询
```

```
< BR >输入产品号:< Input type = text name = "number">
< Input type = submit name = "g" value = "提交">
</Form >
< FORM action = "byPrice.jsp" Method = "post">
  根据价格查询
  < BR > 价格在< Input type = text name = "priceMin" size = 5 >至
        < Input type = text name = "priceMax" size = 5 >之间
        < Input type = submit value = "提交">
</Form >
</Font ></BODY ></HTML >
```

byNumber.jsp

```
<%@ page contentType = "text/html;charset = GB2312" %>
<%@ taglib tagdir = "/WEB-INF/tags" prefix = "inquire" %>
< HTML >< BODY >< Font size = 2 >
<% String number = request.getParameter("number");
   if(number == null)
      number = "";
   byte [] bb = number.getBytes("iso-8859-1");
   number = new String(bb);
%>
< inquire:NumberConditon number = "<% = number %>" />
根据产品号<% = number %>查询到的记录:
< BR ><% = queryResultByNumber %><% -- queryResultByNumber 是 Tag 文件返回的对象 -- %>
</Font ></BODY ></HTML >
```

byPrice.jsp

```
<%@ page contentType = "text/html;charset = GB2312" %>
<%@ taglib tagdir = "/WEB-INF/tags" prefix = "inquire" %>
< HTML >< BODY >< Font size = 2 >
<% String min = request.getParameter("priceMin");
   String max = request.getParameter("priceMax");
%>
< inquire:PriceConditon priceMin = "<% = min %>" priceMax = "<% = max %>"/>
价格在<% = min %>至 <% = max %>之间的记录:
< BR ><% = queryResultByPrice %><% -- queryResultByPrice 是 Tag 文件返回的对象 -- %>
</Font ></BODY ></HTML >
```

NumberCondition.tag

```
<%@ tag pageEncoding = "GB2312" %>
<%@ tag import = "java.sql.* " %>
<%@ attribute name = "number" required = "true" %>
<%@ variable name-given = "queryResultByNumber" scope = "AT_END" %>
<%   StringBuffer result;
     result = new StringBuffer();
     try{  Class.forName("com.mysql.jdbc.Driver");
     }
     catch(Exception e){}
     Connection con;
     Statement sql;
```

```
        ResultSet rs;
        int n = 0;
        try{   result.append("< table border = 1 >");
               String uri =
               "jdbc:mysql://127.0.0.1/warehouse?" +
               "user = root&password = &characterEncoding = gb2312";
               con = DriverManager.getConnection(uri);
               DatabaseMetaData metadata = con.getMetaData();
               ResultSet rs1 = metadata.getColumns(null,null,"product",null);
               int 字段个数 = 0;
               result.append("< tr >");
               while(rs1.next()){
                  字段个数++;
                  String clumnName = rs1.getString(4);
                  result.append("< td >" + clumnName + "</td >");
               }
               result.append("</tr >");
               sql = con.createStatement();
               String condition =
               "SELECT * FROM product Where number = '" + number + "'";
               rs = sql.executeQuery(condition);
               while(rs.next()){
                  result.append("< tr >");
                  for(int k = 1;k <= 字段个数;k++)
                    result.append("< td >" + rs.getString(k) + "</td >");
                    result.append("</tr >");
               }
               result.append("</table >");
               con.close();
        }
        catch(SQLException e){
               result.append(e);
        }
        jspContext.setAttribute("queryResultByNumber",new String(result));
%>
```

PriceConditon.tag

```
<%@ tag pageEncoding = "GB2312" %>
<%@ tag import = "java.sql.*" %>
<%@ attribute name = "priceMax" required = "true" %>
<%@ attribute name = "priceMin" required = "true" %>
<%@ variable name - given = "queryResultByPrice" scope = "AT_END" %>
<%    float max = Float.parseFloat(priceMax);
      float min = Float.parseFloat(priceMin);
      StringBuffer result;
      result = new StringBuffer();
      try{  Class.forName("com.microsoft.sqlserver.jdbc.SQLServerDriver");
      }
      catch(Exception e){}
```

```
        Connection con;
        Statement sql;
        ResultSet rs;
        int n = 0;
        try{  result.append("< table border = 1 >");
              String uri = "jdbc:mysql://127.0.0.1/warehouse";
              con = DriverManager.getConnection(uri,"root","");
              DatabaseMetaData metadata = con.getMetaData();
              ResultSet rs1 = metadata.getColumns(null,null,"product",null);
              int 字段个数 = 0;
              result.append("< tr >");
              while(rs1.next()){
                 字段个数++;
                 String clumnName = rs1.getString(4);
                 result.append("< td >" + clumnName + "</td >");
              }
              result.append("</tr >");
              sql = con.createStatement();
              String condition =
             "SELECT * FROM product Where price < = " +
              max + " AND " + "price > = " + min;
              rs = sql.executeQuery(condition);
              while(rs.next()){
                 result.append("< tr >");
                 for(int k = 1;k < = 字段个数;k++)
                    result.append("< td >" + rs.getString(k) + "</td >");
                 result.append("</tr >");
              }
              result.append("</table >");
              con.close();
        }
        catch(SQLException e){
              result.append(e);
        }
        jspContext.setAttribute("queryResultByPrice",new String(result));
%>
```

6.4.4 排序查询

可以在 SQL 语句中使用 ORDER BY 子语句，对记录排序。例如，按 price 排序查询的 SQL 语句：

```
SELECT * FROM product ORDER BY price。
```

例 6-5 是查询数据库 warehouse 中的 product 表(见 6.1.2 节创建的数据库)。通过 JSP 页面 example6_5.jsp 可以选择价格从低到高排序记录、按生产日期排序记录。Tag 文件 SortTag.tag 根据 JSP 页面选择的排序方式排序记录。example6_5.jsp 的效果如图 6-22(a)、(b)所示。

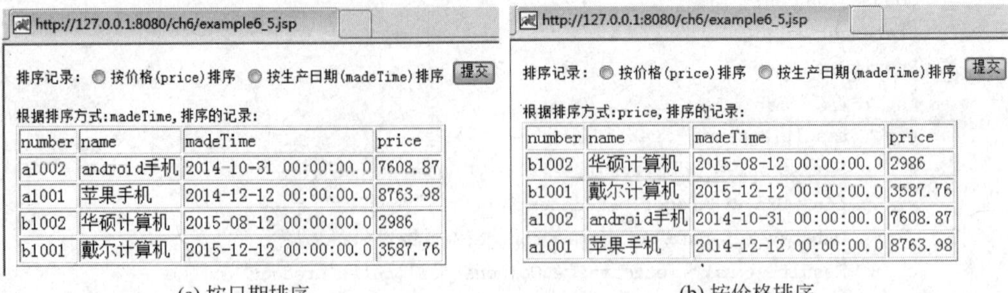

(a) 按日期排序　　(b) 按价格排序

图 6-22　两个排序查询页面效果图

例 6-5

example6_5.jsp

```
<%@ page contentType = "text/html;charset = GB2312" %>
<%@ taglib tagdir = "/WEB-INF/tags" prefix = "inquire" %>
<HTML><BODY><Font size = 2>
<FORM action = "" method = post name = form>
排序记录:<INPUT type = "radio" name = "orderType" value = "price">按价格(price)排序
<INPUT type = "radio" name = "orderType" value = "madeTime">按生产日期(madeTime)排序
   <Input type = submit name = "g" value = "提交">
</Form>
<% String orderType = request.getParameter("orderType");
    if(orderType == null)
      orderType = "";
%>
  <inquire:SortTag orderType = "<% = orderType %>" />
  根据排序方式:<% = orderType %>,排序的记录:
<BR><% = orderResult %><% -- orderResult 是 Tag 文件返回的对象 -- %>
</Font></BODY></HTML>
```

SortTag.tag

```
<%@ tag pageEncoding = "GB2312" %>
<%@ tag import = "java.sql.*" %>
<%@ attribute name = "orderType" required = "true" %>
<%@ variable name-given = "orderResult" scope = "AT_END" %>
<%   String orderCondition =
              "SELECT * FROM product ORDER BY " + orderType;
     StringBuffer result;
     result = new StringBuffer();
     try{  Class.forName("com.mysql.jdbc.Driver");
     }
     catch(Exception e){}
     Connection con;
     Statement sql;
     ResultSet rs;
     int n = 0;
     try{  result.append("<table border = 1>");
           String uri = "jdbc:mysql://127.0.0.1/warehouse";
           con = DriverManager.getConnection(uri,"root","");
```

```
            DatabaseMetaData metadata = con.getMetaData();
            ResultSet rs1 = metadata.getColumns(null,null,"product",null);
            int 字段个数 = 0;
            result.append("<tr>");
            while(rs1.next()){
               字段个数++;
               String clumnName = rs1.getString(4);
               result.append("<td>" + clumnName + "</td>");
            }
            result.append("</tr>");
            sql = con.createStatement();
            rs = sql.executeQuery(orderCondition);
            while(rs.next()){
               result.append("<tr>");
               for(int k = 1;k <= 字段个数;k++)
                 result.append("<td>" + rs.getString(k) + "</td>");
               result.append("</tr>");
            }
            result.append("</table>");
            con.close();
      }
      catch(SQLException e){
            result.append("");
      }
       //返回 orderResult 对象
      jspContext.setAttribute("orderResult",new String(result));
%>
```

6.4.5 模糊查询

可以用 SQL 语句操作符 LIKE 进行模式般配，使用“%”表示零个或多个字符，用一个下划线“_”表示任意一个字符。比如，下述语句查询产品名称中含有“戴”的记录：

```
rs = sql.executeQuery("SELECT * FROM product WHERE name LIKE '%戴%'");
```

例 6-6 是查询数据库 warehouse 中的 product 表(见 6.1.2 创建的数据库)。JSP 页面 example6_6.jsp 负责选择模糊查询条件，Tag 文件负责连接数据库查询记录。example6_6.jsp 效果如图 6-23 所示。

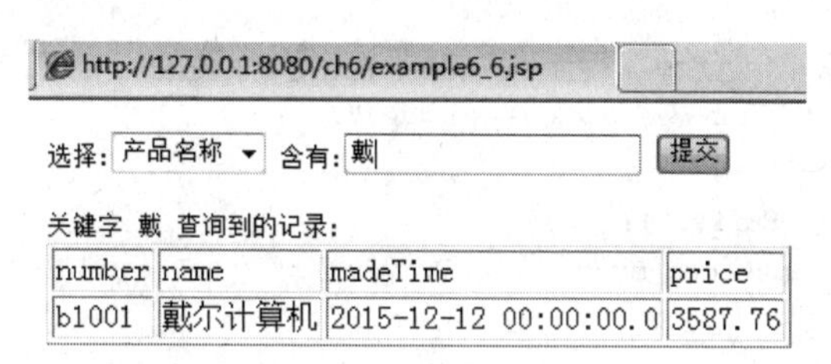

图 6-23 模糊查询

为了避免出现中文乱码(比较字段值中可能有中文)，例子中数据库的连接方式采用(见 6.3.3 节)：

```
"jdbc:mysql://地址/数据库? user = 用户 &password = 密码 &characterEncoding = gb2312";
```

例 6-6

example6_6.jsp

```
<%@ page contentType = "text/html;charset = GB2312" %>
<%@ taglib tagdir = "/WEB-INF/tags" prefix = "inquire" %>
<HTML><BODY color = cyan><Font size = 2>
<FORM action = "" method = post name = form>
   选择:<Select name = "ziduan">
         <Option value = "name">产品名称
         <Option value = "number">产品号
    </Select>
   含有:<INPUT type = "text" name = "keyWord" >
        <Input type = submit name = "g" value = "提交">
</Form>
<%  String ziduan = request.getParameter("ziduan");
    String keyWord = request.getParameter("keyWord");
    if(ziduan == null||keyWord == null){
       ziduan = "";
       keyWord = "";
    }
    byte bb[] = keyWord.getBytes("iso-8859-1");
    keyWord = new String(bb);
    out.println("关键字 " + keyWord);
%>
<inquire:KeyWord ziduan = "<% = ziduan %>" keyWord = "<% = keyWord %>" />
查询到的记录:
<BR><% = foundResult %><% --foundResult 是 Tag 文件返回的对象 -- %>
</Font></BODY></HTML>
```

KeyWord.tag

```
<%@ tag pageEncoding = "GB2312" %>
<%@ tag import = "java.sql.*" %>
<%@ attribute name = "keyWord" required = "true" %>
<%@ attribute name = "ziduan" required = "true" %>
<%@ variable name-given = "foundResult" scope = "AT_END" %>
<%
    String condition = "SELECT * FROM product Where " + ziduan +
                      " like '%" + keyWord + "%'";
    StringBuffer result;
    result = new StringBuffer();
    try{  Class.forName("com.mysql.jdbc.Driver");
    }
    catch(Exception e){}
    Connection con;
    Statement sql;
    ResultSet rs;
    int n = 0;
    try{  result.append("<table border = 1>");
          String uri =
```

```
            "jdbc:mysql://127.0.0.1/warehouse?" +
            "user = root&password = &characterEncoding = gb2312";
            con = DriverManager.getConnection(uri);
            DatabaseMetaData metadata = con.getMetaData();
            ResultSet rs1 = metadata.getColumns(null,null,"product",null);
            int 字段个数 = 0;
            result.append("<tr>");
            while(rs1.next()){
               字段个数++;
               String clumnName = rs1.getString(4);
               result.append("<td>" + clumnName + "</td>");
            }
            result.append("</tr>");
            sql = con.createStatement();
            rs = sql.executeQuery(condition);
            while(rs.next()){
               result.append("<tr>");
               for(int k = 1;k <= 字段个数;k++)
                  result.append("<td>" + rs.getString(k) + "</td>");
               result.append("</tr>");
            }
            result.append("</table>");
            con.close();
        }
        catch(SQLException e){
           result.append("请输入模糊查询条件");
        }
        //返回 foundResult 对象
        jspContext.setAttribute("foundResult",new String(result));
%>
```

6.5 更新记录

可以使用 SQL 语句更新记录中字段的值。Statement 对象调用方法：

```
public int executeUpdate(String sqlStatement);
```

通过参数 sqlStatement 指定的方式实现对数据库表中记录的字段值的更新。例如，下述语句将表 product 中 name 字段值是“海尔电视机”的记录的“price”字段的值更新为 6866：

```
executeUpdate("UPDATE product SET price  =  6866 WHERE name = '海尔电视机'");
```

注意：可以使用一个 Statement 对象进行更新和查询操作，但需要注意的是，当查询语句返回结果集后，如果没有立即输出结果集的记录，而接着执行了更新语句，那么结果集就不能输出记录了。要想输出记录就必须重新返回结果集。

例 6-7 是更新数据库 warehouse 中的 product 表（见 6.1.2 创建的数据库）。例 6-7 根据 number 字段，更新 product 表中记录的 name、madeTime、price 字段的值。在 example6_7.jsp

页面提交某个产品的产品号以及该产品的新名称、生产日期和价格到 newResult.jsp 页面，newResult.jsp 页面调用 Tag 文件 NewRecord.tag 更新记录的字段值。例 6-7 中用到了例 6-2 中的 QueryTag.tag（负责查询 product 表中的记录）。example6_7.jsp 和 newResult.jsp 的效果如图 6-24(a)、(b)所示。

为了避免出现中文乱码，例子中数据库的连接方式采用(见 6.3.3 节)：

```
"jdbc:mysql://地址/数据库?user=用户&password=密码&characterEncoding=gb2312";
```

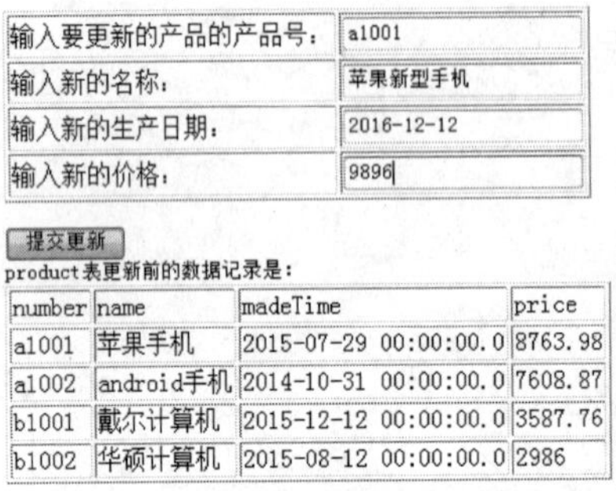

输入要更新的产品的产品号：	a1001
输入新的名称：	苹果新型手机
输入新的生产日期：	2016-12-12
输入新的价格：	9896

提交更新

product表更新前的数据记录是：

number	name	madeTime	price
a1001	苹果手机	2015-07-29 00:00:00.0	8763.98
a1002	android手机	2014-10-31 00:00:00.0	7608.87
b1001	戴尔计算机	2015-12-12 00:00:00.0	3587.76
b1002	华硕计算机	2015-08-12 00:00:00.0	2986

(a) 输入更新数据

product表更新后的数据记录是：

number	name	madeTime	price
a1001	苹果新型手机	2016-12-12 00:00:00.0	9896
a1002	android手机	2014-10-31 00:00:00.0	7608.87
b1001	戴尔计算机	2015-12-12 00:00:00.0	3587.76
b1002	华硕计算机	2015-08-12 00:00:00.0	2986

(b) 更新后的记录

图 6-24　两个更新记录页面效果图

例 6-7

example6_7.jsp

```
<%@ page contentType="text/html;charset=GB2312" %>
<%@ taglib tagdir="/WEB-INF/tags" prefix="inquire" %>
<HTML><BODY><FONT size=2>
<FORM action="newResult.jsp" method=post>
<table border=1>
<tr><td> 输入要更新的产品的产品号：</td>
<td><Input type="text" name="number"></td></tr>
<tr><td>输入新的名称：</td><td><Input type="text" name="name"></td></tr>
<tr><td>输入新的生产日期：</td><td><Input type="text" name="madeTime"></td></tr>
<tr><td>输入新的价格：</td><td><Input type="text" name="price"></td></tr>
</table>
<BR><Input type="submit" name="b" value="提交更新">
<BR>product 表更新前的数据记录是：
<inquire:QueryTag dataBaseName="warehouse"
                  tableName="product"
                  user="root" password=""/>
<BR><%=queryResult %>
</Font></BODY></HTML>
```

newResult.jsp

```
<%@ page contentType="text/html;charset=GB2312" %>
<%@ taglib tagdir="/WEB-INF/tags" prefix="renew" %>
<%@ taglib tagdir="/WEB-INF/tags" prefix="inquire" %>
<HTML><BODY bgcolor=pink><Font size=2>
<% String nu=request.getParameter("number");
```

```
    String na = request.getParameter("name");
    String mT = request.getParameter("madeTime");
    String pr = request.getParameter("price");
    byte bb[] = na.getBytes("iso-8859-1");
    na = new String(bb);
 %>
<renew:NewRecord number = "<% = nu %>" name = "<% = na %>"
                    madeTime = "<% = mT %>" price = "<% = pr %>"/>
<BR>product 表更新后的数据记录是:
<inquire:QueryTag dataBaseName = "warehouse"
                    tableName = "product"
                    user = "root" password = ""/>
<BR><% = queryResult %>
</Font></BODY></HTML>
```

NewRecord.tag

```
<%@ tag pageEncoding = "GB2312" %>
<%@ tag import = "java.sql.*" %>
<%@ attribute name = "number" required = "true" %>
<%@ attribute name = "name" required = "true" %>
<%@ attribute name = "madeTime" required = "true" %>
<%@ attribute name = "price" required = "true" %>
<% float p = Float.parseFloat(price);
    String condition1 = "UPDATE product SET name = '" + name +
                    "' WHERE number = " + "'" + number + "'" ,
           condition2 = "UPDATE product SET madeTime = '" + madeTime +
                    "' WHERE number = " + "'" + number + "'",
           condition3 = "UPDATE product SET price = " + price +
                    " WHERE number = " + "'" + number + "'" ;
    try{ Class.forName("com.mysql.jdbc.Driver");
    }
    catch(Exception e){}
    Connection con;
    Statement sql;
    ResultSet rs;
  try{ String uri =
        "jdbc:mysql://127.0.0.1/warehouse?" +
         "user = root&password = &characterEncoding = gb2312";
         con = DriverManager.getConnection(uri);
         sql = con.createStatement();
         sql.executeUpdate(condition1);
         sql.executeUpdate(condition2);
         sql.executeUpdate(condition3);
         con.close();
  }
  catch(Exception e){
         out.print("" + e);
  }
%>
```

6.6 添加记录

可以使用 SQL 语句添加新的记录。Statement 对象调用方法:

```
public int executeUpdate(String sqlStatement);
```

通过参数 sqlStatement 指定的方式实现向数据库表中添加新的记录。例如,下述语句将向表 product 中添加一条新的记录('012','神通手机','2015-2-26',2687)。

```
executeUpdate("INSERT INTO product VALUES ('012','神通手机','2015-2-26',2687)");
```

注意:可以使用一个 Statement 对象进行添加和查询操作,但需要注意的是,当查询语句返回结果集后,如果没有立即输出结果集的记录,而接着执行了添加语句,那么结果集就不能输出记录了。要想输出记录就必须重新返回结果集。

例 6-8 是向数据库 warehouse 中的 product 表添加新的记录(见 6.1.2 节创建的数据库)。example6_8.jsp 页面提交新的记录到 newDatabase.jsp 页面,该页面调用 Tag 文件 AddRecord.tag 添加新记录到 product 表。例 6-8 中用到了例 6-2 中的 QueryTag.tag(负责查询 product 表中的记录)。example6_8.jsp 和 newDatabase.jsp 的效果如图 6-25(a)、(b)所示。

为了避免出现中文乱码,例子中数据库的连接方式采用(见 6.3.3):

```
"jdbc:mysql://地址/数据库?user=用户&password=密码&characterEncoding=gb2312";
```

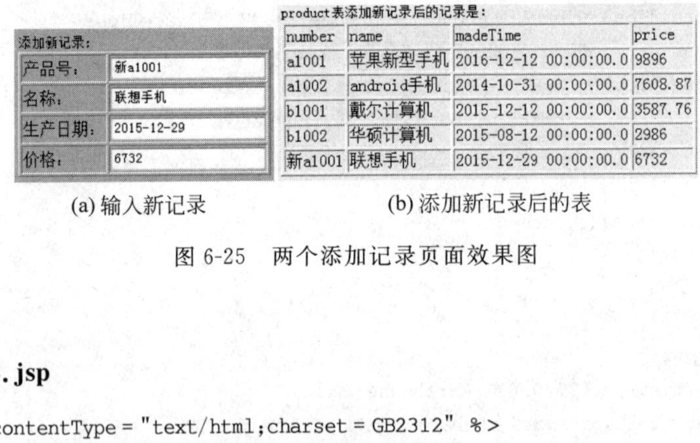

添加新记录:

产品号:	新a1001
名称:	联想手机
生产日期:	2015-12-29
价格:	6732

(a) 输入新记录

product表添加新记录后的记录是:

number	name	madeTime	price
a1001	苹果新型手机	2016-12-12 00:00:00.0	9896
a1002	android手机	2014-10-31 00:00:00.0	7608.87
b1001	戴尔计算机	2015-12-12 00:00:00.0	3587.76
b1002	华硕计算机	2015-08-12 00:00:00.0	2986
新a1001	联想手机	2015-12-29 00:00:00.0	6732

(b) 添加新记录后的表

图 6-25 两个添加记录页面效果图

例 6-8

example6_8.jsp

```
<%@ page contentType="text/html;charset=GB2312" %>
<%@ taglib tagdir="/WEB-INF/tags" prefix="inquire" %>
<HTML><BODY bgcolor=pink><FONT size=2>
<FORM action="newDatabase.jsp" method=post>
添加新记录:
<table border=1>
<tr><td>产品号:</td><td><Input type="text" name="number"></td></tr>
<tr><td>名称:</td><td><Input type="text" name="name"></td></tr>
<tr><td>生产日期:</td><td><Input type="text" name="madeTime"></td></tr>
<tr><td>价格:</td><td><Input type="text" name="price"></td></tr>
</table>
<BR><Input type="submit" name="b" value="提交">
```

```
<BR>product 表添加新记录前的记录是:
<inquire:QueryTag dataBaseName = "warehouse"
                  tableName = "product"
                  user = "root" password = ""/>
<BR><% = queryResult %><% -- queryResult 是 QueryTag 文件返回的对象 -- %>
</Font></BODY></HTML>
```

newDatabase. jsp

```
<%@ page contentType = "text/html;charset = GB2312" %>
<%@ taglib tagdir = "/WEB - INF/tags" prefix = "inquire" %>
<HTML><BODY bgcolor = yellow><Font size = 2>
<%!
    String handleStr(String s) {
        try {
          byte bb[] = s.getBytes("iso - 8859 - 1");
          return new String(bb);
        }
        catch(Exception exp){}
        return s;
    }
%>
<% String nu = handleStr(request.getParameter("number"));
   String na = handleStr(request.getParameter("name"));
   String mT = handleStr(request.getParameter("madeTime"));
   String pr = handleStr(request.getParameter("price"));
%>
<inquire:AddRecord number = "<% = nu %>" name = "<% = na %>"
                   madeTime = "<% = mT %>" price = "<% = pr %>"/>
<BR>product 表添加新记录后的记录是:
<inquire:QueryTag dataBaseName = "warehouse"
                  tableName = "product"
                  user = "root" password = ""/>
<BR><% = queryResult %><% -- queryResult 是 QueryTag 返回的对象 -- %>
</Font></BODY></HTML>
```

AddRecord. tag

```
<%@ tag pageEncoding = "GB2312" %>
<%@ tag import = "java.sql. * " %>
<%@ attribute name = "number" required = "true" %>
<%@ attribute name = "name" required = "true" %>
<%@ attribute name = "madeTime" required = "true" %>
<%@ attribute name = "price" required = "true" %>
<% float p = Float.parseFloat(price);
   String condition =
   "INSERT INTO product VALUES" +
              "(" + "'" + number + "','" + name + "','" + madeTime + "'," + p + ")";
   try{ Class.forName("com.mysql.jdbc.Driver");
   }
   catch(Exception e){}
   Connection con;
   Statement sql;
```

```
        ResultSet rs;
        try{ String uri =
              "jdbc:mysql://127.0.0.1/warehouse?" +
              "user = root&password = &characterEncoding = gb2312";
               con = DriverManager.getConnection(uri);
               sql = con.createStatement();
               sql.executeUpdate(condition);
               con.close();
        }
        catch(Exception e){
              out.print("" + e);
        }
    %>
```

6.7 删除记录

可以使用 SQL 语句删除记录。Statement 对象调用方法:

```
public int executeUpdate(String sqlStatement);
```

删除数据库表中的记录。例如,下述语句将删除产品号是 888 的记录:

```
executeUpdate("DELETE FROM product WHERE number = '888' ");
```

例 6-9 是删除数据库 warehouse 中的 product 表的某项记录(见 6.1.2 节创建的数据库)。example6_9.jsp 页面提交被删除的记录的 number 字段的值到 delete.jsp 页面,delete.jsp 页面调用 Tag 文件 DelRecord.tag 删除相应的记录。例 6-9 中还用到了例 6-2 中的 QueryTag.tag(负责查询 product 表中的记录)。example6_9.jsp 和 delete.jsp 的效果如图 6-26(a)、(b)所示。

为了避免出现中文乱码,例子中数据库的连接方式采用(见 6.3.3 节):

```
"jdbc:mysql://地址/数据库?user = 用户 &password = 密码 &characterEncoding = gb2312";
```

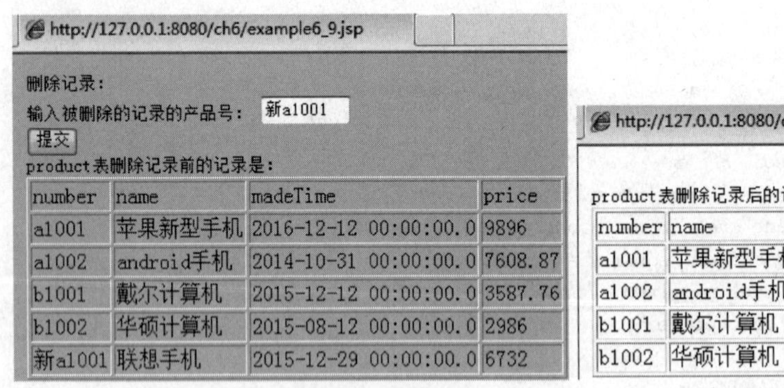

number	name	madeTime	price
a1001	苹果新型手机	2016-12-12 00:00:00.0	9896
a1002	android手机	2014-10-31 00:00:00.0	7608.87
b1001	戴尔计算机	2015-12-12 00:00:00.0	3587.76
b1002	华硕计算机	2015-08-12 00:00:00.0	2986
新a1001	联想手机	2015-12-29 00:00:00.0	6732

(a) 输入待删除记录的产品号

number	name	madeTime	price
a1001	苹果新型手机	2016-12-12 00:00:00.0	9896
a1002	android手机	2014-10-31 00:00:00.0	7608.87
b1001	戴尔计算机	2015-12-12 00:00:00.0	3587.76
b1002	华硕计算机	2015-08-12 00:00:00.0	2986

(b) 删除记录后的表

图 6-26　两个删除记录页面效果图

注意:可以使用一个 Statement 对象进行删除和查询操作,但需要注意的是,当查询语句返回结果集后,如果没有立即输出结果集的记录,而接着执行了删除语句,那么结果集就不能输出记录了。要想输出记录就必须重新返回结果集。

例 6-9

example6_9.jsp

```
<%@ page contentType="text/html;charset=GB2312" %>
<%@ taglib tagdir="/WEB-INF/tags" prefix="inquire" %>
<HTML><BODY bgcolor=pink><FONT size=2>
<FORM action="delete.jsp" method=post>
删除记录：<BR>输入被删除的记录的产品号：
<Input type="text" name="number" size=8>
<BR><Input type="submit" name="b" value="提交">
<BR>product 表删除记录前的记录是：
<inquire:QueryTag dataBaseName="warehouse"
                  tableName="product"
                  user="root" password=""/>
<BR><%=queryResult %><%-- queryResult 是 QueryTag.tag 返回的对象 --%>
</Font></BODY></HTML>
```

delete.jsp

```
<%@ page contentType="text/html;charset=GB2312" %>
<%@ taglib tagdir="/WEB-INF/tags" prefix="inquire" %>
<HTML><BODY><Font size=2>
<% String num=request.getParameter("number");
   if(num==null)
      num="";
   byte [] bb=num.getBytes("iso-8859-1");
   num= new String(bb);
%>
<inquire:DelRecord number="<%=num%>" />
<BR>product 表删除记录后的记录是：
<inquire:QueryTag dataBaseName="warehouse"
                  tableName="product"
                  user="root" password=""/>
<BR><%=queryResult %><%--queryResult 是 QueryTag.tag 返回的对象 --%>
</Font></BODY></HTML>
```

DelRecord.tag

```
<%@ tag pageEncoding="GB2312" %>
<%@ tag import="java.sql.*" %>
<%@ attribute name="number" required="true" %>
<% String condition =
        "DELETE FROM product WHERE number = '"+number+"'";
    try{ Class.forName("com.mysql.jdbc.Driver");
    }
    catch(Exception e){}
    Connection con;
    Statement sql;
    ResultSet rs;
  try{ String uri=
       "jdbc:mysql://127.0.0.1/warehouse?"+
```

```
            "user = root&password = &characterEncoding = gb2312";
            con = DriverManager.getConnection(uri);
            con = DriverManager.getConnection(uri,"root","");
            sql = con.createStatement();
            sql.executeUpdate(condition);
            con.close();
        }
        catch(Exception e){
            out.print("" + e);
        }
    %>
```

6.8 用结果集操作数据库中的表

尽管可以用 SQL 语句对数据库中表进行更新、插入操作，但也可以使用内存中的 ResultSet 对象对底层数据库表进行更新和插入操作(这些操作由系统自动转化为相应的 SQL 语句)，优点是不必熟悉有关更新、插入的 SQL 语句，而且方便编写代码，缺点是必须要事先返回结果集。为了避免出现中文乱码，数据库的连接方式采用(见 6.3.3)：

```
"jdbc:mysql://地址/数据库?user = 用户 &password = 密码 &characterEncoding = gb2312";
```

首先，必须得到一个可滚动的 ResultSet 对象。例如：

```
String uri = "jdbc:mysql://127.0.0.1/warehouse?" +
             "user = root&password = &characterEncoding = gb2312";
con = DriverManager.getConnection(uri);;
Statement sql =
con.createStatement(ResultSet.TYPE_SCROLL_SENSITIVE,ResultSet.CONCUR_UPDATABLE);
ResultSet rs = sql.executeQuery("SELECT * FROM product");
```

6.8.1 更新记录中的列值

使用结果集更新数据库表中第 n 行记录中某列的值的步骤是：

1. 结果集 rs 的游标移动到第 n 行

让结果集调用 absolue()方法将游标移到第 n 行：

```
rs.absolute(n);
```

2. 结果集将第 n 行的 p 列的列值更新

结果集可以使用下列方法更新列值：

```
updateInt(String columnName, int x),updateInt(int columnIndex int x)
updateLong(String columnName, long x), updateLong(int columnIndex, long x)
updateDouble(String columnName, double x),updateDouble(int columnIndex, double x)
updateString(String columnName, String x) , updateString(int columnIndex,String x)
updateBoolean(String columnName, boolean x),updateBoolean(int columnIndex, boolean x)
updateDate(String columnName, Date x) ,updateDate(int columnIndex, Date x)
```

3. 更新数据库中的表

最后，结果集调用 updateRow()方法，用结果集中的第 n 行更新数据库表中的第 n 行记录。以下代码片段更新 product 表中的第 3 行记录的 name 列(字段)的值。

```
rs.absolute(3);
rs.updateString(2, "IBM PC");     //也可以写成 rs.updateString("name", "IBM PC");
rs.updateRow();
```

6.8.2 插入记录

使用结果集向数据库表中插入(添加)一行记录步骤是：

1. 结果集 rs 的游标移动到插入行

结果集中有一个特殊区域，用作构建要插入的行的暂存区域(staging area)，习惯上将该区域位置称作结果集的插入行。为了向数据库表中插入一行新的记录，必须首先将结果集的游标移动到插入行，代码如下：

```
rs.moveToInsertRow();
```

2. 更新插入行的列值

结果集可以用 updateXxx()方法更新插入行的列值。例如：

```
rs.updateString(1, "c002");
rs.updateString(2, "IBM iPad");
rs.updateDate(3,Date());
rs.updateDouble(4, 5356);
```

3. 插入记录

最后，结果集调用 insertRow()方法用结果集中的插入行向数据库表中插入一行新记录。

例 6-10 是向数据库 warehouse 中的 product 表插入记录(见 6.1.2 节创建的数据库)。example6_10.jsp 提交新的记录到 newRecord.jsp 页面，该页面调用 Tag 文件 InsertRecord.tag 插入新记录到 product 表。例 6-10 还使用了例 6-2 中的 QueryTag.tag(负责查询 product 表中的记录)。example6_10.jsp 和 newRecord.jsp 的效果如图 6-27(a)、(b)所示。

number	name	madeTime	price
a1001	iPhone 4	2015-07-29 00:00:00.0	8768
a1002	android手机	2014-10-31 00:00:00.0	7608.87
b1001	戴尔计算机	2015-12-12 00:00:00.0	3587.76
b1002	华硕计算机	2015-08-12 00:00:00.0	2986

(a) 输入要插入的记录

http://127.0.0.1:8080/ch6/newRecord.jsp

product表添加新记录后的记录是：

number	name	madeTime	price
a1001	iPhone 4	2015-07-29 00:00:00.0	8768
a1002	android手机	2014-10-31 00:00:00.0	7608.87
b1001	戴尔计算机	2015-12-12 00:00:00.0	3587.76
b1002	华硕计算机	2015-08-12 00:00:00.0	2986
c001	IBM iPad	2015-06-01 00:00:00.0	5671

(b) 插入记录后的表

图 6-27 两个插入记录页面效果图

例 6-10

example6_10.jsp

```
<%@ page contentType = "text/html;charset = GB2312" %>
<%@ taglib tagdir = "/WEB-INF/tags" prefix = "inquire" %>
<HTML><BODY bgcolor = white ><FONT size = 2 >
<FORM action = "newRecord.jsp" method = post >
添加新记录：
<table border = 1 >
<tr><td>产品号：</td><td><Input type = "text" name = "number"></td></tr>
<tr><td>名称：</td><td><Input type = "text" name = "name"></td></tr>
<tr><td>生产日期(**** - ** - **)：
    </td><td><Input type = "text" name = "madeTime"></td></tr>
<tr><td>价格：</td><td><Input type = "text" name = "price"></td></tr>
</table>
<BR><Input type = "submit" name = "b" value = "提交">
<BR>product 表添加新记录前的记录是：
<inquire:QueryTag dataBaseName = "warehouse"
                  tableName = "product"
                  user = "root" password = ""/>
<BR><% = queryResult %><%-- queryResult 是 QueryTag.tag 返回的对象 --%>
</Font></BODY></HTML>
```

newRecord.jsp

```
<%@ page contentType = "text/html;charset = GB2312" %>
<%@ taglib tagdir = "/WEB-INF/tags" prefix = "inquire" %>
<HTML><BODY bgcolor = cyan><Font size = 2>
<% String nu = request.getParameter("number");
   String na = request.getParameter("name");
   String mT = request.getParameter("madeTime");
   String pr = request.getParameter("price");
   byte bb[] = na.getBytes("iso-8859-1");
   na = new String(bb);
%>
<inquire:InsertRecord number = "<% = nu %>" name = "<% = na %>"
madeTime = "<% = mT %>" price = "<% = pr %>"/>
<BR>product 表添加新记录后的记录是：
<inquire:QueryTag dataBaseName = "warehouse"
                  tableName = "product"
                  user = "root" password = ""/>
<BR><% = queryResult %><%-- queryResult 是 QueryTag.tag 返回的对象 --%>
</Font></BODY></HTML>
```

InsertRecord.tag

```
<%@ tag pageEncoding = "GB2312" %>
<%@ tag import = "java.sql.*" %>
<%@ tag import = "java.util.*" %>
<%@ attribute name = "number" required = "true" %>
<%@ attribute name = "name" required = "true" %>
<%@ attribute name = "madeTime" required = "true" %>
```

```
<%@ attribute name="price" required="true" %>
<% float p=Float.parseFloat(price);
    try{  Class.forName("com.mysql.jdbc.Driver");
    }
    catch(Exception e){}
    Connection con;
    Statement sql;
    ResultSet rs;
    Calendar calendar=Calendar.getInstance();
    try{
         String uri= "jdbc:mysql://127.0.0.1/warehouse";
         con=DriverManager.getConnection(uri,"root","");
         sql=con.createStatement
          (ResultSet.TYPE_SCROLL_SENSITIVE,ResultSet.CONCUR_UPDATABLE);
         rs=sql.executeQuery("SELECT * FROM product");
         rs.moveToInsertRow();   //rs的游标移动到插入行
         rs.updateString(1,number);
         rs.updateString(2, name);
         String [] str=madeTime.split("[-/]");
         int year=Integer.parseInt(str[0]);
         int month=Integer.parseInt(str[1]);
         int day=Integer.parseInt(str[2]);
         calendar.set(year,month-1,day);
         java.sql.Date date=new java.sql.Date(calendar.getTimeInMillis());
         rs.updateDate(3,date);
         rs.updateDouble(4,p);
         rs.insertRow();                                    //向prodcut表插入一行记录
         con.close();
    }
    catch(Exception e){
        out.print(""+e);
    }
%>
```

6.9 预处理语句

Java提供了更高效率的数据库操作机制，就是PreparedStatement对象，该对象被习惯地称作预处理语句对象。本节学习怎样使用预处理语句对象操作数据库中的表。

6.9.1 预处理语句的优点

当向数据库发送一个SQL语句，比如“Select * From product”，数据库库中的SQL解释器负责将把SQL语句生成底层的内部命令，然后执行该命令，完成有关的数据操作。如果不断地向数据库提交SQL语句势必增加数据库中SQL解释器的负担，影响执行的速度。如果应用程序能针对连接的数据库，事先就将SQL语句解释为数据库底层的内部命令，然后直接让数据库去执行这个命令，显然不仅减轻了数据库的负担，而且也提高了访问数据库的速度。

Connection 连接对象 con 调用 prepareStatement(String sql)方法：

```
PreparedStatement pre = con.prepareStatement(String sql);
```


对参数 sql 指定的 SQL 语句进行预编译处理，生成该数据库底层的内部命令，并将该命令封装在 PreparedStatement 对象 pre 中，那么该对象调用下列方法都可以使得该底层内部命令被数据库执行：

```
ResultSet executeQuery()
boolean execute()
int executeUpdate()
```

只要编译好了 PreparedStatement 对象 pre，那么 pre 可以随时地执行上述方法，提高了访问数据库的速度。

在下面的例 6-11 使用预处理语句来查询 warehouse 数据库中 product 表的全部记录(有关 product 表见 6.1.2 节)。

例 6-11

example6_11.jsp

```
<%@ page contentType = "text/html;charset = GB2312" %>
<%@ taglib tagdir = "/WEB-INF/tags" prefix = "inquire" %>
<HTML><Body bgcolor = cyan><Font size = 2>
<inquire:PrepareTag dataBaseName = "warehouse"
                    tableName = "product"
                    user = "root" password = ""/>
在<% = biao %>表查询到记录：<% -- biao 是 Tag 文件返回的对象 -- %>
<BR><% = queryResult %><% -- queryResult 是 Tag 文件返回的对象 -- %>
</Font></Body></HTML>
```

PrepareTag.tag

```
<%@ tag pageEncoding = "GB2312" %>
<%@ tag import = "java.sql.*" %>
<%@ attribute name = "dataBaseName" required = "true" %>
<%@ attribute name = "tableName" required = "true" %>
<%@ attribute name = "user" required = "true" %>
<%@ attribute name = "password" required = "true" %>
<%@ variable name-given = "biao" scope = "AT_END" %>
<%@ variable name-given = "queryResult" scope = "AT_END" %>
<% StringBuffer result;
   result = new StringBuffer();
   try{ Class.forName("com.mysql.jdbc.Driver");
   }
   catch(Exception e){}
   Connection con;
   PreparedStatement pre;                                  //预处理语句 pre
   ResultSet rs;
   try{ result.append("<table border = 1>");
        String uri = "jdbc:mysql://127.0.0.1/" + dataBaseName;
        con = DriverManager.getConnection(uri,user,password);
        DatabaseMetaData metadata = con.getMetaData();
```

```
            ResultSet rs1 = metadata.getColumns(null,null,tableName,null);
            int 字段个数 = 0;
            result.append("<tr>");
            while(rs1.next()){
                字段个数++;
                String clumnName = rs1.getString(4);
                result.append("<td>" + clumnName + "</td>");
            }
            result.append("</tr>");
            pre = con.prepareStatement("SELECT * FROM " + tableName);           //预处理语句 pre
            rs = pre.executeQuery();
            while(rs.next()){
                  result.append("<tr>");
                  for(int k = 1;k <= 字段个数;k++)
                      result.append("<td>" + rs.getString(k) + "</td>");
                  result.append("</tr>");
            }
            result.append("</table>");
            con.close();
        }
        catch(SQLException e){
              result.append("用户名或密码错误");
        }
        jspContext.setAttribute("queryResult",new String(result));
        jspContext.setAttribute("biao",tableName);                  //返回 biao 对象
    %>
```

6.9.2 使用通配符

在对 SQL 进行预处理时可以使用通配符"?"来代替字段的值，只要在预处理语句执行之前再设置通配符所表示的具体值即可。例如：

```
prepareStatement pre = con.prepareStatement("SELECT * FROM product WHERE price < ? ");
```

那么在 sql 对象执行之前，必须调用相应的方法设置通配符"?"代表的具体值。比如：

```
pre.setDouble(1,6565);
```

指定上述预处理语句 pre 中通配符"?"代表的值是 6565。通配符按着它们在预处理的"SQL 语句"中从左至右依次出现的顺序分别被称作第 1 个、第 2 个…… 第 m 个通配符。比如，下列方法：

```
void setDouble(int parameterIndex,int x);
```

用来设置通配符的值，其中参数 parameterIndex 用来表示 SQL 语句中从左到右的第 parameterIndex 个通配符号，x 是该通配符所代表的具体值。

尽管

```
pre = con.prepareStatement("SELECT * FROM product WHERE price < ? ");
pre.setDouble(1,6565);
```

的功能等同于

```
pre = con.prepareStatement("SELECT * FROM message WHERE price < 6565 ");
```

但是,使用通配符可以使得应用程序更容易动态地改变 SQL 语句中关于字段值的条件。

预处理语句设置通配符“?”的值的常用方法有:

```
void setDate(int parameterIndex,Date x)
void setDouble(int parameterIndex,double x)
void setFloat(int parameterIndex,float x)
void setInt(int parameterIndex,int x)
void setLong(int parameterIndex,long x)
void setString(int parameterIndex,String x)
```

在下面的例 6-12 中,example6_12.jsp 提交新的记录到 insertRecord.jsp 页面,该页面调用 Tag 文件 PrepareInsert.tag 添加新记录到 product 表(有关 product 表见 6.1.2 节)。另外例 6-12 中还用到了例 6-11 中的 PrepareTag.tag 文件,该文件负责查询 product 表中的记录。

例 6-12

example6_12.jsp

```
<%@ page contentType = "text/html;charset = GB2312" %>
<%@ taglib tagdir = "/WEB-INF/tags" prefix = "inquire" %>
<HTML><BODY bgcolor = orange><FONT size = 2>
<FORM action = "insertRecord.jsp" method = post>
添加新记录:
<table border = 1>
<tr><td>产号:</td><td><Input type = "text" name = "number"></td></tr>
<tr><td>名称:</td><td><Input type = "text" name = "name"></td></tr>
<tr><td>日期:</td><td><Input type = "text" name = "madeTime"></td></tr>
<tr><td>价格:</td><td><Input type = "text" name = "price"></td></tr>
</table>
<BR><Input type = "submit" name = "b" value = "提交">
<BR>product 表添加新记录前的记录是:
<inquire:PrepareTag dataBaseName = "warehouse"
                     tableName = "product"
                     user = "root" password = ""/>
<BR><% = queryResult %><% -- queryResult 是 PrepareTag 文件返回的对象 -- %>
</Font></BODY></HTML>
```

insertRecord.jsp

```
<%@ page contentType = "text/html;charset = GB2312" %>
<%@ taglib tagdir = "/WEB-INF/tags" prefix = "ok" %>
<HTML><BODY bgcolor = yellow><Font size = 2>
<%!
    String handleStr(String s) {
        try {
            byte bb[] = s.getBytes("iso-8859-1");
```

```
            return new String(bb);
        }
        catch(Exception exp){}
        return s;
    }
%>
<% String nu = handleStr(request.getParameter("number"));
   String na = handleStr(request.getParameter("name"));
   String mT = handleStr(request.getParameter("madeTime"));
   String pr = handleStr(request.getParameter("price"));
%>
<ok:PrepareInsert number = "<% = nu %>" name = "<% = na %>"
                   madeTime = "<% = mT %>" price = "<% = pr %>"/>
<BR>product 表添加新记录后的记录是:
<ok:PrepareTag dataBaseName = "warehouse"
               tableName = "product"
              user = "root" password = ""/>
<BR><% = queryResult %><% -- queryResult 是 Tag 文件返回的对象 -- %>
</Font></BODY></HTML>
```

PrepareInser. tag

```
<% @ tag pageEncoding = "GB2312" %>
<% @ tag import = "java.sql. * " %>
<% @ tag import = "java.util. * " %>
<% @ attribute name = "number" required = "true" %>
<% @ attribute name = "name" required = "true" %>
<% @ attribute name = "madeTime" required = "true" %>
<% @ attribute name = "price" required = "true" %>
<% float p = Float.parseFloat(price);
    try{  Class.forName("com.mysql.jdbc.Driver");
    }
    catch(Exception e){}
    Connection con;
    PreparedStatement pre;              //预处理语句
    ResultSet rs;
    Calendar calendar = Calendar.getInstance();
    try{ String uri =
        "jdbc:mysql://127.0.0.1/warehouse?" +
         "user = root&password = &characterEncoding = gb2312";
         con = DriverManager.getConnection(uri);
          String insertCondition = "INSERT INTO product VALUES (?,?,?,?)";
          pre = con.prepareStatement(insertCondition);
          pre.setString(1,number);     //设置的第 1 个? 的值是字符串 number
          pre.setString(2, name);      //设置的第 2 个? 的值是字符串 name
          pre.setString(3,madeTime);   //设置的第 3 个? 的值是 madeTime
          pre.setDouble(4,p);          //设置的第 4 个? 的值是 p
          int m = pre.executeUpdate();
          if(m! = 0){
```

```
                out.println("对表中插入" + m + "条记录成功");
            }
        }
        catch(SQLException exp){
            out.print(exp);
        }
    %>
```


6.10 事　　务

事务由一组 SQL 语句组成,所谓“事务处理”是指:应用程序保证事务中的 SQL 语句要么全部都执行,要么一个都不执行。

事务是保证数据库中数据完整性与一致性的重要机制。应用程序和数据库建立连接之后,可能使用多个 SQL 语句操作数据库中的一个表或多个表。比如,一个管理资金转账的应用程序为了完成一个简单的转账业务可能需要两个 SQL 语句,即需要将数据库 user 表中 id 号是 0001 的记录的 userMoney 字段的值由原来的 100 更改为 50,然后将 id 号是 0002 的记录的 userMoney 字段的值由原来的 20 更新为 70。应用程序必须保证这 2 个 SQL 语句要么全都执行,要么全都不执行。

JDBC 事务处理步骤如下:

1. setAutoCommit(boolean autoCommit)方法

事务处理的第一步骤是使用 setAutoCommit(boolean autoCommit)方法关闭自动提交模式,这样做的理由是和数据库建立的连接对象 con 的提交模式是自动提交模式,即该连接对象 con 产生的 Statement 或 PreparedStatement 对象对数据库提交任何一个 SQL 语句操作都会立刻生效,使得数据库中的数据发生变化,这显然不能满足事物处理的要求。比如,在转账操作时,将用户“geng”的 userMoney 的值由原来的 100 更改为 50(减去 50 的操作)的操作不应当立刻生效,而应等到“zhang”的用户的 userMoney 的值由原来的 20 更新为 70(增加 50 的操作)后一起生效,如果第 2 个语句 SQL 语句操作未能成功,第一个 SQL 语句操作就不应当生效。因此,为了能进行事务处理,必须关闭 con 的自动提交模式(自动提交模式是连接对象 con 的默认设置)。

con 对象首先调用 setAutoCommit(boolean autoCommit)方法,将参数 autoCommit 取值为 false 来关闭自动提交模式:

```
con.setAutoCommit(false);
```

2. commit()方法

con 调用 setAutoCommit(false)后,con 产生的 Statement 对象对数据库提交任何一个 SQL 语句操作都不会立刻生效,这样一来,就有机会让 Statement 对象(PreparedStatement 对象)提交多个 SQL 语句,这些 SQL 语句就是一个事务。事务中的 SQL 语句不会立刻生效,直到连接对象 con 调用 commit()方法。con 调用 commit()方法就是让事务中的 SQL 语句全部生效。

3. rollback()方法

con 调用 commit()方法进行事务处理时,只要事务中任何一个 SQL 语句没有生效,就抛出 SQLException 异常。在处理 SQLException 异常时，必须让 con 调用 rollback()方法，其作用是：撤销事务中成功执行过的 SQL 语句对数据库数据所做的更新、插入或删除操作，即撤销引起数据发生变化的 SQL 语句操作，将数据库中的数据恢复到 commi()方法执行之前的状态。

为了下列例 6-13 的需要，在 bank 数据库中创建了表 user 表(有关建数据库和表的操作见 6.1 节)，表的字段及属性如下：

name(文本) userMoney(双精度型)

在下面的例 6-13 使用了事务处理，将 user 表中 name 字段是“geng”的 userMoney 的值减少 50，并将减少的 50 增加到 name 字段是“zhang”的 userMony 属性值上。运行效果如图 6-28 所示。

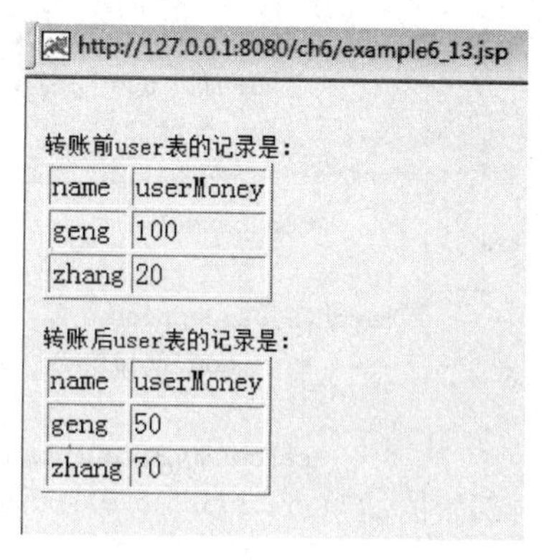

图 6-28 事务处理

例 6-13

example6_13.jsp

```
<%@ page contentType = "text/html;charset = GB2312" %>
<%@ page import = "java.sql.*" %>
<%@ taglib tagdir = "/WEB-INF/tags" prefix = "inquire" %>
<HTML><BODY bgcolor = yellow><FONT size = 2>
<BR>转账前 user 表的记录是：
<inquire:PrepareTag dataBaseName = "bank"
                    tableName = "user"
                    user = "root"
                    password = ""/>
<BR><% = queryResult %><%-- queryResult 是例 6-12 中 Tag 文件返回的对象 --%>
<% Connection con = null;
   Statement sql;
   ResultSet rs;
   try { Class.forName("com.mysql.jdbc.Driver");
   }
   catch(ClassNotFoundException e){}
   try{ int n = 50;
         String uri =
        "jdbc:mysql://127.0.0.1/bank?" +
        "user = root&password = &characterEncoding = gb2312";
         con = DriverManager.getConnection(uri);
         con.setAutoCommit(false);     //关闭自动提交模式
         sql = con.createStatement();
         rs = sql.executeQuery("SELECT * FROM user WHERE name = 'geng'");
         rs.next();
         double gengMoney = rs.getDouble("userMoney");
         gengMoney = gengMoney - n;
         if(gengMoney >= 0) {
           rs = sql.executeQuery("SELECT * FROM user WHERE name = 'zhang'");
           rs.next();
           double zhangMoney = rs.getDouble("userMoney");
```

```
            zhangMoney = zhangMoney + n;
            sql.executeUpdate
            ("UPDATE user SET userMoney = " + gengMoney + " WHERE name = 'geng'");
            sql.executeUpdate
            ("UPDATE user SET userMoney = " + zhangMoney + " WHERE name = 'zhang'");
            con.commit();              //开始事务处理
          }
          con.close();
      }
      catch(SQLException e){
          try{ con.rollback();          //撤销事务所做的操作
          }
          catch(SQLException exp){}
          out.println(e);
      }
%>
<BR>转账后 user 表的记录是:
<inquire:PrepareTag dataBaseName = "bank"
                    tableName = "user"
                    user = "root"
                    password = ""/>
<BR><% = queryResult %><% -- queryResult 是例 6-12 中 Tag 文件返回的对象 -- %>
</Font></BODY></HTML>
```

6.11 常见数据库连接

6.11.1 连接 Microsoft SQL Server 数据库

本节讲解怎样连接 SQL Server 2012 管理的数据库,内容同样适合于 SQL Server 2005 和 SQL Server 2008。

1. Microsoft SQL Server 2012

SQL Server 2012 是一个功能强大且可靠的免费数据库管理系统,它为轻量级(lightweight)网站和桌面应用程序提供丰富和可靠的数据存储。登录 http://www.microsoft.com,比如,登录 http://www.microsoft.com/zh-cn/download/default.aspx(微软的下载中心),然后在热门下载里选择选项“服务器”,然后选择下载 Microsoft SQL Server 2012 Express 以及相应的管理工具 Microsoft SQL Server 2008 Management Studio Express 或 Microsoft SQL Server Management Studio Express。

SQL Server 2012 Express 是免费的,对于 64 位操作系统可下载 SQLEXPR_x64_CHS.exe,对于 32 位操作系统可下载 SQLEXPR32_x86_CHS.exe。

Microsoft SQL Server Management Studio Express(SSMS)是微软提供的免费的数据库端管理软件,用于访问、配置、管理和开发 SQL Server 的所有组件,同时它还合并了多种图形工具和丰富的脚本编辑器,利用它们,技术水平各不相同的开发人员和管理员都可以使用 SQL Server。对于 64 位操作系统可下载 CHS\x64\SQLManagementStudio_x64_CHS.exe,对于 32 位操作系统可下载 CHS\x86\SQLManagementStudio_x86_CHS.exe。

安装好 SQL Server 2012 后，需启动 SQL Server 2012 提供的数据库服务器(数据库引擎)，以便使远程的计算机访问它所管理的数据库。在安装 SQL Server 2012 时如果选择的是自动启数据库服务器，数据库服务器会在开机后自动启动，否则需手动启动 SQL Server 2012 服务器，可以单击“开始”→“程序”→Microsoft SQL Server，启动 SQL Server 2012 服务器。

在本章中，为了便于调试程序，在同一台计算机同时安装了 Tomcat 服务器和 SQL Server 2012 数据库服务器，即使这样，为了能让 Tomcat 服务器管理的 Web 应用程序访问 SQL Server 2012 管理的数据库，也必须要启动 SQL Server 2012 提供的数据库服务器。

2. 建立数据库

打开 SSMS 提供的“对象资源管理器”，将出现相应的操作界面，如图 6-29 所示。

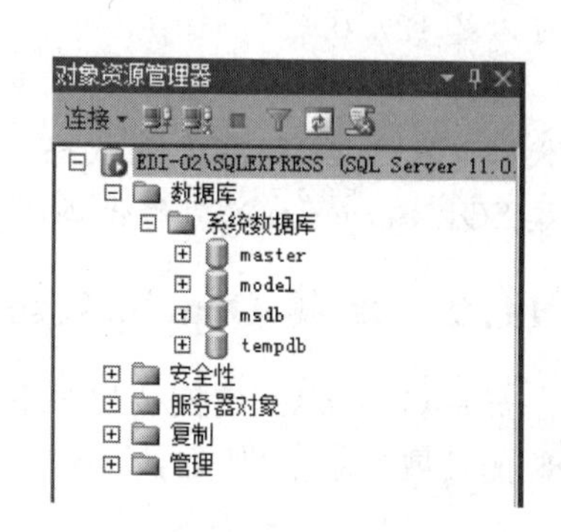

图 6-29　SQL Server 对象资源管理器

图 6-29 所示意的界面上的“数据库”目录下是已有的数据库的名称，在“数据库”目录上单击鼠标右键可以建立新的数据库，比如建立名称是 warehouse 的数据库。

创建好数据库后，就可以建立若干个表。如果准备在 warehouse 数据库中创建名字为 product 的表，那么可以用鼠标单击“数据库”下的 warehouse 数据库，在 warehouse 管理的“表”的选项上右击鼠标，选择“新建表”，将出现相应的建表界面。

3. 加载针对 SQL Server 的 JDBC 数据库驱动程序

可以登录 www.micsosoft.com 下载 Microsoft JDBC Driver 4.0 for SQL Server 即下载 sqljdbc_1.1.1501.101_enu.exe。安装 sqljdbc_1.1.1501.101_enu.exe 后，在安装目录的 enu 子目录中可以找到驱动程序文件：sqljdbc.jar，将该驱动程序复制到 Tomcat 服务器所使用的 JDK 的扩展目录中，即 JDK 安装目录下的“\jre\lib\ext”文件夹中，比如：D:\jdk1.7\jre\lib\ext，或复制到 Tomcat 服务器安装目录的\common\lib 文件夹中，比如：D:\apache-tomcat-8.0.3\common\lib。

应用程序加载 SQL Server 驱动程序代码如下：

```
try {  Class.forName("com.microsoft.sqlserver.jdbc.SQLServerDriver");
}
catch(Exception e){
}
```

4. 建立连接

假设 SQL Server 数据库服务器所驻留的计算机的 IP 地址是 192.168.100.1，SQL Server 数据库服务器占用的端口是 1433。应用程序要和 SQL Server 数据库服务器管理的数据库 warehouse 建立连接，而有权访问数据库 warehouse 的用户的 ID 和密码分别是 sa、dog123456，那么建立连接的代码如下：

```
try{  String uri = "jdbc:sqlserver://192.168.100.1:1433;DatabaseName = warehouse";
      String user = "sa";
```

```
        String password = "dog123456";
        con = DriverManager.getConnection(uri,user,password);
    }
catch(SQLException e){
        System.out.println(e);
}
```

应用程序一旦和某个数据库建立连接，就可以通过 SQL 语句和该数据库中的表交互信息，比如查询、修改、更新表中的记录。

注意：如果用户和要连接 SQL Server 2000 服务器驻留在同一计算机上，使用的 IP 地址可以是 127.0.0.1。

注意：对于例 6-1 至例 6-13 中，只要将例子中加载针对 MySQL 的 JDBC 数据库驱动程序代码以及连接 MySQL 的代码更换成加载针对 SQL Server 的 JDBC 数据库驱动程序和连接 SQL Server 数据库的代码，就可以让 Web 页面访问 SQL Server 数据库。

6.11.2 连接 Oracle 数据库

安装 Oracle 后，将安装目录下的 jdbc\lib\classes12.jar 复制到 Tomcat 引擎所使用的 JDK 的扩展目录中，比如：

```
D:/jdk1.7/jre/lib/ext
```

或复制到 Tomcat 服务器安装目录的\common\lib 文件夹中，比如：

```
D:\apache-tomcat-8.0.3\common\lib
```

通过如下的两个步骤和一个 Oracle 数据库建立连接：

1. 加载驱动程序

```
Class.forName("oracle.jdbc.driver.OracleDriver").newInstance();
```

2. 建立连接

```
Connection con =
DriverManager.getConnection( "jdbc:oracle:thin:@主机:端口号:数据库名","用户名","密码");
```

例如：

```
String user = "scott";
String password = "tiger";
con = DriverManager.getConnection
    ("jdbc:oracle:thin:@192.168.96.1:1521:oracle9i",user,password);
```

6.11.3 连接 Microsoft Access 数据库

1. 建立数据库

使用 Microsoft Access 可以建立多个数据库。操作步骤如下：

单击“开始”→“所有程序”→Microsoft Office→Microsoft Access，在新建数据库界面选择“空 Access 数据库”，然后命名、保存新建的数据库，比如建立名字是 shop 的数据库，保存在 D:\1000 中。

2. 创建表

创建好数据库后，就可以在该数据库下建立若干个表。为了在 shop 数据库中创建名字为 goods 的表，需在 shop 管理的“表”的界面上选择“使用设计器创建表”，单击界面上的“设计”菜单，弹出建表界面，如图 6-30 所示。利用建表界面建立 goods 表，该表的字段(属性)为：

```
number(文本) name(文本) madeTime(日期) price(数字,双精度)
```

其中，number 字段为主键(在字段上右击鼠标来设置字段是否是主键)，如图 6-30 所示。

goods

字段名称	数据类型
number	文本
name	文本
madeTime	日期/时间
price	数字

图 6-30 goods 表及字段属性

在 shop 管理的“表”的界面上，用鼠标双击已创建的 goods 表可以为该表添加记录。

3. 连接数据库

目前官方没有提供针对 Access 数据库的 JDBC 驱动程序(但可以购买一些公司开发的针对 Access 数据库的 JDBC 驱动程序)。应用程序可以通过 Microsoft 提供的 ODBC 来连接访问 Access 数据库，即使用 JDBC-ODBC 桥接器方式连接访问 Access 数据库。使用 JDBC-ODBC 桥接器方式的机制是，应用程序只需建立 JDBC 和 ODBC 之间的连接，即所谓的建立 JDBC-ODBC 桥接器，而和数据库的连接由 ODBC 去完成。

1) 建立 JDBC-ODBC 桥接器

JDBC 使用 java.lang 包中的 Class 类建立 JDBC-ODBC 桥接器。Class 类通过调用它的静态方法 forName 加载 sun.jdbc.odbc 包中的 JdbcOdbcDriver 类来建立 JDBC-ODBC 桥接器。建立桥接器时可能发生异常，必须捕获这个异常，建立桥接器的代码是：

```
try{ Class.forName("sun.jdbc.odbc.JdbcOdbcDriver");
}
catch(ClassNotFoundException e) {
     System.out.println(e);
}
```

2) ODBC 数据源

应用程序所在的计算机负责创建数据源，即使用 ODBC 数据源管理工具创建数据源，因此，必须保证应用程序所在计算机有 ODBC 数据源管理工具。Windows 7 以及 Windows XP 都有 ODBC 数据源管理工具。

对于 Windows 7 需要通过运行 C:/Windows/SysWOW64/odbcad32.exe 启动 ODBC 管理工具，否则无法看到有关的数据库驱动程序，对于 Windows XP，选择“控制面板”→“管理工具”→“ODBC 数据源”(某些 Windows XP 系统,需选择“控制面板”→“性能和维护”→“管理工具”→“ODBC 数据源”)。启动 ODBC 管理工具后,在其界面上显示了用户已有的数据源的名称(如图 6-31 所示)。选择“用户 DSN”或“系统 DSN”,单击“添加”按钮,可以创建新的数据源;单击“配置”按钮,可以重新配置已有的数据源;单击“删除”按钮,可以删除已有的数据源。

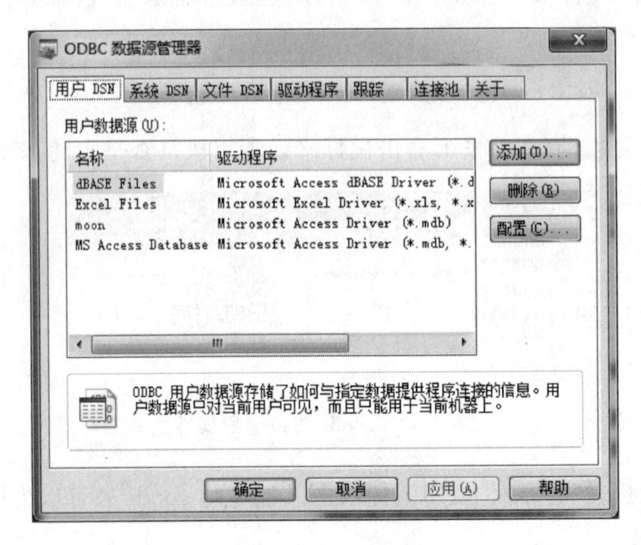

图 6-31　添加、修改或删除数据源

在图 6-31 所示的界面上选择单击“添加”按钮,出现为新增的数据源选择驱动程序界面,如图 6-32 所示。如果要访问的 Access 数据库版本是 Access 2003 版之前的,可以选择 Microsoft Access Driver(* . mdb),否则选择 Microsoft Access Driver(* . mdb, * accdb)。单击“完成”按钮。

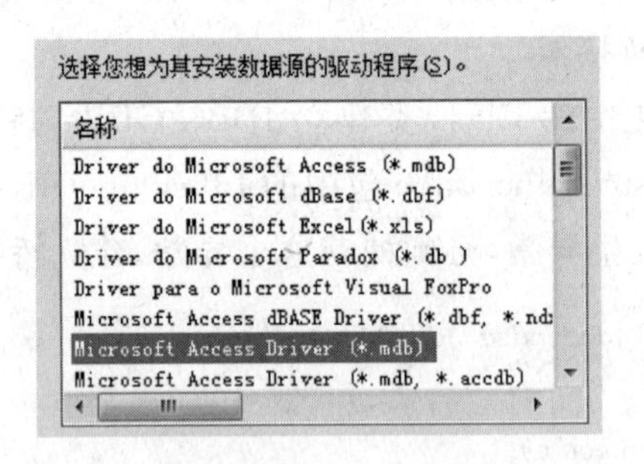

图 6-32　为新增的数据源选择驱动程序

在图 6-32 中为数据源选择相应的驱动程序后,将出现设置数据源具体项目的对话框,如图 6-33 所示。在名称栏里为数据源起一个自己喜欢的名字,这里起的名字是 myData。这个 myData 数据源就是指某个数据库。在“数据库选择”栏中选择一个数据库,这里选择的是 D:\1000 下的 shop 数据库(见前面所建立的数据库)。需要注意的是,在设置数据源时,请关闭 Microsoft Access 打开的 shop. mdb 数据库,否则系统将提示“非法路径”。

3) 建立连接

编写连接数据库代码不会出现数据库的名称,只能出现数据源的名字。首先使用 java.

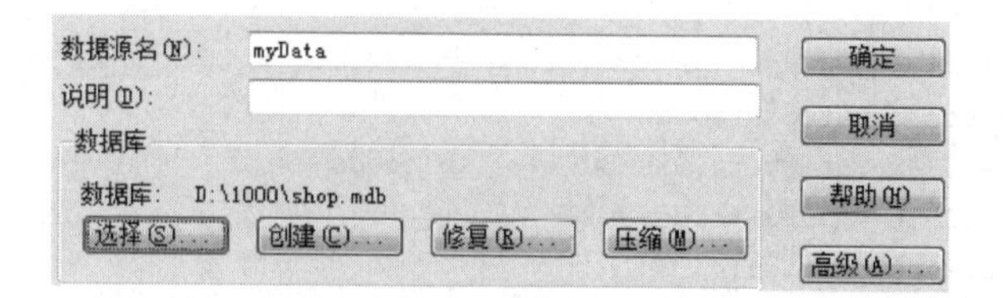

图 6-33　设置数据源的名字和对应的数据库

sql 包中的 Connection 类声明一个对象，然后再使用类 DriverManager 调用它的静态方法 getConnection 创建这个连接对象：

```
Connection con = DriverManager.getConnection("jdbc:odbc:数据源名字",
                                              "loginName"," password ");
```

假如没有为数据源设置 loginName 和 password，那么连接形式是：

```
Connection con = DriverManager. getConnection("jdbc:odbc:数据源名字", "","");
```

为了能和数据源 myData 交换数据，建立连接时应捕获 SQLException 异常：

```
try{ Connection con = DriverManager.getConnection("jdbc:odbc:myData", "","");
}
catch(SQLException e){}
```

程序一旦和某个数据源建立连接，就可以通过 SQL 语句和该数据源所指定的数据库中的表交互信息，比如查询、修改、更新表中的记录。

下面例 6-14 连接 Access 数据库 shop，查询其中的 goods 表，效果如图 6-34 所示。

http://127.0.0.1:8080/ch6/example6_14.jsp

产品号	名称	生产日期	价格
a1001	洗衣机	2015-12-12	3567.0
a1002	烤箱	2015-09-23	987.0
b1001	双人沙发	2015-09-12	1098.0

图 6-34　查询 Access 数据库

例 6-14

example6_14.jsp

```
<%@ page contentType = "text/html;charset = GB2312" %>
<%@ page import = "java.sql. * " %>
<HTML><BODY bgcolor = cyan>
<% Connection con;
    Statement sql;
    ResultSet rs;
    try{ Class.forName("sun.jdbc.odbc.JdbcOdbcDriver");
    }
    catch(ClassNotFoundException e){
            out.print(e);
```

```
    }
    try {   con = DriverManager.getConnection("jdbc:odbc:myData","","");
          sql = con.createStatement();
          rs = sql.executeQuery("SELECT * FROM goods ");
          out.print("<table border = 2 >");
          out.print("<tr>");
            out.print("<th width = 100 >" + "产品号");
            out.print("<th widLh = 100 >" + "名称");
            out.print("<th width = 50 >" + "生产日期");
            out.print("<th width = 50 >" + "价格");
         out.print("</TR>");
         while(rs.next()){
           out.print("<tr>");
             out.print("<td>" + rs.getString(1) + "</td>");
             out.print("<td>" + rs.getString(2) + "</td>");
             out.print("<td>" + rs.getDate("madeTime") + "</td>");
             out.print("<td>" + rs.getFloat("price") + "</td>");
           out.print("</tr>") ;
         }
         out.print("</table>");
         con.close();
    }
    catch(SQLException e){
         out.print(e);
    }
%>
</BODY></HTML>
```

6.12 实验 1：查询记录

要求在 webapps 目录下新建一个 Web 服务目录：practice6。除特别要求外，实验所涉及的 JSP 页面均保存在 practice6 中；Tag 文件保存在 practice6\WEB-INF\tags 目录中。

实验中所用的数据库为 Student，其中的表为 message 表。在进行实验之前首先完成下列任务：

(1) 使用 MySQL 数据库管理工具创建一个数据库 Student。

(2) 在数据库中 Student 中创建名字为 message 的表，表的各个字段及属性如图 6-35 所示。

列名	数据类型	长度	允许空
number	char	50	
name	char	50	
birthday	datetime	8	✓
email	char	100	✓

图 6-35　message 表

假设访问数据库的用户名为 root，该用户的密码是空(没有密码)。

1. 相关知识点

1）加载 mysql 的 JDBC-数据库驱动

应用程序加载 mysql 的 JDBC-数据库驱动程序代码如下：

```
try{ Class.forName("com.mysql.jdbc.Driver");
}
catch(Exception e){}
```

2）连接数据库

为了能和 MySQL 数据库服务器管理的数据库建立连接，必须保证该 MySQL 数据库服务器已经启动，如果没有更改过 MySQL 数据库服务器的配置，有权访问数据库的用户的 ID 是 root，密码是空（没有密码），那么建立连接的代码如下：

```
try{   String uri = " jdbc:mysql:// 192.168.100.1:3306/Student";
       String user = "root";
       String password = "";
       con = DriverManager.getConnection(uri,user,password);
   }
catch(SQLException e){
       System.out.println(e);
}
```

其中 root 用户有权访问数据库 Student，root 用户的密码是空。

2. 实验目的

本实验的目的是让学生掌握使用 JDBC 查询数据库中表的记录。

3. 实验要求

编写三个 JSP 页面：inputCondition.jsp、byNumber.jsp 和 byName.jsp 页面。编写两个 Tag 文件：NumberCondtion.tag 和 NameConditon.tag。

1）inputCondition.jsp 的具体要求

inputCondition.jsp 页面提供两个表单。其中一个表单允许用户输入要查询的学生的学号，即输入 message 表中 number 字段的查询条件，然后将查询条件提交给要 byNumber.jsp，另一个表单允许用户输入要查询的学生的姓名，即输入 message 表中 name 字段的查询条件，然后将查询条件提交给要 byName.jsp。inputCondition.jsp 页面的效果如图 6-36 所示。

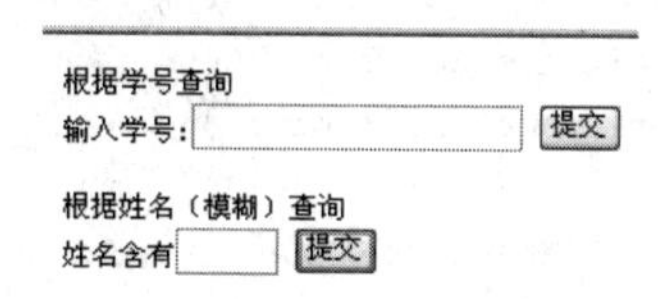

图 6-36　inputCondition.jsp 页面的效果

2）byNumber.jsp 的具体要求

byNumber.jsp 页面首先获得 inputCondition.jsp 页面提交的关于 number 字段的查询条件，然后使用 Tag 标记调用 Tag 文件 ByNumber.tag，并将 number 字段的查询条件传递给 ByNumber.tag。

3) byName. jsp 的具体要求

byName. jsp 页面首先获得 inputCondition. jsp 页面提交的关于 name 字段的查询条件,然后使用 Tag 标记调用 Tag 文件 ByName. tag,并将 name 字段的查询条件传递给 ByName. tag。

4) NumberCondtion. tag 的具体要求

NumberCondtion. tag 文件使用 attribute 指令获得 byNumber. jsp 页面传递过来的 number 字段的查询条件,然后和数据源 redsun 建立连接,根据得到的查询条件查询 message 表。NumberCondtion. tag 文件使用 variable 指令将查询结果返回给 byNumber. jsp 页面。

5) NameConditone. tag 的具体要求

NameConditon. tag 文件使用 attribute 指令获得 byName. jsp 页面传递过来的 name 字段的查询条件,然后和数据源 redsun 建立连接,根据得到的查询条件查询 message 表。NameConditon. tag 文件使用 variable 指令将查询结果返回给 byName. jsp 页面。

4. 参考代码

代码仅供参考,学生可按着实验要求,参考本代码编写代码。

JSP 页面参考代码如下:

inputCondition. jsp

```
<%@ page contentType = "text/html;charset = GB2312" %>
<HTML><BODY><Font size = 2>
<FORM action = "byNumber.jsp" Method = "post">
   根据学号查询
   <BR>输入学号:<Input type = text name = "number">
   <Input type = submit name = "g" value = "提交">
</Form>
<FORM action = "byName.jsp" Method = "post">
  根据姓名(模糊)查询
  <BR> 姓名含有<Input type = text name = "name" size = 5>
  <Input type = submit value = "提交">
</Form>
</Font></BODY></HTML>
```

byNumber. jsp

```
<%@ page contentType = "text/html;charset = GB2312" %>
<%@ taglib tagdir = "/WEB-INF/tags" prefix = "look" %>
<HTML><BODY bgcolor = cyan><Font size = 2>
<% String number = request.getParameter("number");
 %>
<look:NumberCondition number = "<% = number %>" />
根据学号<% = number %>查询到的记录:
<BR><% = queryResultByNumber %>
</Font></BODY></HTML>
```

byName. jsp

```
<%@ page contentType = "text/html;charset = GB2312" %>
<%@ taglib tagdir = "/WEB-INF/tags" prefix = "look" %>
<HTML><BODY bgcolor = pink><Font size = 2>
```

```
<% String name = request.getParameter("name");
    byte bb[] = name.getBytes("iso-8859-1");
    name = new String(bb);
%>
<look:NameCondition name = "<% = name %>" />
姓名含有"<% = name %>"的记录:
<BR><% = queryResultByName %>
</Font></BODY></HTML>
```

Tag 文件参考代码如下：

NumberCondition.tag

```
<%@ tag pageEncoding = "GB2312" %>
<%@ tag import = "java.sql.*" %>
<%@ attribute name = "number" required = "true" %>
<%@ variable name-given = "queryResultByNumber" scope = "AT_END" %>
<% StringBuffer result;
    result = new StringBuffer();
    try{  Class.forName("com.mysql.jdbc.Driver");
    }
    catch(ClassNotFoundException e){
      out.print(e);
    }
    Connection con;
    Statement sql;
    ResultSet rs;
    int n = 0;
    try{  result.append("<table border = 1>");
          String uri = "jdbc:mysql://127.0.0.1/Student";
          String user = "root";
          String password = "";
          con = DriverManager.getConnection(uri,user,password);
          DatabaseMetaData metadata = con.getMetaData();
          ResultSet rs1 = metadata.getColumns(null,null,"message",null);
          int 字段个数 = 0;
          result.append("<tr>");
          while(rs1.next()){
             字段个数++;
             String clumnName = rs1.getString(4);
             result.append("<td>" + clumnName + "</td>");
          }
          result.append("</tr>");
          sql = con.createStatement();
          String condition = "SELECT * FROM message Where number = '" + number + "'";
          rs = sql.executeQuery(condition);
          while(rs.next()){
result.append("<tr>");
             for(int k = 1;k <= 字段个数;k++)
                result.append("<td>" + rs.getString(k) + "</td>");
             result.append("</tr>");
          }
```

```
            result.append("</table>");
            con.close();
        }
        catch(SQLException e){
            esult.append(e);
        }
        jspContext.setAttribute("queryResultByNumber",new String(result));
%>
```

NameCondition.tag

```
<%@ tag pageEncoding = "GB2312" %>
<%@ tag import = "java.sql.*" %>
<%@ attribute name = "name" required = "true" %>
<%@ variable name-given = "queryResultByName" scope = "AT_END" %>
<% StringBuffer result;
    result = new StringBuffer();
    try{ Class.forName("com.mysql.jdbc.Driver");
    }
    catch(ClassNotFoundException e){
        out.print(e);
    }
    Connection con;
    Statement sql;
    ResultSet rs;
    int n = 0;
    try{ result.append("<table border = 1>");
            String uri = "jdbc:mysql://127.0.0.1/Student";
            String user = "root";
            String password = "";
            con = DriverManager.getConnection(uri,user,password);
            DatabaseMetaData metadata = con.getMetaData();
            ResultSet rs1 = metadata.getColumns(null,null,"message",null);
            int 字段个数 = 0;
            result.append("<tr>");
            while(rs1.next()){
                字段个数++;
                String clumnName = rs1.getString(4);
                result.append("<td>" + clumnName + "</td>");
            }
            result.append("</tr>");
            sql = con.createStatement();
            String condition = "SELECT * FROM message Where name Like '%" + name + "%'";
            rs = sql.executeQuery(condition);
            while(rs.next()){
                result.append("<tr>");
                for(int k = 1;k <= 字段个数;k++)
                    result.append("<td>" + rs.getString(k) + "</td>");
                result.append("</tr>");
            }
            result.append("</table>");
```

```
            con.close();
        }
        catch(SQLException e){
            result.append(e);
        }
        jspContext.setAttribute("queryResultByName",new String(result));
%>
```

6.13 实验2：更新记录

要求在webapps目录下新建一个Web服务目录：practice6。除特别要求外，实验所涉及的JSP页面均保存在practice6中；Tag文件保存在practice6\WEB-INF\tags目录中。

实验2中所用的数据库为实验1中的Student，其中的表为message表。

1. 相关知识点

1）返回Statement对象

假设建立的数据库连接对象是con，那么首先要返回一个Statement对象，代码如下：

```
Statement sql = con.createStatement();
```

2）向数据库发送关于更新记录的SQL语句

```
int m = sql.executeUpdate(更新记录的SQL语句);
```

更新成功 m 的值为1否则为0。

2. 实验目的

本实验的目的是让学生掌握使用JDBC更新数据库中表的记录。

3. 实验要求

编写两个JSP页面：inputNew.jsp和newResult.jsp页面。编写一个NewRecord.tag文件：NewRecord.tag。另外，本实验2还用到实验1中的NameCondition.tag文件。

1）inputNew.jsp的具体要求

inputNew.jsp页面提供一个表单，该表单允许用户将某个学生的新的姓名、出生日期和email提交到newResult.jsp页面。inputNew.jsp页面效果如图6-37所示。

输入要更新的学生的学号：	
输入新的姓名：	
输入新的出生日期：	
输入新的email：	

图6-37　inputNew.jsp页面的效果

2）newResult.jsp的具体要求

newResult.jsp页面首先获得inputNew.jsp页面提交的关于name字段、birthday字段和email字段的更新条件，然后使用Tag标记调用Tag文件NewRecord.tag更新记录的字段值。

3）NewRecord.tag的具体要求

NewRecord.tag文件使用attribute指令获得newResult.jsp页面传递过来的name字

段、birthday 字段和 email 字段的更新条件,然后和数据源 redsun 建立连接,更新数据库表中的相应记录。

4. 参考代码

代码仅供参考,学生可按着实验要求,参考本代码编写代码。

JSP 页面参考代码如下:

inputNew.jsp

```
<%@ page contentType = "text/html;charset = GB2312" %>
<%@ taglib tagdir = "/WEB-INF/tags" prefix = "inquire" %>
<HTML><BODY><FONT size = 2>
<FORM action = "newResult.jsp" method = post>
<table border = 1>
<tr><td> 输入要更新的学生的学号:</td>
<td><Input type = "text" name = "number"></td></tr>
<tr><td>输入新的姓名:</td><td><Input type = "text" name = "name"></td></tr>
<tr><td>输入新的出生日期:</td><td><Input type = "text" name = "birthday"></td></tr>
<tr><td>输入新的 email:</td><td><Input type = "text" name = "email"></td></tr>
</table>
<BR><Input type = "submit" name = "b" value = "提交更新">
<BR>message 表更新前的数据记录是:
<inquire:NameCondition name = ""/>
<BR><% = queryResultByName %>
</Font></BODY></HTML>
```

newResult.jsp

```
<%@ page contentType = "text/html;charset = GB2312" %>
<%@ taglib tagdir = "/WEB-INF/tags" prefix = "renew" %>
<%@ taglib tagdir = "/WEB-INF/tags" prefix = "inquire" %>
<HTML><BODY bgcolor = cyan>
<Font size = 2>
<% String nu = request.getParameter("number");
   String na = request.getParameter("name");
   String bd = request.getParameter("birthday");
   String em = request.getParameter("email");
   byte bb[] = na.getBytes("iso-8859-1");
   na = new String(bb);
 %>
<renew:NewRecord number = "<% = nu %>" name = "<% = na %>"
   birthday = "<% = bd %>" email = "<% = em %>"/>
  message 表更新后的数据记录是:
<inquire:NameCondition name = ""/>
<BR><% = queryResultByName %>
</Font></BODY></HTML>
```

Tag 文件参考代码如下:

NewRecord.tag

```
<%@ tag pageEncoding = "GB2312" %>
<%@ tag import = "java.sql.*" %>
```

```
<%@ attribute name = "number" required = "true" %>
<%@ attribute name = "name" required = "true" %>
<%@ attribute name = "birthday" required = "true" %>
<%@ attribute name = "email" required = "true" %>
<%
   String condition1 = "UPDATE message SET name = '" + name + "' WHERE number = " + "'" + number + "
'" ,
   condition2 = "UPDATE message SET birthday = '" + birthday + "' WHERE number = " + "'" + number
+ "'",
   condition3 = "UPDATE message SET email = '" + email + "' WHERE number = " + "'" + number + "'" ;
   try { Class.forName("com.mysql.jdbc.Driver");
   }
   catch(ClassNotFoundException e){
       out.print(e);
    }
    Connection con;
    Statement sql;
    ResultSet rs;
   try{ String uri =
        "jdbc:mysql://127.0.0.1/Student?" +
         "user = root&password = &characterEncoding = gb2312";
         con = DriverManager.getConnection(uri);
          sql = con.createStatement();
          sql.executeUpdate(condition1);
          sql.executeUpdate(condition2);
          sql.executeUpdate(condition3);
          con.close();
    }
    catch(Exception e){
         out.print("" + e);
    }
%>
```

6.14 实验 3：删除记录

要求在 webapps 目录下新建一个 Web 服务目录：practice6。除特别要求外，实验所涉及的 JSP 页面均保存在 practice6 中；Tag 文件保存在 practice6\WEB-INF\tags 目录中。

实验 3 中所用的数据库为实验 1 中的 Student，其中的表为 message 表。

1. 相关知识点

删除数据库中表的记录的步骤如下。

1）与数据库 Student 建立连接

```
String uri = "jdbc:mysql://127.0.0.1/Student";
String user = "root";
String password = "";
con = DriverManager.getConnection(uri,user,password);;
```

2) 返回 Statement 对象

```
Statement sql = con.createStatement();
```

3) 向数据库发送关于删除记录的 SQL 语句

```
int m = sql.executeUpdate(删除记录的 SQL 语句);
```

删除成功 m 的值为 1 否则为 0。

2. 实验目的

本实验的目的是让学生掌握使用 JDBC 删除数据库中表的记录。

3. 实验要求

编写两个 JSP 页面：inputNumber.jsp 和 delete.jsp 页面。编写一个 Tag 文件：DelRecord.tag。另外,本实验 3 还用到实验 1 中的 NameCondition.tag 文件。

1) inputNumber.jsp 的具体要求

inputNumber.jsp 页面提供一个表单,该表单允许用户将某个学生的学号提交到 delete.jsp 页面。inputNumber.jsp 页面效果如图 6-38 所示。

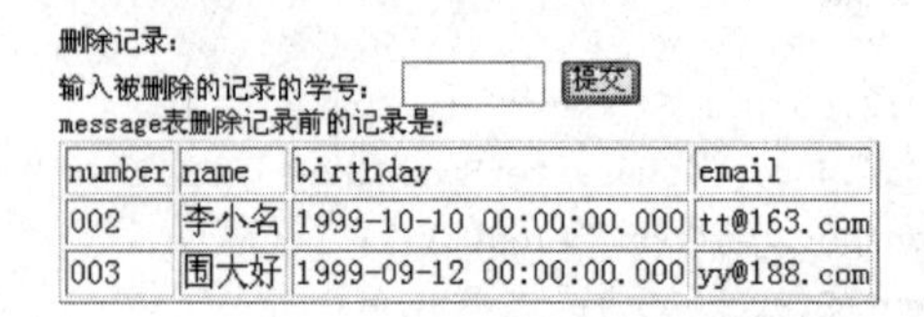

number	name	birthday	email
002	李小名	1999-10-10 00:00:00.000	tt@163.com
003	围大好	1999-09-12 00:00:00.000	yy@188.com

图 6-38　inpuNumber.jsp 页面的效果

2) delete.jsp 的具体要求

delete.jsp 页面首先获得 inputNumber.jsp 页面提交的关于 number 字段,即学生的学号,然后使用 Tag 标记调用 Tag 文件 DelRecord.tag 删除数据库表中相应的记录。

3) DelRecord.tag 的具体要求

DelRecord.tag 文件使用 attribute 指令获得 delete.jsp 页面传递过来的 numbere 字段的删除条件,然后和数据源 redsun 建立连接,删除数据库表中的相应记录。

4. 参考代码

代码仅供参考,学生可按着实验要求,参考本代码编写代码。

JSP 页面参考代码如下：

inputNumber.jsp

```
<%@ page contentType = "text/html;charset = GB2312" %>
<%@ taglib tagdir = "/WEB-INF/tags" prefix = "inquire" %>
<HTML><BODY><FONT size = 2>
<FORM action = "delete.jsp" method = post>
删除记录：
<br>输入被删除的记录的学号：
    <Input type = "text" name = "number" size = 8>
    <Input type = "submit" name = "b" value = "提交">
<br>message 表删除记录前的记录是：
  <inquire:NameCondition name = ""/>
<br><% = queryResultByName %>
```

```
</Font></BODY></HTML>
```

delete.jsp

```
<%@ page contentType = "text/html;charset = GB2312" %>
<%@ taglib tagdir = "/WEB-INF/tags" prefix = "inquire" %>
<%@ taglib tagdir = "/WEB-INF/tags" prefix = "del" %>
<HTML><BODY bgcolor = cyan><Font size = 2>
<% String nu = request.getParameter("number");
%>
  <del:DelRecord number = "<% = nu %>" />
   message 表删除记录后的记录是：
  <inquire:NameCondition name = ""/>
  <br><% = queryResultByName %>
</Font></BODY></HTML>
```

Tag 文件参考代码如下：

DelRecord.tag

```
<%@ tag pageEncoding = "GB2312" %>
<%@ tag import = "java.sql.*" %>
<%@ attribute name = "number" required = "true" %>
<% String condition = "DELETE FROM message WHERE number = '" + number + "'";
    try{
          Class.forName("com.mysql.jdbc.Driver");
    }
    catch(ClassNotFoundException e){
          out.print(e);
    }
    Connection con;
    Statement sql;
    ResultSet rs;
    try{ String uri =
         "jdbc:mysql://127.0.0.1/Student?" +
          "user = root&password = &characterEncoding = gb2312";
          con = DriverManager.getConnection(uri);
           sql = con.createStatement();
           sql.executeUpdate(condition);
           con.close();
    }
    catch(Exception e){
         out.print("" + e);
    }
%>
```

习 题 6

1. 首先使用 MySQL 数据库在客户端建立一个数据库，并在数据库中创建表(见 6.1)。然后编写一个 JSP 页面 a.jsp，要求 a.jsp 调用 Tag 文件 GetRecord.tag 查询数据库中表的全部记录。a. jsp 调用 Tag 文件时,使用 Tag 标记将数据库表名传递给 Tag 文件。

2. 编写一个 JSP 页面 b. jsp,要求 b. jsp 调用 Tag 文件 AddRecord. tag 向数据库的表中添加一条记录(事先建立数据库和表见 6.1 节)。

3. 编写一个 JSP 页面 c. jsp,要求 c. jsp 调用 Tag 文件 RenewRecord. tag 更新 warehouse 数据库中 product 表中的一条记录(有关 product 表见 6.1 节)。c. jsp 调用 Tag 文件时,使用 Tag 标记将表名 product 和更新条件传递给 Tag 文件。

4. 编写一个 JSP 页面 d. jsp,要求 d. jsp 调用 Tag 文件 DelRecord. tag 删除 warehouse 数据库中 product 表中的记录(有关 product 表见 6.1 节)。d. jsp 调用 Tag 文件时,使用 Tag 标记将表名 product 和删除条件传递给 Tag 文件。

第7章 JSP与JavaBean

本章导读

主要内容

- 编写JavaBean和使用JavaBean
- 获取和修改JavaBean的属性
- 使用JavaBean的简单例子
- JavaBean与文件操作
- JavaBean与数据库操作

难点

- JavaBean与文件操作
- JavaBean与数据库操作

关键实践

- 有效期限为request的JavaBean
- 有效期限为session的JavaBean
- 有效期限为application的JavaBean

第3章介绍了Tag文件,其核心内容是JSP页面可以将数据的处理过程指派给一个或几个Tag文件来完成,而且通过第5章和第6章的学习认识到了Tag文件的关键作用:代码复用。本节将学习JSP提供的另一项技术:JavaBean,该技术不仅能实现代码的复用,而且是MVC模式中的重用成员之一(有关MVC模式将在第9章讲述),而MVC模式无疑是Web设计中最重要的模式之一,因此JavaBean技术是JSP的重要组成部分。

Tag文件和本章介绍的JavaBean技术各有特点,应针对具体的Web应用综合使用,只有真正熟悉了这两项技术才会在具体的Web设计中灵活地使用它们。

在谈论JavaBean之前先看一个通俗的事情:组装电视机。组装一台电视机时,人们可以选择多个组件,例如电阻、电容、显像管等。一个组装电视机的人不必关心显像管是怎么研制的,只要根据说明书了解其中的属性和功能就可以了。不同的电视机可以安装相同的显像管,显像管的功能完全相同,但是在不同的电视机里面,一台电视机的显像管发生了故障并不影响其他的电视机,也可能两台电视机安装了一个共享的组件——天线,如果天线发生了故障,两台电视机都受到同样的影响。

JavaBean是一种Java类,通过封装属性和方法成为具有某种功能或者处理某个业务的对象,简称bean。按着Sun公司的定义,JavaBean是一个可重复使用的软件组件,由于JavaBean是基于Java语言的,因此JavaBean不依赖平台,具有以下特点:

- 可以实现代码的重复利用。
- 易编写、易维护、易使用。
- 可以在任何安装了 Java 运行环境的平台上使用,而不需要重新编译。

已经知道,一个基本的 JSP 页面就是由普通的 HTML 标记和 Java 程序片组成,如果程序片和 HTML 大量交互在一起,就显得页面混杂,不易维护。JSP 页面应当将数据的处理过程指派给一个或几个 bean 来完成,只需在 JSP 页面中调用这个 bean 即可。不提倡大量的数据处理都用 Java 程序片来完成。在 JSP 页面中调用 bean,可有效地分离静态工作部分和动态工作部分。

在本章中,在 webapps 目录下新建一个 Web 服务目录: ch7,因此,除非特别约定,本章例子中涉及的 JSP 页面均保存在 ch7 中。

7.1 编写 JavaBean 和使用 JavaBean

7.1.1 bean 的编写与保存

JavaBean 分为可视组件和非可视组件。在 JSP 中主要使用非可视组件,对于非可视组件,不必去设计它的外观,主要关心它的属性和方法。

1. 编写 bean

编写 JavaBean 就是编写一个 Java 的类,所以只要会写类就能编写一个 bean,这个类创建的一个对象称作一个 bean。为了能让使用这个 bean 的应用程序构建工具(例如 JSP 引擎)知道这个 bean 的属性和方法,只需在类的方法命名上遵守以下规则:

(1) 如果类的成员变量的名字是 xxx,那么为了更改或获取成员变量的值,即更改或获取属性的值,在类中必须提供两个方法:

- getXxx(),用来获取属性 xxx 的值。
- setXxx(),用来修改属性 xxx 的值。

即方法的名字用 get 或 set 为前缀,后缀是将成员变量名字的首字母大写的字符序列。对于 boolean 类型的成员变量,即布尔逻辑类型的属性,允许使用 is 代替上面的 get 和 set。

(2) 类中方法的访问权限都必须是 public 的。

(3) 类中如果有构造方法,那么这个构造方法的访问权限也是 public 的,并且是无参数的。

下面编写一个简单的 bean,并说明在 JSP 中怎样使用这个 bean。如果使用 Tomcat 5.0 版本后的引擎服务器,bean 必须带有包名。使用 package 语句给 bean 一个包名。包名可以是一个合法的标识符,也可以是若干个标识符加“.”分割而成。例如:

```
package gping;
package tom.jiafei;
```

以下是用来创建 bean 的 Java 源文件。

Circle.java

```
package tom.jiafei;
public class Circle{
```

```
    int radius;
    public Circle(){
      radius = 1;
    }
    public int getRadius(){
      return radius;
    }
    public void setRadius(int newRadius){
      radius = newRadius;
    }
    public double circleArea(){
      return Math.PI * radius * radius;
    }
    public double circleLength(){
      return 2.0 * Math.PI * radius;
    }
}
```

2. 保存 bean

为了让 JSP 页面使用 bean，Tomcat 服务器必须使用相应的字节码创建一个对象，即创建一个 bean。为了让 Tomcat 服务器能找到字节码，字节码文件必须保存在特定的目录中。

将上述 Java 源文件保存为 Circle.java，例如保存在 D:\1000 中(源文件可以保存在任意目录中)，并编译通过，得到字节码文件 Circle.class。为使 JSP 页面使用 bean，Tomcat 服务器必须使用相应的字节码创建一个对象，即创建一个 bean。为了让 Tomcat 服务器能找到字节码，字节码文件必须保存在特定的目录中。

首先，在当前 Web 服务目录下建立子目录结构：\WEB-INF\classes，为了让 Tomcat 服务器启用上述\WEB-INF\classes 目录，必须重新启动 Tomcat 服务器。WEB-INF\classes 被启用后，就可以根据类的包名，在 classes 下再建立相应的子目录，例如类的包名为 tom.jiafei，那么在 classes 下建立子目录结构：tom\jiafei，如图 7-1 所示。

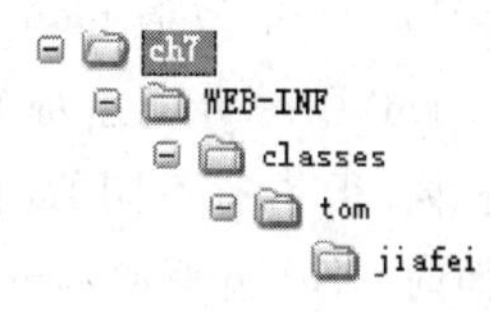

图 7-1　bean 的存放目录

把创建 bean 的字节码文件复制到 ch7\WEB-INF\classes\tom\jiafei 中。

7.1.2　使用 bean

在使用 bean 的 JSP 页面中，首先必须有相应的 page 指令。例如：

```
<%@ page import = "tom.jiafei.*" %>
```

然后在 JSP 页面中再使用 JSP 动作标记：useBean，来加载使用 bean。

1. useBean 标记的格式

JSP 页面通过使用 JSP 动作标记：useBean，来加载使用 bean，useBean 动作标记的格式如下：

```
<jsp:useBean id = "名字" class = "创建 bean 的类" scope = "bean 有效期限"> </jsp:useBean>
```

或

```
<jsp:useBean id="名字" class="创建 bean 的类" scope="bean 有效期限"/>
```

需要特别注意的是：其中的"创建 bean 的类"要带有包名。例如：

```
class="tom.jiafei.Circle"
```


当 JSP 引擎上某个含有 useBean 动作标记的 JSP 页面被加载执行时，JSP 引擎将首先在一个同步块中查找内置 pageContent 对象中是否含有名字 id 和作用域 scope 的对象。如果这个对象存在，JSP 引擎就分配一个这样的对象给用户，这样，用户就在服务器端获得了一个作用域是 scope、名字是 id 的 bean（就像组装电视机时获得了一个有一定功能和使用范围的电子元件）。如果在 pageContent 对象中没有查找到指定作用域是 scope、名字是 id 的对象，就根据 class 指定的类创建一个名字是 id 的对象，即创建了一个名字是 id 的 bean，并添加到 pageContent 内置对象中，并指定该 bean 的作用域是 scope，同时 JSP 引擎在服务器端分配给用户一个作用域是 scope、名字是 id 的 bean。

JSP 引擎的内置 pageContent 对象用来存储供 JSP 引擎使用的数据对象，即通过该内置对象向用户提供不同类型的各种数据对象。当含有 useBean 标记的 JSP 页面被执行后，bean 就被存放在 pageContent 对象中，如果更改了创建 bean 的 Java 类文件后，pageContent 对象中的 bean 并不能被更新，这是因为任何 JSP 页面再次被访问执行时，总是先到 pageContent 中查找 bean，而 pageContent 对象直到 JSP 引擎关闭才释放它存储的数据对象。因此，如果修改了创建 bean 的字节码，必须重新启动 JSP 引擎。

2. bean 的有效期限

下面就 useBean 标记中 scope 取值的不同情况进行阐述：

1）scope 取值 page

JSP 引擎分配给每个用户的 bean 是互不相同的，也就是说，尽管每个用户的 bean 的功能相同，但占有不同的内存空间。该 bean 的有效期限是当前页面，当 JSP 引擎执行完这个页面时，JSP 引擎取消分配给该用户的 bean。

2）scope 取值 request

JSP 引擎分配给每个用户的 bean 是互不相同的，该 bean 的有效期限是 request 期间。用户在网站的访问期间可能请求过多个页面，如果这些页面含有 scope 取值是 request 的 useBean 标记，那么 pageContent 对象在每个页面分配给用户的 bean 也是互不相同的。JSP 引擎对请求做出响应之后，取消分配给用户的这个 bean。

3）scope 取值 session

该 bean 的有效期限是用户的会话期间，也就是说，如果用户在某个 Web 服务目录多个页面中相互连接，每个页面都含有一个 useBean 标记，而且各个页面的 useBean 标记中 id 的值相同、scope 的值都是 session，那么，该用户在这些页面得到的 bean 是相同的一个（占有相同的内存空间）。如果用户在某个页面更改了这个 bean 的属性，其他页面的这个 bean 的属性也将发生同样的变化。当用户的会话（session）消失，JSP 引擎取消分配的 bean，即释放 bean 所占有的内存空间。需要注意的是，不同用户的 scope 取值是 session 的 bean 是互不相同的（占有不同的内存空间），也就是说，当两个用户同时访问一个 JSP 页面时，一个用户对自己 bean 的属性的改变，不会影响到另一个用户。

4）scope 取值 application

JSP 引擎为 Web 服务目录下所有的 JSP 页面分配一个共享的 bean，不同用户的 scope 取值是 application 的 bean 也都是相同的一个，也就是说，当多个用户同时访问一个 JSP 页面时，任何一个用户对自己 bean 的属性的改变，都会影响到其他的用户。

从上面的叙述可知道，有效期限最长的 bean 是 scope 取值为 application 的 bean，最短的是 scope 取值为 page 的 bean。例如 scope 取值为 page 的 bean 的有效期限就小于 scope 取值为 request 的 bean，这是因为对于 scope 取值为 page 的 bean，当 JSP 引擎执行完页面，在做出响应之前就取消了分配用户的 bean，而对于 scope 取值为 request 的 bean，在执行完页面，做出响应之后才取消分配用户的 bean。图 7-2 给出了 bean 的有效期限的比较示意图。

注意：当使用作用域是 session 的 bean 时，要保证用户端支持 Cooker。

例 7-1 中，负责创建 bean 的类是上述的 Circle 类（Circle. class 保存在 ch7\WEB-INF\classes\tom\jiafei 目录中），创建的 bean 的 id 是 girl、scope 是 page。example 7_1. jsp 的效果如图 7-3 所示。

例 7-1

example7_1. jsp

```
<%@ page contentType = "text/html; charset = GB2312" %>
<%@ page import = "tom.jiafei.*" %>
<HTML><BODY bgcolor = cyan><Font size = 3>
    <jsp:useBean id = "girl" class = "tom.jiafei.Circle" scope = "page" >
    </jsp:useBean>
  <%-- 通过上述 useBean 标记,用户获得了一个 scope 是 page,id 是 girl 的 bean --%>
    <% girl.setRadius(100);
     %>
      圆的半径是:<% = girl.getRadius() %>
<br>圆的周长是:<% = girl.circleLength() %>
<br>圆的面积是:<% = girl.circleArea() %>
</BODY></HTML>
```

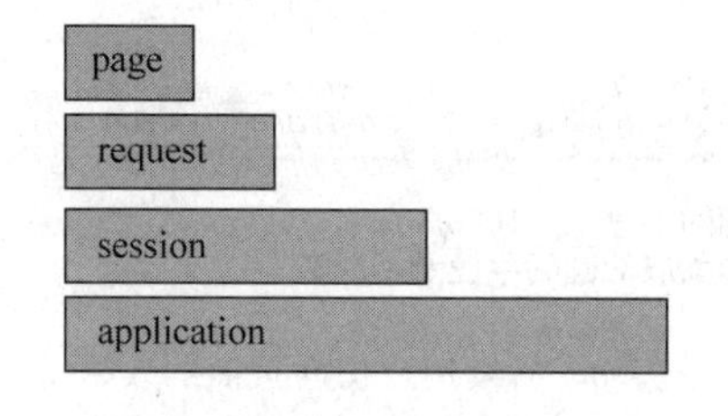

图 7-2　bean 的有效期限之比较

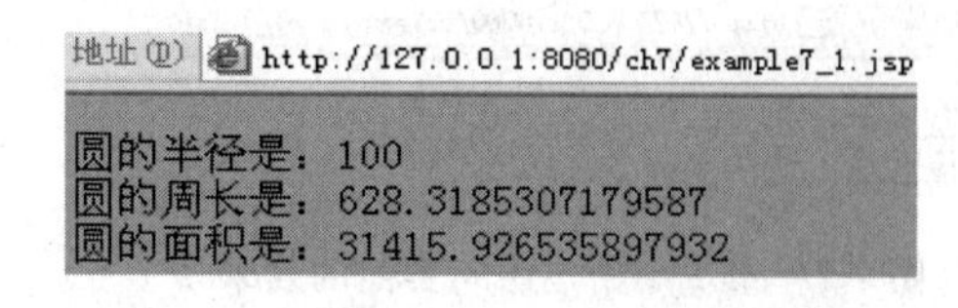

图 7-3　使用 scope 是 page 的 bean

例 7-2 中，将 bean 的 scope 设为 session，bean 的 id 仍然是 girl，创建该 bean 的类文件仍然是上述的 Circle. class。在 bean1. jsp 页面，girl 的半径 radius 的值是 1，然后链接到 bean2. jsp 页面，显示半径 radius 的值，然后将 girl 的半径 radius 的值更改为 600，当再刷新 bean1. jsp 时会发现 radius 的值已经变成了 600。bean1. jsp 的效果如图 7-4 和图 7-5 所示。

例 7-2

bean1. jsp

```
<%@ page contentType = "text/html; charset = GB2312" %>
<%@ page import = "tom.jiafei.*" %>
```

```
<HTML><BODY bgcolor = cyan><Font size = 2>
    <jsp:useBean id = "girl" class = "tom.jiafei.Circle" scope = "session">
    </jsp:useBean>
<P>圆的半径是:
  <% = girl.getRadius() %>
  <A href = "bean2.jsp"><BR>bean2.jsp</A>
</BODY></HTML>
```

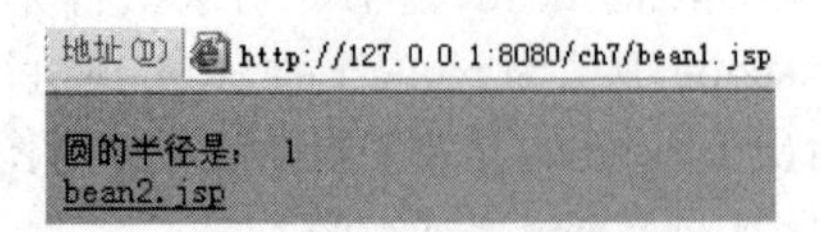

图 7-4　在 bean1.jsp 查看 radius 初始值

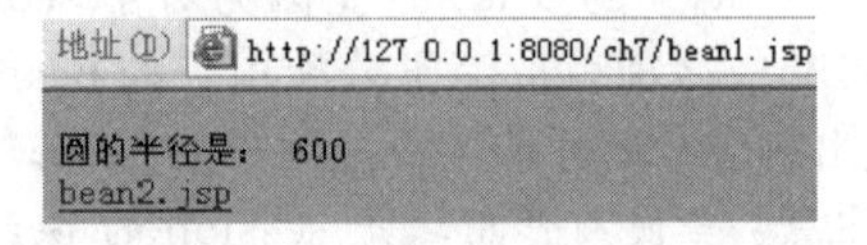

图 7-5　在 bean1.jsp 重新查看 radius 的值

bean2.jsp(效果如图 7-6 所示)

```
<%@ page contentType = "text/html; charset = GB2312" %>
<%@ page import = "tom.jiafei.*" %>
<HTML><BODY bgcolor = yellow><Font size = 2>
  <jsp:useBean id = "girl" class = "tom.jiafei.Circle" scope = "session">
    </jsp:useBean>
<P>圆的半径是:
    <% = girl.getRadius() %>
  <% girl.setRadius(600); %>
<P>修改后的圆的半径是:
    <% = girl.getRadius() %>
</BODY></HTML>
```

图 7-6　在 bean2.jsp 修改 radius 的值

例 7-3 中,将 bean 的 scope 设为 application。当第一个用户访问这个页面时,显示 bean 的 radius 的值,然后把 radius 的值修改为 1000(如图 7-7 所示),当其他用户访问这个页面时,看到的 radius 的值都是 1000(如图 7-8 所示)。

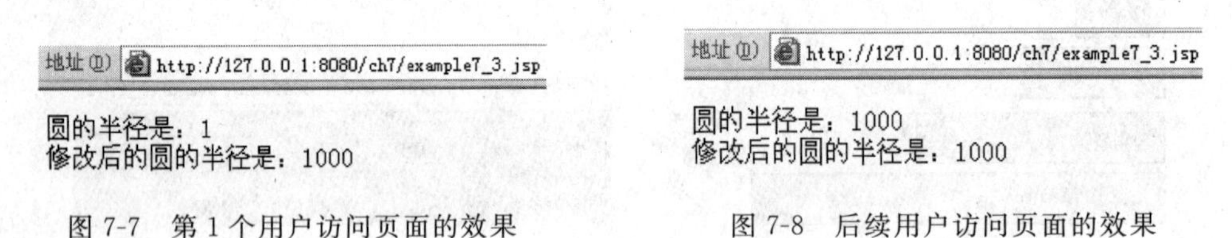

图 7-7　第 1 个用户访问页面的效果　　图 7-8　后续用户访问页面的效果

例 7-3

example7_3.jsp

```
<%@ page contentType = "text/html; charset = GB2312" %>
<%@ page import = "tom.jiafei.*" %>
  <jsp:useBean id = "girl" class = "tom.jiafei.Circle" scope = "application"/>
<HTML><BODY>
      圆的半径是:<% = girl.getRadius() %>
    <% girl.setRadius(1000); %>
    <br>修改后的圆的半径是:<% = girl.getRadius() %>
</BODY></HTML>
```

7.2 获取和修改 bean 的属性

当 JSP 页面使用 useBean 动作标记创建一个 bean 后，就可在 Java 程序片中让这个 bean 调用方法产生行为，如例 7-1、例 7-2 和例 7-3 所示。JSP 页面还可以使用动作标记 getProperty、setProperty 获取和修改 bean 的属性。下面讲述怎样使用 JSP 的动作标记去获取和修改 bean 的属性。

7.2.1 getProperty 动作标记

使用该标记可以获得 bean 的属性值，并将这个值以字符串的形式显示给用户。使用这个标记之前，必须使用 useBean 标记获取一个 bean。

getProperty 动作标记的格式：

```
<jsp:getProperty name = "bean 的 id" property = "bean 的属性"/>
```

或

```
<jsp:getProperty name = "bean 的 id" property = "bean 的属性"/>
</jsp:getProperty>
```

其中，name 取值是 bean 的 id，用来指定要获取哪个 bean 的属性的值，property 取值是该 bean 的一个属性的名字。该指令的作用相当于 Java 表达式：

```
<% = bean 的 id.getXxx() %>
```

因此，bean 必须保证有相应的 getXxx 方法，否则 JSP 页面将无法使用 getProperty 标记。

例 7-4 中，创建 bean 的 Java 源文件 Circle2.java 将前面的 Circle.java 给予改进，增加 circleArea 和 circleLength 两个属性，例 7-4 中的 JSP 页面 example7_4.jsp 使用 useBean 标记得到 id 是 apple 的 bean，并使用 getProperty 标记获取 apple 的各个属性的值。

例 7-4

Circle2.java

```
package tom.jiafei;
public class Circle2{
    double radius = 1;
    double circleArea = 0;
    double circleLength = 0;
    public double getRadius(){
      return radius;
    }
    public void setRadius(double newRadius){
      radius = newRadius;
    }
    public double getCircleArea(){
      circleArea = Math.PI * radius * radius;
      return circleArea;
```

```
    }
    public double getCircleLength(){
      circleLength = 2.0 * Math.PI * radius;
      return circleLength;
    }
}
```

注意：Circle2 类中的 circleArea 和 circleLength 属性是关联属性，即二者的值依赖于 radius 属性的值，所以不应该提供针对它们的 set 方法，只需提供 get 方法即可。

example7_4.jsp(效果如图 7-9 所示)

```
<%@ page contentType = "text/html; charset = GB2312" %>
<%@ page import = "tom.jiafei.Circle2" %>
<HTML><BODY bgcolor = cyan><FONT size = 4>
    <jsp:useBean id = "apple" class = "tom.jiafei.Circle2" scope = "page"/>
      圆的半径是：<jsp:getProperty name = "apple" property = "radius"/>
<br>圆的面积是：<jsp:getProperty name = "apple" property = "circleArea"/>
<br>圆的周长是：<jsp:getProperty name = "apple" property = "circleLength"/>
</FONT></BODY></HTML>
```

地址(D) http://127.0.0.1:8080/ch7/example7_4.jsp

圆的半径是：1.0
圆的面积是：3.141592653589793
圆的周长是： 6.283185307179586

图 7-9　使用 getProperty 标记

7.2.2　setProperty 动作标记

使用该标记可以设置 bean 的属性值。使用这个标记之前，必须使用 useBean 标记得到一个可操作的 bean，而且 bean 必须保证有相应的 setXxx 方法。

setProperty 动作标记可以通过两种方式设置 bean 属性的值。

1. 将 bean 属性的值设置为一个表达式的值或字符串

这种方式不如后种方式方便，但当涉及属性值是汉字时，使用这种方式更好一些。

将 bean 属性的值设置为一个表达式的值的使用格式：

```
<jsp:setProperty name = "bean 的 id" property = "bean 的属性" value = "<% = expression %>"/>
```

将 bean 属性的值设置为一个字符串的使用格式：

```
<jsp:setProperty name = "bean 的 id" property = "bean 的属性" value = 字符串 />
```

如果将表达式的值设置为 bean 属性的值，表达式值的类型必须和 bean 的属性的类型一致。如果将字符串设置为 bean 的属性的值，这个字符串会自动被转化为 bean 的属性的类型。在 Java 中，将数字字符构成的字符串转化为其他数值类型的方法如下：

```
Integer.parseInt(String s),转化到 int。
Long.parseLong(String s),转化到 long。
Float.parseFloat(String s),转化到 float。
```

Double.parseDouble(String s),转化到 double。

例 7-5 中,写了一个描述学生的 bean。在一个 JSP 页面中获得 scope 是 page 的 bean,并使用动作标记设置,获取该 bean 的属性。

例 7-5

Student.java

```
package tom.jiafei;
public class Student{
    String name = null;
    long number;
    double height,weight;
    public String getName(){
      return name;
    }
    public void setName(String newName){
      name = newName;
    }
    public long getNumber(){
      return number;
    }
    public void setNumber(long newNumber){
      number = newNumber;
    }
    public double getHeight(){
      return height;
    }
    public void setHeight(double newHeight){
      height = newHeight;
    }
    public double getWeight(){
      return weight;
    }
    public void setWeight(double newWeight){
      weight = newWeight;
    }
}
```

example7_5.jsp(效果如图 7-10 所示)

```
<%@ page contentType = "text/html; charset = GB2312" %>
<%@ page import = "tom.jiafei.Student" %>
<jsp:useBean id = "zhang" class = "tom.jiafei.Student" scope = "page"/>
<HTML><BODY bgcolor = cyan><FONT size = 4>
<jsp:setProperty name = "zhang" property = "name" value = "张小三"/>
    名字是:<jsp:getProperty name = "zhang" property = "name" />
<jsp:setProperty name = "zhang" property = "number" value = "1999001"/>
<br>学号是:<jsp:getProperty name = "zhang" property = "number"/>
<jsp:setProperty name = "zhang" property = "height" value = "<% = 1.78 %>"/>
<br>身高是:<jsp:getProperty name = "zhang" property = "height"/>米
<jsp:setProperty name =  "zhang" property = "weight" value = "67.65"/>
```

```
<br>体重是：<jsp:getProperty name="zhang" property="weight"/>公斤
</FONT></BODY></HTML>
```

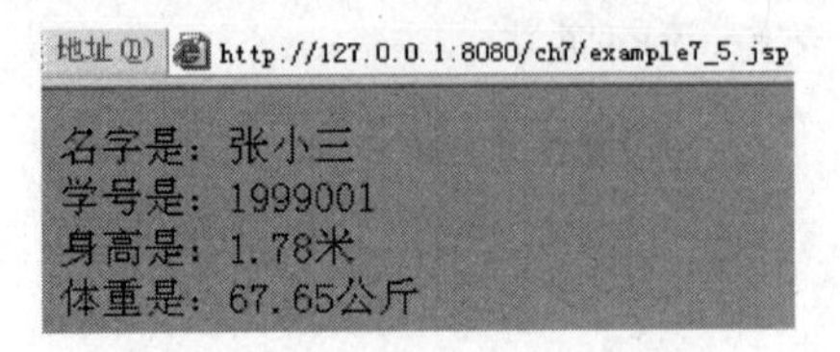

图 7-10 使用 setProperty 标记

2. 通过 HTTP 表单的参数的值来设置 bean 的相应属性的值

使用 setProperty 设置 bean 属性值的第 2 种方式是通过 HTTP 表单的参数的值来设置 bean 的相应属性的值，JSP 引擎会自动将参数的字符串值转换为 bean 相对应的属性的值。

如果使用 HTTP 表单的参数的值来设置 bean 中相对应的属性的值可以使用如下 setProperty 标记：

```
<jsp:setProperty name= "bean 的 id" property=" * " />
```

使用上述标记设置 bean 的属性值，要求 bean 的"属性名"和表单中所对应的"参数名"相同，该标记不用再具体指定 beans 属性的值将对应表单中哪个参数指定的值，系统会自动根据名字进行匹配对应。

如果需要明确 bean 的某个属性的值设置为表单中的对应的参数值，需使用如下 setProperty 标记：

```
<jsp:setProperty name= "bean 的 id" property="属性名" param="参数名"/>
```

使用上述方式设置 bean 的属性值，不要求 bean 的"属性名"和表单中所对应的"参数名"相同。

例 7-6 通过表单来指定 bean 的属性值。由于用户可能提交汉语的姓名，所以例 7-6 中的 StudentTwo.java 文件对例 7-5 中的 Student.java 进行了改动，将其中的 getName 方法进行如下的改动：

```
public String getName(){
    try{
        byte [] b = name.getBytes("ISO-8859-1");
        name = new String(b);
        return name;
    }
    catch(Exception e){
        return name;
    }
}
```

例 7-6

StudentTwo.java

```
package tom.jiafei;
public class StudentTwo{
```

```
    String name = null;
    long number;
    double height,weight;
    public String getName(){
        try{
            byte [] b = name.getBytes("ISO - 8859 - 1");
            name = new String(b);
            return name;
        }
        catch(Exception e){
            return name;
        }
    }
    public void setName(String newName){
        name = newName;
    }
    public long getNumber(){
        return number;
    }
    public void setNumber(long newNumber){
        number = newNumber;
    }
    public double getHeight(){
        return height;
    }
    public void setHeight(double newHeight){
        height = newHeight;
    }
    public double getWeight(){
        return weight;
    }
    public void setWeight(double newWeight){
        weight = newWeight;
    }
}
```

example7_6.jsp（效果如图 7-11 所示）

```
<% @ page contentType = "text/html; charset = GB2312" %>
<% @ page import = "tom.jiafei.StudentTwo" %>
<jsp: useBean id = "zhang" class = "tom.jiafei.StudentTwo" scope = "page"/>
<HTML><BODY><FONT size = 2>
<FORM action = "" Method = "post">
    输入姓名：<Input type = text name = "xingming">
<br>输入学号：<Input type = text name = "xuehao">
<br>输入身高：<Input type = text name = "shenggao">
<br>输入体重：<Input type = text name = "tizhong">
    <Input type = submit value = "提交">
</FORM>
<jsp: setProperty name = "zhang" property = "name" param = "xingming"/>
<jsp: setProperty name = "zhang" property = "number" param = "xuehao"/>
```

```
<jsp:setProperty name="zhang" property="height" param="shenggao"/>
<jsp:setProperty name="zhang" property="weight" param="tizhong"/>
    名字是:<jsp:getProperty name="zhang" property="name"/>
<br>学号是:<jsp:getProperty name="zhang" property="number"/>
<br>身高是:<jsp:getProperty name="zhang" property="height"/>米
<br>体重是:<jsp:getProperty name="zhang" property="weight"/>公斤
```

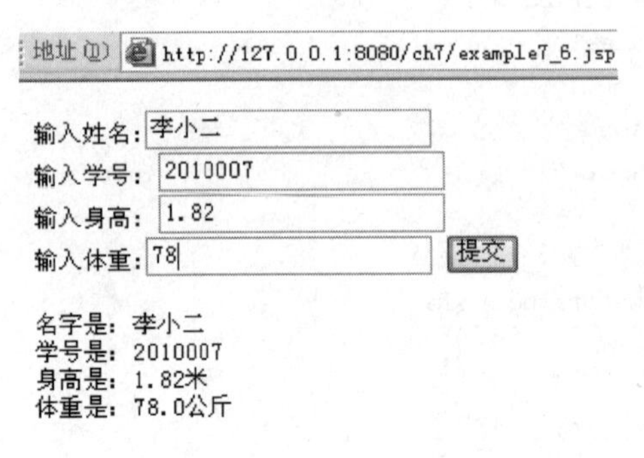

图 7-11 使用 setProperty 标记

7.3 bean 的辅助类

通过上面的学习,已经知道怎样使用一个简单的 bean。有时在写一个 bean 的时候,除了需要用 import 语句引入 Java 的内置包中的类,还可能需要其他自己写的一些类,那么只要将这些类的字节码文件和 bean 的字节码放在同一目录中即可。

例 7-7 中,使用一个 bean 列出 JSP 页面所在目录中特定扩展名的文件。在写 bean 的类文件 ListFile 时,需要一个实现 FilenameFilter 接口的辅助类 FileName,该类可以帮助 bean 列出指定扩展名的文件。例 7-7 中 Java 源文件 ListFile.java 编译通过后,会生成两个字节码文件:ListFile.class 和 FileName.class。需要将这两个字节码文件都复制到 ch7\WEB-INF\classes\tom\jiafei 中。

例 7-7

ListFile.java

```
package tom.jiafei;
import java.io.*;
class FileName implements FilenameFilter{
    String str = null;
    FileName (String s){
        str = "." + s;
    }
    public boolean accept(File dir,String name){
        return name.endsWith(str);
    }
}
public class ListFile{
    String extendsName = null,allFileName = null;
    public void setExtendsName(String s){
```

```
            extendsName = s;
        }
        public String getExtendsName(){
            return extendsName;
        }
        public String getAllFileName(){
            StringBuffer str = new StringBuffer();
            File dir = new File("D: /apache - tomcat - 6.0.13/webapps/ch7");
            FileName help = new FileName(extendsName);
            String file_name[] = dir.list(help);
            for(int i = 0; i < file_name.length; i++ ){
                str.append(file_name[i] + " ");
            }
            allFileName = new String(str);
            return allFileName;
        }
}
```

example7_7.jsp（效果如图 7-12 所示）

```
<%@ page contentType = "text/html; charset = GB2312" %>
<%@ page import = "tom.jiafei.ListFile" %>
<jsp: useBean id = "file" class = "tom.jiafei.ListFile" scope = "request"/>
<HTML><BODY><FONT size = 2>
<FORM action = "" Method = "post" >
    输入文件的扩展名：<Input type = text name = "extendsName">
  <Input type = submit value = "提交">
</FORM>
  <jsp: setProperty name = "file" property = " * "/>
  当前 JSP 页面所在目录中，扩展名是：
<jsp: getProperty name = "file" property = "extendsName"/>
  的文件有：
  <br><jsp: getProperty name = "file" property = "allFileName"/>
</FONT></BODY></HTML>
```

图 7-12　使用 bean 的辅助类

7.4　使用 bean 的简单例子

7.4.1　三角形

JSP 页面通过表单输入三角形三边的长度并提交给该页面，表单提交后，JSP 页面将计算三角形面积的任务交给一个 bean 去完成。

1. 三角形 bean

Triangle.java

```
package tom.jiafei;
public class Triangle{
    double sideA,sideB,sideC;
    double area;
    boolean isTriangle;
    public void setSideA(double a){
        sideA = a;
    }
    public double getSideA(){
        return sideA;
    }
    public void setSideB(double b){
        sideB = b;
    }
    public double getSideB(){
        return sideB;
    }
    public void setSideC(double c){
        sideC = c;
    }
    public double getSideC(){
        return sideC;
    }
    public double getArea(){
        double p = (sideA + sideB + sideC)/2.0;
        area = Math.sqrt(p * (p - sideA) * (p - sideB) * (p - sideC));
        return area;
    }
    public boolean isTriangle(){
        if(sideA < sideB + sideC&&sideB < sideA + sideC&&sideC < sideA + sideB)
            isTriangle = true;
        else
            isTriangle = false;
        return isTriangle;
    }
}
```

2. JSP 页面

triangle.jsp(效果如图 7-13 所示)

```
<%@ page contentType = "text/html; charset = GB2312" %>
<%@ page import = "tom.jiafei.Triangle" %>
<jsp:useBean id = "triangle" class = "tom.jiafei.Triangle" scope = "page"/>
<HTML><BODY><FONT size = 2>
<FORM action = "" Method = "post">
<P>输入三角形的边 A:<Input type = text name = "sideA" value = 0>
<P>输入三角形的边 B:<Input type = text name = "sideB" value = 0>
<P>输入三角形的边 C:<Input type = text name = "sideC" value = 0>
```

```
<Input type=submit value="提交">
<P>你给出三角形的三边是：
<jsp:setProperty name= "triangle" property=" * " />
<BR>边 A 是：<jsp:getProperty name="triangle" property="sideA"/>
<BR>边 B 是：<jsp:getProperty name="triangle" property="sideB"/>
<BR>边 C 是：<jsp:getProperty name= "triangle" property="sideC"/>
<P>这三个边能构成一个三角形吗？
<jsp:getProperty name="triangle" property="triangle"/>
<P>面积是：<jsp:getProperty name="triangle" property="area"/>
</FONT></BODY></HTML>
```

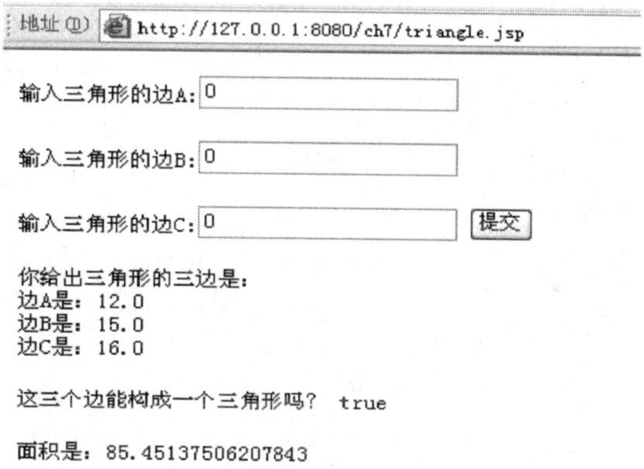

地址(D) http://127.0.0.1:8080/ch7/triangle.jsp

输入三角形的边A: 0

输入三角形的边B: 0

输入三角形的边C: 0 提交

你给出三角形的三边是:
边A是: 12.0
边B是: 15.0
边C是: 16.0

这三个边能构成一个三角形吗? true

面积是: 85.45137506207843

图 7-13　使用 bean 计算三角形面积

7.4.2　猜数字

当用户访问 getNumber.jsp 页面时，随机给用户一个 1～100 之间的整数，然后用户使用该页面提供的表单输入自己的猜测，并提交给 guess.jsp 页面，guess.jsp 页面使用一个 bean 来处理用户的猜测，该 bean 负责判断用户的猜测是否正确。

1. 猜数字 bean

GuessNumber.java

```
package tom.jiafei;
public class GuessNumber{
    int answer=0,                    //待猜测的整数
    guessNumber=0,                   //用户的猜测
    guessCount=0;                    //用户猜测的次数
    String result=null;
    boolean right=false;
    public void setAnswer(int n){
        answer=n;
        guessCount=0;
    }
    public int getAnswer(){
        return answer;
    }
    public void setGuessNumber(int n){
```

```
        guessNumber = n;
        guessCount ++ ;
        if(guessNumber == answer){
            result = "恭喜,猜对了";
            right = true;
        }
        else if(guessNumber > answer){
            result = "猜大了";
            right = false;
        }
        else if(guessNumber < answer){
            result = "猜小了";
            right = false;
        }
    }
    public int getGuessNumber(){
        return guessNumber;
    }
    public int getGuessCount(){
        return guessCount;
    }
    public String getResult(){
        return result;
    }
    public boolean isRight(){
        return right;
    }
}
```

2. JSP 页面

getNumber.jsp(效果如图 7-14 所示)

```
<%@ page contentType = "text/html; charset = GB2312" %>
<%@ page import = "tom.jiafei.GuessNumber" %>
<jsp:useBean id = "guess" class = "tom.jiafei.GuessNumber" scope = "session"/>
<HTML><BODY bgcolor = cyan><FONT size = 2>
<% int n = (int)(Math.random() * 100) + 1;
   String str = response.encodeRedirectURL("guess.jsp");
%>
<jsp:setProperty name = "guess" property = "answer" value = "<% = n %>"/>
   随机给你一个 1 到 100 之间的数,请猜测这个数是多少?
<Form action = "<% = str %>" method = post >
   输入你的猜测:<Input type = text name = "guessNumber">
<Input type = submit value = "提交">
</FONT></FORM></BODY></HTML>
```

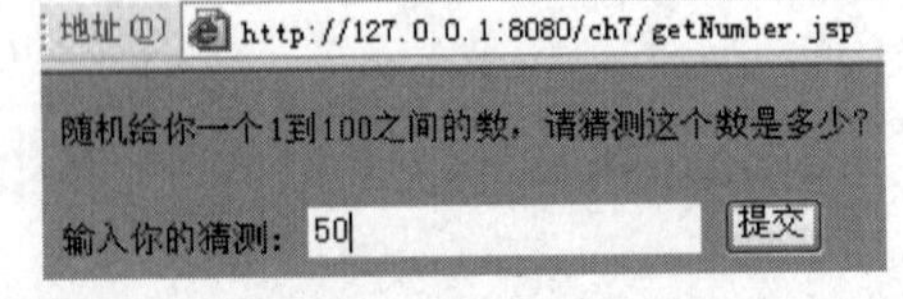

图 7-14　输入猜测

guess.jsp（效果如图 7-15 所示）

```
<%@ page contentType = "text/html; charset = GB2312" %>
<%@ page import = "tom.jiafei.GuessNumber" %>
<jsp:useBean id = "guess" class = "tom.jiafei.GuessNumber" scope = "session" />
<% String strGuess = response.encodeRedirectURL("guess.jsp"),
          strGetNumber = response.encodeRedirectURL("getNumber.jsp");
%>
<HTML><BODY>
<jsp:setProperty name = "guess" property = "guessNumber" param = "guessNumber"/>
   这是第<jsp:getProperty name = "guess" property = "guessCount" />猜,
<jsp:getProperty name = "guess" property = "result"/>。
   你给出的数是 <jsp:getProperty name = "guess" property = "guessNumber"/>。
<% if(guess.isRight() == false){
%><FORM action = "<% = strGuess %>" method = post >
      再输入你的猜测：<Input type = text name = "guessNumber">
      <Input type = submit value = "提交">
   </FORM>
<% }
%>
<BR><A href = "<% = strGetNumber %>">链接到 getNumber.jsp 重新玩猜数。</A>
</BODY></HTML>
```

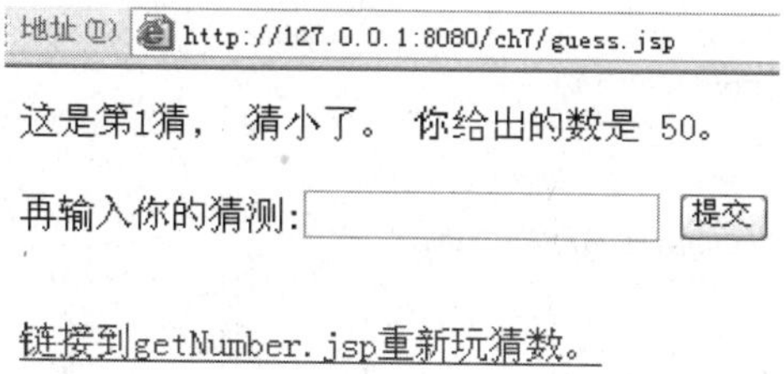

地址(D) http://127.0.0.1:8080/ch7/guess.jsp

这是第1猜，猜小了。你给出的数是 50。

再输入你的猜测: 提交

链接到getNumber.jsp重新玩猜数。

(a) 猜小了

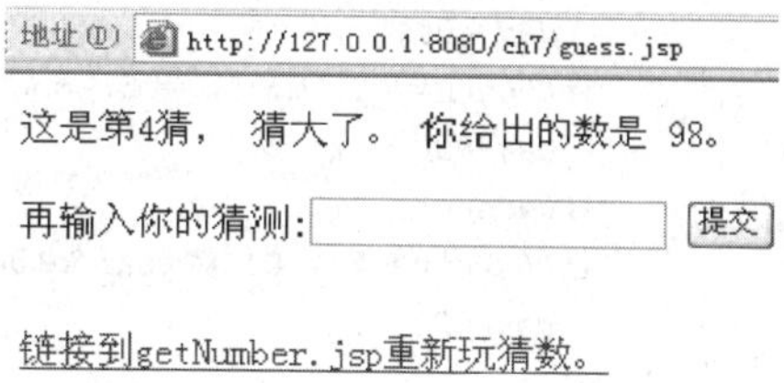

地址(D) http://127.0.0.1:8080/ch7/guess.jsp

这是第4猜，猜大了。你给出的数是 98。

再输入你的猜测: 提交

链接到getNumber.jsp重新玩猜数。

(b) 猜大了

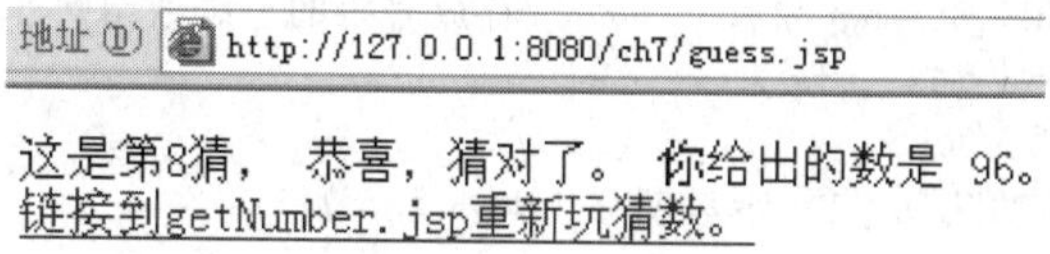

地址(D) http://127.0.0.1:8080/ch7/guess.jsp

这是第8猜，恭喜，猜对了。你给出的数是 96。
链接到getNumber.jsp重新玩猜数。

(c) 猜对了

图 7-15　guess.jsp 的效果图

7.4.3　日历

JSP 页面通过表单选择年和月，然后调用 bean，bean 负责 JSP 页面选择的年、月的“日历”。

1. 日历 bean

CalendarBean.java

```
package tom.jiafei;
import java.util.*;
public class CalendarBean{
```

```
String calendar = null;
int year = 2005,month = 0;
public void setYear(int year){
  this.year = year;
}
public int getYear(){
  return year;
}
public void setMonth(int month){
  this.month = month;
}
public int getMonth(){
  return month;
}
public String getCalendar(){
  StringBuffer buffer = new StringBuffer();
  Calendar 日历 = Calendar.getInstance();
  日历.set(year,month - 1,1);          //将日历翻到 year 年 month 月 1 日,注意 0 表示一月
  //依次类推,11 表示 12 月
  //获取 1 日是星期几(get 方法返回的值是 1 表示星期日,返回的值是 7 表示星期六)
    int 星期几 = 日历.get(Calendar.DAY_OF_WEEK) - 1;
    int day = 0;
    if(month == 1||month == 3||month == 5||month == 7||month == 8||month == 10||month == 12)
      day = 31;
    if(month == 4||month == 6||month == 9||month == 11)
      day = 30;
    if(month == 2){
    if(((year % 4 == 0)&&(year % 100!= 0))||(year % 400 == 0))
      day = 29;
    else
      day = 28;
    }
    String a[] = new String[42];        //存放号码的一维数组
    for(int i = 0; i < 42; i++ )
        a[i] = " ";
    for(int i = 星期几,n = 1; i <星期几 + day; i++ ){
      if(n <= 9)
        a[i] = String.valueOf(n) + " ";
      else
        a[i] = String.valueOf(n);
      n++ ;
    }
    //用表格显示数组
    buffer.append("< table border = 3 >");
    buffer.append("< tr >") ;
    String xingqi[] = {"星期日","星期一","星期二","星期三","星期四","星期五",
"星期六"};
    for(int k = 0; k < 7; k++ )
       buffer.append("< td >" + xingqi[k] + "</td >");
    buffer.append("</tr >");
    for(int k = 0; k < 42; k = k + 7){
```

```
            buffer.append("<tr>") ;        //换行
            for(int j=k; j<7+k; j++)
                buffer.append("<td>" + a[j] + "</td>");
            buffer.append("</tr>");
        }
        buffer.append("</table");
        calendar = new String(buffer);
        return calendar;
    }
}
```

2. JSP 页面

showcalendar.jsp(效果如图 7-16 所示)

```
<%@ page contentType="text/html; charset=GB2312" %>
<%@ page import="tom.jiafei.CalendarBean" %>
<HTML><BODY bgcolor=cyan><FONT size=4>
  <jsp:useBean id="rili" class="tom.jiafei.CalendarBean" scope="request"/>
  <FORM action="" method=post name=form>
    选择日历的年份：
  <Select name="year">
            <Option value="2009">2009 年
            <Option value="2010">2010 年
            <Option value="2011">2011 年
            <Option value="2012">2012 年
  </Select>
  选择日历的月份：
            <Select name="month">
            <Option value="1">1 月
            <Option value="2">2 月
            <Option value="3">3 月
            <Option value="4">4 月
            <Option value="5">5 月
            <Option value="6">6 月
            <Option value="7">7 月
            <Option value="8">8 月
            <Option value="9">9 月
            <Option value="10">10 月
            <Option value="11">11 月
            <Option value="12">12 月
            </Select>
            <BR><BR>
            <INPUT TYPE="submit" value="提交你的选择" name="submit">
            </FORM>
            <jsp:setProperty name="rili" property="*"/>
  <FONT color="red"><jsp:getProperty name="rili" property="year"/></FONT>年
  <FONT color="blue"><jsp:getProperty name="rili" property="month"/></FONT>月
  的日历：
  <jsp:getProperty name= "rili" property="calendar" />
</FONT></BODY></HTML>
```

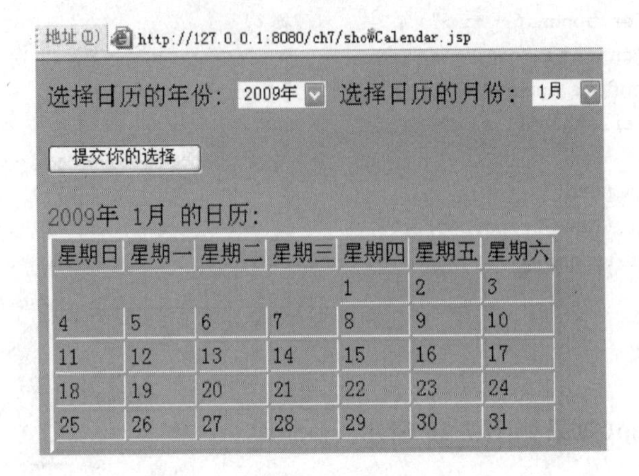

图 7-16　日历

7.4.4　四则运算

JSP 页面通过表单输入两个数和四则运算符号提交给该页面,表单提交后,JSP 页面将计算任务交给一个 bean 去完成。

1. 四则运算 bean

ComputerBean.java

```
package tom.jiafei;
public class ComputerBean{
    double numberOne,numberTwo,result;
    String operator = "+";
    public void setNumberOne(double n){
      numberOne = n;
    }
    public double getNumberOne(){
      return numberOne;
    }
    public void setNumberTwo(double n){
      numberTwo = n;
    }
    public double getNumberTwo(){
      return numberTwo;
    }
    public void setOperator(String s){
      operator = s.trim();
    }
    public String getOperator(){
      return operator;
    }
    public double getResult(){
      if(operator.equals("+"))
        result = numberOne + numberTwo;
      else if(operator.equals("-"))
```

```
        result = numberOne - numberTwo;
      else if(operator.equals(" * "))
        result = numberOne * numberTwo;
      else if(operator.equals("/"))
        result = numberOne/numberTwo;
      return result;
    }
}
```

2. JSP 页面

computer.jsp(效果如图 7-17 所示)

```
<%@ page contentType = "text/html; charset = GB2312" %>
<%@ page import = "tom.jiafei. * " %>
  <jsp: useBean id = "computer" class = "tom.jiafei.ComputerBean" scope = "session"/>
<HTML><BODY bgcolor = yellow><Font size = 2>
  <jsp: setProperty name = "computer" property = " * "/>
  <FORM action = "" method = post name = form>
      <Input type = text name = "numberOne"
                value = <jsp: getProperty name = "computer" property = "numberOne"/> size = 5>
          <Select name = "operator">
                <Option value = " + "> +
                <Option value = " - "> -
                <Option value = " * "> *
                <Option value = "/">/
          </Select>
        <Input type = text name = "numberTwo"
                value = <jsp: getProperty name = "computer" property = "numberTwo"/> size = 5>
        = <jsp: getProperty name = "computer" property = "result"/>
      <BR> <INPUT TYPE = "submit" value = "提交你的选择" name = "submit">
    </FORM>
</BODY></HTML>
```

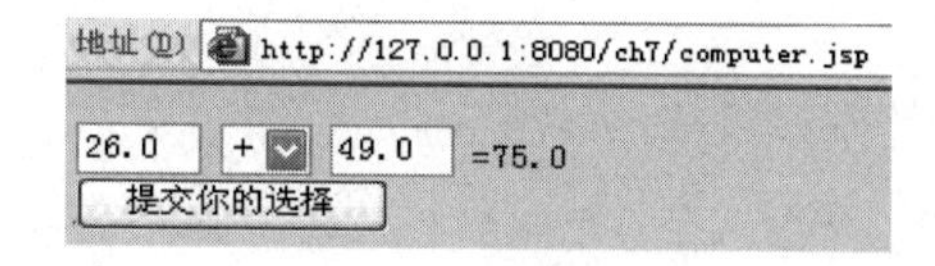

图 7-17　四则运算

7.4.5　浏览图片

图片是扩展名为.jpg 的图像文件，存放在当前 Web 服务目录 ch7 的子目录 image 中，要求图像文件的名字中不能含有汉字和空格。JSP 页面调用一个负责浏览图像的 bean，效果如图 7-18 所示。

1. 浏览图片 bean

Play.java

```
package tom.jiafei;
import java.io. * ;
class FileName implements FilenameFilter{
```

```
        public boolean accept(File dir,String name){
            boolean boo = false;
              if(name.endsWith(".jpg")||name.endsWith(".JPG"))
                  boo = true;
              return boo;
        }
}
public class Play{
    int imageNumber = 0,max;
    String pictureName[],playImage;
    public Play(){
        File f = new File("."); //该文件被认为在 Web 引擎的/bin 目录中
        String path = f.getAbsolutePath();
        path = path.substring(0,path.indexOf("bin") - 1);
        File dir = new File(path + "/webapps/ch7/image");
        pictureName = dir.list(new FileName());
        max = pictureName.length;
    }
    public void setImageNumber(int n){
       if(n < 0)
          n = max - 1;
       if(n == max)
          n = 0;
       imageNumber = n;
    }
    public int getImageNumber(){
       return imageNumber;
    }
    public String getPlayImage(){
       playImage = new String("< image src = image/" + pictureName[imageNumber] + " " +
                " width = 260 height = 200 ></image >");
       return playImage;
    }
}
```

2. JSP 页面

showPic.jsp(效果如图 7-18 所示)

```
<%@ page contentType = "text/html;charset = GB2312" %>
<%@ page import = "tom.jiafei.*" %>
<jsp:useBean id = "play" class = "tom.jiafei.Play" scope = "session"/>
<jsp:setProperty name = "play" property = "imageNumber" param = "imageNumber"/>
<jsp:getProperty name = "play" property = "playImage"/>
<HTML><BODY bgcolor = cyan><Font size = 2>
<Table><FORM action = "" method = post>
  <tr>
  <td><Input type = submit name = "ok" value = "上一张"></td>
   <Input type = "hidden" name = "imageNumber" value = "<% = play.getImageNumber() - 1 %>">
   </FORM>
   <FORM action = "" method = post>
   <td><Input type = submit name = "ok" value = "下一张"></td>
```

```
    <Input type = "hidden" name = "imageNumber" value = "<% = play.getImageNumber() + 1 %>">
    </tr>
    </FORM>
   </Table>
</Font></BODY></HTML>
```

图 7-18 浏览图片效果图

7.5 JavaBean 与文件操作

7.5.1 读文件

bean 可以列出用户指定的目录中的文件并可读取目录中文件的内容，bean 的 scope 取值为 session。有两个 JSP 页面：select.jsp 和 read.jsp，其中，select.jsp 页面调用 bean 选择目录；read.jsp 页面调用 bean 读取相应目录下的文件。

1. 读文件 bean

ReadFile.java

```
package tom.jiafei;
import java.io.*;
public class ReadFile{
    String fileDir = "c:/",fileName = "";
    String listFile,readContent;
    public void setFileDir(String s){
        fileDir = s;
    }
    public String getFileDir(){
        return fileDir;
    }
    public void setFileName(String s){
        fileName = s;
    }
    public String getFileName(){
        return fileName;
    }
    public String getListFile(){
```

```
            File dir = new File(fileDir);
            File file_name[] = dir.listFiles();
            StringBuffer list = new StringBuffer();
            for(int i = 0; i < file_name.length; i++ ){
                if ((file_name[i]!= null)&&(file_name[i].isFile())){
                  String temp = file_name[i].toString();
                  int n = temp.lastIndexOf("\\");
                  temp = temp.substring(n + 1);
                  list.append(" " + temp);
                }
            }
            listFile = new String(list);
            return listFile;
        }
        public String getReadContent(){                        //读取文件
            try{  File file = new File(fileDir,fileName);
                  FileReader in = new FileReader(file);
                  BufferedReader inTwo = new BufferedReader(in);
                  StringBuffer stringbuffer = new StringBuffer();
                  String s = null;
                  while ((s = inTwo.readLine())!= null){
                      byte bb[] = s.getBytes();
                      s = new String(bb);
                      stringbuffer.append("\n" + s);
                  }
                  String temp = new String(stringbuffer);
                  readContent = "< TextArea rows = 10 cols = 62 >" + temp + "</TextArea >";
            }
            catch(IOException e){
                  readContent = "< TextArea rows = 8 cols = 62 ></TextArea >";

            }
            return readContent;
        }
    }
```

2. JSP 页面

select.jsp(效果如图 7-19 所示)

```
<%@ page contentType = "text/html; charset = GB2312" %>
<HTML><BODY bgcolor = yellow><Font size = 2>
<P>选择一个目录:
<FORM action = "read.jsp" method = post>
    <Select name = "fileDir" >
        <Option value = "D:/1000"> D:/1000
        <Option value = "D:/apache - tomcat - 6.0.13/webapps/ch6"> Web 服务目录 ch6
        <Option value = "D:/apache - tomcat - 6.0.13/webapps/ch7"> Web 服务目录 ch7
    </Select>
<Input type = submit value = "提交">
</FORM>
</FONT></BODY></HTML>
```

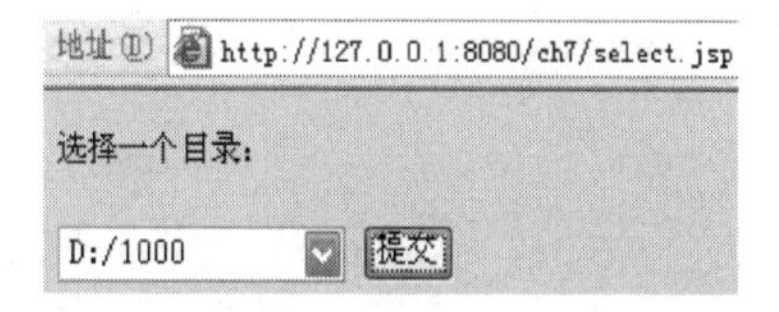

图 7-19　使用 bean 列出目录

read.jsp(效果如图 7-20 所示)

```
<%@ page contentType = "text/html; charset = GB2312" %>
<%@ page import = "tom.jiafei.ReadFile" %>
<HTML><BODY bgcolor = cyan><Font size = 2>
<jsp:useBean id = "file" class = "tom.jiafei.ReadFile" scope = "session"/>
<jsp:setProperty name = "file" property = "fileDir" param = "fileDir"/>
<P>该目录<jsp:getProperty name = "file" property = "fileDir"/>
        有如下文件:<BR>
    <jsp:getProperty name = "file" property = "listFile"/>
<FORM action = "" method = post name = form1>
    输入一个文件名字:<input type = text name = "fileName">
    <Input type = submit value = "提交">
</FORM>
    <jsp:setProperty name = "file" property = "fileName" param = "fileName"/>
        文件:<jsp:getProperty name = "file" property = "fileName"/>
        内容如下:<BR>
    <jsp:getProperty name = "file" property = "readContent"/>
<BR>
        <A href = "select.jsp">重新选择目录</A>
</Body></HTML></HTML>
```

图 7-20　使用 bean 读取文件

7.5.2　写文件

bean 可以将 JSP 页面 write.jsp 提交的文本信息以文件格式保存到服务器，文件所在的目录以及名字由 write.jsp 负责指定。

1. 写文件 bean

WriteFile. java

```
package tom.jiafei;
import java.io.*;
public class WriteFile{
    String filePath = null,
           fileName = null,
           fileContent = null;
    boolean success;
    public void setFilePath(String s){
        filePath = s;
        try{  File path = new File(filePath);
              path.mkdir();
           }
        catch(Exception e){}
    }
    public String getFilePath(){
        return filePath;
    }
    public void setFileName(String s){
        fileName = s;
    }
    public String getFileName(){
        return fileName;
    }
    public void setFileContent(String s){
        fileContent = s;
        byte content[] = fileContent.getBytes();
        try{  File file = new File(filePath,fileName);
              FileOutputStream in = new FileOutputStream(file);
              in.write(content,0,content.length);
              in.close();
              success = true;
        }
        catch(Exception e){
             success = false;
        }
    }
    public boolean isSuccess(){
        return success;
    }
}
```

2. JSP 页面

write.jsp(效果如图 7-21 所示)

```
<%@ page contentType = "text/html; Charset = GB2312" %>
<%@ page import = "tom.jiafei.WriteFile" %>
<jsp:useBean id = "ok" class = "tom.jiafei.WriteFile" scope = "page"/>
<jsp:setProperty name = "ok" property = "filePath" param = "filePath"/>
```

```
<jsp:setProperty name = "ok" property = "fileName" param = "fileName"/>
<jsp:setProperty name = "ok" property = "fileContent" param = "fileContent"/>
<HTML><BODY bgcolor = cyan><Font size = 2>
  <FORM action = "" method = post>
    请选择一个目录:<Select name = "filePath">
                      <Option value = "C:/1000">C:/1000
                      <Option value = "D:/2000">D:/2000
                  </Select>
<BR>输入保存文件的名字:<Input type = text name = "fileName">
<BR>输入文件的内容:<BR>
  <TextArea name = "fileContent" Rows = "10" Cols = "40"></TextArea>
<BR><Input type = submit value = "提交">
</FORM>
<% if(ok.isSuccess() == true){
 %>  你写文件成功,文件所在目录:
        <jsp:getProperty name = "ok" property = "filePath" />
        <BR>文件名字:
        <jsp:getProperty name = "ok" property = "fileName"/>
<%
    }
 %>
</FONT></BODY></HTML>
```

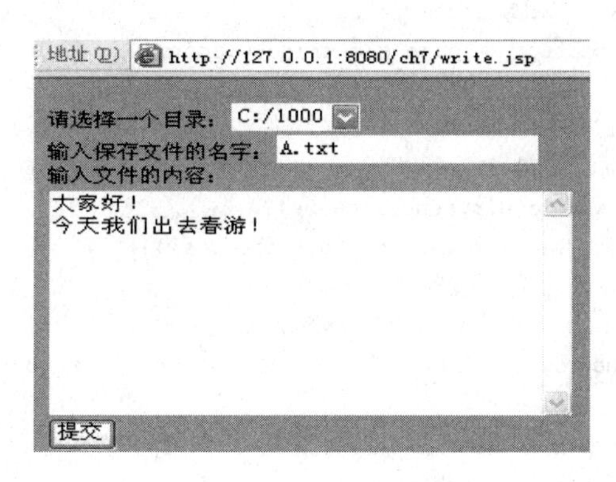

图 7-21　使用 bean 写文件

7.5.3　上传文件

JSP 页面 upfile.jsp 调用 bean 将用户选择的文件上传至服务器(上传至当前 Web 服务目录的 image 子目录中),如果上传的是图像文件,就显示上传的图像,但是需要注意的是,如果上传的图像的名字中含有汉字或空格,那么图像可以成功上传,但是上传页面不能显示该图像。

上传文件 bean 使用了 javax.servlet.http 包中的类,javax.servlet.http 包不在 JDK 的核心类库中,因此需要将 Tomcat 安装目录 lib 子目录中的 servlet-api.jar 文件复制到 Tomcat 服务器所使用的 JDK 的扩展目录中。例如,复制到 D:\jdk1.6\jre\lib\ext 中。

1. 上传文件 bean

UpFile.java

```
package tom.jiafei;
```

```
import java.io.*;
import javax.servlet.http.*;
public class UpFile{
    HttpServletRequest request;
    HttpSession session;
    String upFileMessage = "",showImage;
    public void setRequest(HttpServletRequest request){
        this.request = request;
    }
    public void setSession(HttpSession session){
        this.session = session;
    }
    public String getShowImage(){
      return showImage;
    }
public String getUpFileMessage(){
        String fileName = null;
        try{ String tempFileName = (String)session.getId();                //用户的 session 的 id
             String webDir = request.getContextPath();
             webDir = webDir.substring(1);
             File f = new File("");
             String path = f.getAbsolutePath();
             int index = path.indexOf("bin");
             String tomcatDir = path.substring(0,index);                  //tomcat 的安装目录
             File dir = new File(tomcatDir + "/webapps/" + webDir + "/image");
             dir.mkdir();
              //建立临时文件 f1
           File f1 = new File(dir,tempFileName);
           FileOutputStream o = new FileOutputStream(f1);
            //将客户上传的全部信息存入 f1
           InputStream in = request.getInputStream();
           byte b[] = new byte[10000];
           int n;
           while( (n = in.read(b))! = -1){
              o.write(b,0,n);
           }
           o.close();
           in.close();
            //读取临时文件 f1,从中获取上传文件的名字和上传文件的内容
           RandomAccessFile randomRead = new RandomAccessFile(f1,"r");
            //读出 f1 的第 2 行,析取出上传文件的名字
           int second = 1;
           String secondLine = null;
           while(second <= 2) {
                secondLine = randomRead.readLine();
                second++;
           }
           //获取 f1 中第 2 行中"filename"之后" = "出现的位置:
           int position = secondLine.lastIndexOf(" = ");
            //客户上传的文件的名字是
           fileName = secondLine.substring(position + 2,secondLine.length() - 1);
```

```
    randomRead.seek(0);                //再定位到文件 f1 的开头
    //获取第 4 行回车符号的位置
    long forthEndPosition = 0;
    int forth = 1;
    while((n = randomRead.readByte())! = - 1&&(forth <= 4)){
       if(n == '\n'){
          forthEndPosition = randomRead.getFilePointer();
          forth++;
       }
    }
    //根据客户上传文件的名字,将该文件存入磁盘
    byte cc[] = fileName.getBytes("ISO - 8859 - 1");
    fileName = new String(cc);
    File f2 =  new File(dir,fileName);
    RandomAccessFile randomWrite = new RandomAccessFile(f2,"rw");
    //确定出文件 f1 中包含客户上传的文件的内容的最后位置,即倒数第 6 行
    randomRead.seek(randomRead.length());
    long endPosition = randomRead.getFilePointer();
    long mark = endPosition;
    int j = 1;
    while((mark >= 0)&&(j <= 6)) {
        mark -- ;
        randomRead.seek(mark);
        n = randomRead.readByte();
        if(n == '\n'){
           endPosition = randomRead.getFilePointer();
           j++;
        }
    }
    //将 randomRead 流指向文件 f1 的第 4 行结束的位置
    randomRead.seek(forthEndPosition);
    long startPoint = randomRead.getFilePointer();
   //从 f1 读出客户上传的文件存入 f2(读取第 4 行结束位置和倒数第 6 行之间的内容)
    while(startPoint < endPosition - 1){
        n = randomRead.readByte();
        randomWrite.write(n);
        startPoint = randomRead.getFilePointer();
    }
    randomWrite.close();
    randomRead.close();

    f1.delete();                       //删除临时文件

    upFileMessage = fileName + " 上传成功(Successfully UpLoad)";
    showImage = fileName;
    return upFileMessage;
 }
 catch(Exception exp){
    if(fileName! = null){
        upFileMessage = fileName + "上传失败( Fail to UpLoad)";
        return upFileMessage;
```

```
                }
                else{
                    upFileMessage = "";
                    return upFileMessage;
                }
            }
        }
}
```

2. JSP 页面

upfile.jsp(效果如图 7-22 所示)

```
<%@ page contentType = "text/html; charset = GB2312" %>
<%@ page import = "tom.jiafei.UpFile" %>
<jsp:useBean id = "upFile" class = "tom.jiafei.UpFile" scope = "session"/>
<HTML><BODY bgcolor = yellow size = 3 >
    <br>选择要上传的文件:<br>
    <FORM action = "" method = "post" ENCTYPE = "multipart/form-data">
        <INPUT type = FILE name = "boy" size = "30">
        <BR><INPUT type = "submit" name = "g" value = "提交">
    </FORM>
    <%   upFile.setRequest(request);
          upFile.setSession(session);
     %>
    <jsp:getProperty name = "upFile" property = "upFileMessage"/>
    <BR><img src = image/<jsp:getProperty name = "upFile" property = "showImage"/>
                                          width = 80 height = 80 ></img>
</BODY></HTML>
```

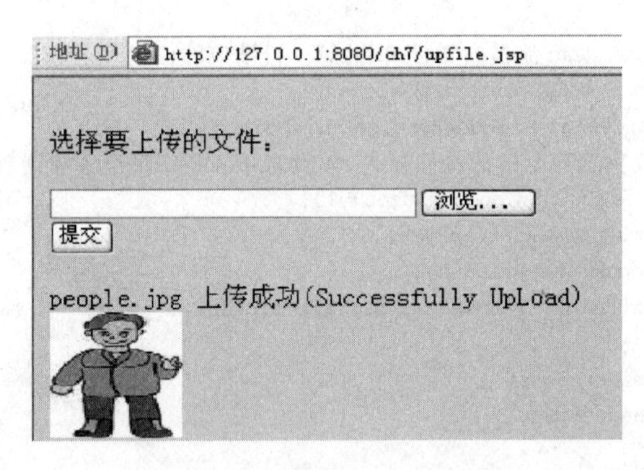

图 7-22　使用 bean 上传文件

7.6　JavaBean 与数据库操作

7.6.1　查询记录

bean 使用加载 MySQL 的 JDBC-数据库驱动程序方式连接数据库、查询表。该 bean 具有很好的通用性，可以查询 MySQL 管理的数据库中的任何表中的记录。JSP 页面 inquire.jsp 只需提供数据库以及相应的表的名字即可。

1. 查询记录 bean

QueryBean.java

```
package tom.jiafei;
import java.sql.*;
public class QueryBean{
    String databaseName = "";              //数据库名
    String tableName = "";                 //表名
    String user = "";                      //用户
    String password = "" ;                 //密码
    StringBuffer queryResult;              //查询结果
    public QueryBean(){
        queryResult = new StringBuffer();
        try{ Class.forName("com.mysql.jdbc.Driver");
        }
        catch(Exception e) {}
    }
    public void setDatabaseName(String s){
        databaseName = s.trim();
        queryResult = new StringBuffer();
    }
    public String getDatabaseName(){
        return databaseName;
    }
    public void setTableName(String s){
        tableName = s.trim();
        queryResult = new StringBuffer();
    }
    public String getTableName(){
        return tableName;
    }
    public void setPassword(String s){
        password = s.trim();
        queryResult = new StringBuffer();
    }
    public String getPassword(){
        return password;
    }
    public void setUser(String s){
        user = s.trim();
        queryResult = new StringBuffer();
    }
    public String getUser(){
        return user;
    }
    public StringBuffer getQueryResult(){
        Connection con;
        Statement sql;
```

```
        ResultSet rs;
        try{ queryResult.append("< table border = 1 >");
             String uri = "jdbc:mysql://127.0.0.1/" + databaseName;
             con = DriverManager.getConnection(uri,user,password);
             DatabaseMetaData metadata = con.getMetaData();
             ResultSet rs1 = metadata.getColumns(null,null,tableName,null);
             int 字段个数 = 0;
             queryResult.append("< tr >");
             while(rs1.next()){
                字段个数++;
                String clumnName = rs1.getString(4);
                queryResult.append("< td >" + clumnName + "</td >");
             }
             queryResult.append("</tr >");
             sql = con.createStatement();
             rs = sql.executeQuery("SELECT * FROM " + tableName);
             while(rs.next()){
                queryResult.append("< tr >");
                for(int k = 1;k <= 字段个数;k++)
                    queryResult.append("< td >" + rs.getString(k) + "</td >");
                queryResult.append("</tr >");
             }
             queryResult.append("</table >");
             con.close();
        }
        catch(SQLException e){
             queryResult.append("请输入正确的用户名和密码");
        }
        return queryResult;
    }
}
```

2. JSP 页面

JSP 页面中使用 JavaBean 查询数据库 book 中的 booklist 表(见 6.1.2 节创建的数据库),效果如图 7-23 所示。

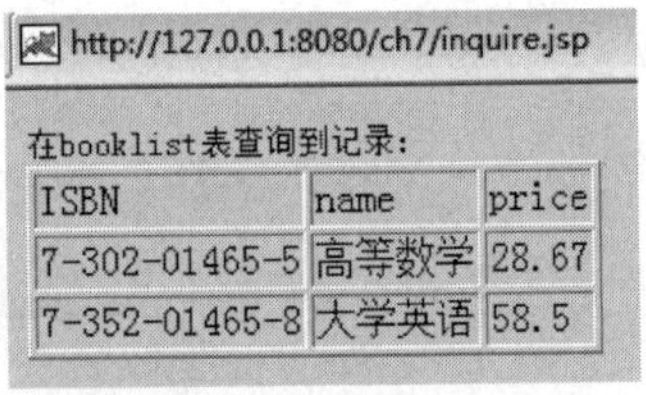

ISBN	name	price
7-302-01465-5	高等数学	28.67
7-352-01465-8	大学英语	58.5

http://127.0.0.1:8080/ch7/inquire.jsp — 在booklist表查询到记录:

图 7-23 使用 bean 查询记录

inquire.jsp(效果如图 7-23 所示)

```
<%@ page contentType = "text/html;charset = GB2312" %>
<%@ page import = "tom.jiafei.QueryBean" %>
< jsp:useBean id = "base" class = "tom.jiafei.QueryBean"
scope = "page"/>
< jsp:setProperty name = "base" property = "databaseName"
value = "book"/>
< jsp:setProperty name = "base" property = "tableName" value = "booklist"/>
< jsp:setProperty name = "base" property = "user" value = "root"/>
< jsp:setProperty name = "base" property = "password" value = ""/>
< HTML >< Body bgcolor = cyan >< Font size = 2 >
在< jsp:getProperty name = "base" property = "tableName"/>表查询到记录:
< BR >< jsp:getProperty name = "base" property = "queryResult"/>
</Font ></Body ></HTML >
```

7.6.2 分页显示记录

bean 在实现分页显示记录时使用了 CachedRowSetImpl 对象,该 bean 具有很好的通用性,可以分页显示 MySQL 管理的数据库中的任何表中的记录。

在第 6 章,学习了 JDBC 操作数据库的有关知识,基本熟悉了 ResultSet 对象。需要特别注意的是,ResultSet 对象和数据库连接对象(Connnection 对象)实现了紧密的绑定,一旦连接对象被关闭,ResultSet 对象中的数据立刻消失。这就意味着,如果用户使用分页方式显示 ResultSet 对象中的数据,就必须始终保持和数据库的连接,直到用户将 ResultSet 对象中的数据查看完毕。每种数据库在同一时刻都有允许的最大连接数目,因此当多个用户同时分页读取数据库表的记录时,应当避免长时间占用数据库的连接资源。

com. sun. rowset 包提供了 CachedRowSetImpl 类,该类实现了 CachedRowSet 接口。CachedRowSetImpl 对象可以保存 ResultSet 对象中的数据,而且 CachedRowSetImpl 对象不依赖 Connnection 对象,这意味着一旦把 ResultSet 对象中的数据保存到 CachedRowSetImpl 对象中后,就可以关闭和数据库的连接。CachedRowSetImpl 继承了 ResultSet 的所有方法,因此可以像操作 ResultSet 对象一样来操作 CachedRowSetImpl 对象。

将 ResultSet 对象 result 中的数据保存到 CachedRowSetImpl 对象 rowSet 中的代码如下:

```
CachedRowSetImpl rowSet = new CachedRowSetImpl();
rowSet.populate(result);
```

注意:JDK1. 5 之后的版本,比如 JDK1. 6 和 JDK1. 7,如果使用了 Sun 公司专用包(包名以 com. sun 为前缀),在编译时将提示使用了 Sun 的专用 API,只要该 API 未废弃,程序可正常运行。

假设 CachedRowSetImpl 对象中有 m 行记录,准备每页显示 n 行,那么,总页数的计算公式是:

- 如果 m 除以 n 的余数大于 0,总页数等于 m 除以 n 的商加 1;
- 如果 m 除以 n 的余数等于 0,总页数等于 m 除以 n 的商。

即

```
总页数 = (m % n) = = 0? (m/n):(m/n + 1);
```

如果准备显示第 p 页的内容,应当把 CachedRowSetImpl 对象中的游标移动到第 $(p-1)*n+1$ 行。

1. 分页显示记录 bean

ShowRecordByPage. java

```
package tom.jiafei;
import java.sql. * ;
import com.sun.rowset. * ;
public class ShowRecordByPage{
    int pageSize = 10;                    //每页显示的记录数
    int pageAllCount = 0;                 //分页后的总页数
```

```
int showPage = 1 ;                    //当前显示页
StringBuffer presentPageResult;       //显示当前页内容
CachedRowSetImpl rowSet;              //用于存储 ResultSet 对象
String databaseName = "";             //数据库名称
String tableName = "";                //表的名字
String user = "" ;                    //用户
String password = "" ;                //密码
String 字段[] = new String[100] ;
int 字段个数 = 0;
public ShowRecordByPage(){
   presentPageResult = new StringBuffer();
   try{ Class.forName("com.mysql.jdbc.Driver");
   }
   catch(Exception e){}
}
public void setPageSize(int size){
   pageSize = size;
   字段个数 = 0;
   String uri = "jdbc:mysql://127.0.0.1/" + databaseName;
   try{ Connection con = DriverManager.getConnection(uri,user,password);
         DatabaseMetaData metadata = con.getMetaData();
         ResultSet rs1 = metadata.getColumns(null,null,tableName,null);
         int k = 0;
         while(rs1.next()){
            字段个数++;
            字段[k] = rs1.getString(4);      //获取字段的名字
            k++;
         }
         Statement sql = con.createStatement(ResultSet.TYPE_SCROLL_SENSITIVE,
                        ResultSet.CONCUR_READ_ONLY);
         ResultSet rs = sql.executeQuery("SELECT * FROM " + tableName);
         rowSet = new CachedRowSetImpl();   //创建行集对象
         rowSet.populate(rs);
         con.close();                      //关闭连接
         rowSet.last();
         int m = rowSet.getRow();   //总行数
         int n = pageSize;
         pageAllCount = ((m % n) == 0)? (m/n):(m/n + 1);
   }
   catch(Exception exp){}
}
public int getPageSize(){
   return pageSize;
}
public int getPageAllCount(){
   return pageAllCount;
}
public void setShowPage(int n){
   showPage = n;
}
public int getShowPage(){
```

```
        return showPage;
    }
    public StringBuffer getPresentPageResult(){
        if(showPage > pageAllCount)
            showPage = 1;
        if(showPage <= 0)
            showPage = pageAllCount;
        presentPageResult = show(showPage);
        return presentPageResult;
    }
    public StringBuffer show(int page){
        StringBuffer str = new StringBuffer();
        str.append("<table border = 1>");
        str.append("<tr>");
        for(int i = 0;i<字段个数;i++)
            str.append("<th>" + 字段[i] + "</th>");
        str.append("</tr>");
        try{ rowSet.absolute((page - 1) * pageSize + 1);
                boolean boo = true;
                for(int i = 1;i <= pageSize&&boo;i++){
                    str.append("<tr>");
                    for(int k = 1;k <= 字段个数;k++)
                        str.append("<td>" + rowSet.getString(k) + "</td>");
                    str.append("</tr>");
                    boo = rowSet.next();
                }
        }
        catch(SQLException exp){}
        str.append("</table>");
        return str;
    }
    public void setDatabaseName(String s){
        databaseName = s.trim();
    }
    public String getDatabaseName(){
        return databaseName;
    }
    public void setTableName(String s){
        tableName = s.trim();
    }
    public String getTableName(){
        return tableName;
    }
    public void setPassword(String s){
        password = s.trim();;
    }
    public void setUser(String s){
        user = s.trim();
    }
    public String getUser(){
        return user;
    }
}
```

2. JSP页面

JSP页面中使用JavaBean分页(每页显示3条记录)显示数据库car中的carlist表(见6.1.2节创建的数据库)中的记录,效果如图7-24所示。

showRecord.jsp(效果如图7-24所示)

```
<%@ page contentType = "text/html;charset = GB2312" %>
<%@ page import = "java.sql.*" %>
<%@ page import = "tom.jiafei.ShowRecordByPage" %>
<jsp:useBean id = "look" class = "tom.jiafei.ShowRecordByPage" scope = "session"/>
<jsp:setProperty name = "look" property = "databaseName" value = "car"/>
<jsp:setProperty name = "look" property = "tableName" value = "carList"/>
<jsp:setProperty name = "look" property = "user" value = "root"/>
<jsp:setProperty name = "look" property = "password" value = ""/>
<jsp:setProperty name = "look" property = "pageSize" value = "3"/>
<HTML><BODY bgcolor = cyan>
    数据库
    <jsp:getProperty name =  "look" property = "databaseName"/>中
    <jsp:getProperty name =  "look" property = "tableName"/>表的记录将被分页显示。
<br>共有 <jsp:getProperty name = "look" property = "pageAllCount"/> 页,
    每页最多显示<jsp:getProperty name = "look" property = "pageSize" />条记录。
  <jsp:setProperty name =  "look" property = "showPage" />
 <jsp:getProperty name =  "look" property = "presentPageResult" />
 <BR>单击"前一页"或"后一页"按钮查看记录(当前显示第
 <jsp:getProperty name =  "look" property = "showPage" /> 页)。
<Table>
 <tr><td><FORM action = "">
       <Input type = hidden name = "showPage" value = "<% = look.getShowPage() - 1 %>" >
       <Input type = submit name = "g" value = "前一页">
       </FORM>
    </td>
    <td><FORM action = "">
       <Input type = hidden name = "showPage" value = "<% = look.getShowPage() + 1 %>" >
       <Input type = submit name = "g" value = "后一页">
       </Form>
    </td>
    <td> <FORM action = "">
          输入页码:<Input type = text name = "showPage" size = 5 >
          <Input type = submit name = "g" value = "提交">
          </FORM>
    </td>
 </tr>
 </Table>
</BODY></HTML>
```

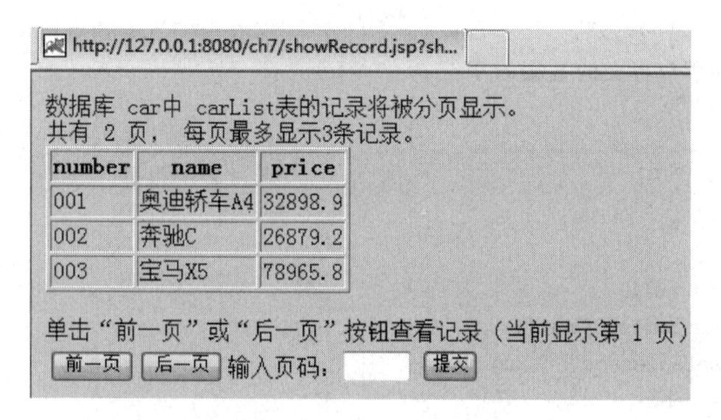

图 7-24　分页显示页面效果图

7.7　标准化考试

试题文件的书写格式要求是：

- 第 1 行必须是全部试题的答案(用来判定考试者的分数)。
- 每道题目之间用 1 行 ***** 分割(至少含有两个 **)。
- 整个试题用 endend 结尾(编写 bean 需要这个标记)。

试题可以是任何一套标准的测试题，例如英语的标准化考试，包括单词测试，阅读理解等。试题存放的目录依赖于 JSP 页面给出的要求，例如这里的 JSP 页面要求试题存放在 D:/2000 或 D:/1000 目录中。试题中每道题目提供 A、B、C、D 4 个选择，如下列 A. txt 所示：

A. txt

```
AC
1. 北京奥运会是什么时间开幕的？<br>
   A. 2008-08-08   B. 2008-08-01<br>   C. 2008-10-01    D. 2008-07-08
******
2. 下列哪个国家不属于亚洲？<br>
   A. 沙特          B. 印度<br>         C. 巴西          D. 越南
*****
endend
```

1. 标准化考试 bean

用一个 bean 每次读取一道试题，实现网络标准化考试。使用网络进行标准化考试是一种常见的考试形式，大部分标准化考试都使用数据库来处理有关的数据。使用数据库可以方便地管理有关的数据，但却降低了系统的效率。基于文件来管理有关的数据，可以提高系统的效率，但要求合理的组织有关数据，以便系统方便地管理数据。

bean 负责每次读入试题的一道题目，用户阅读题目，然后给出自己的选择。

Test. java

```
package tom.jiafei;
import java.io.*;
public class Test{
   String filename = "",                          //存放考题文件名字的字符串
```

```
correctAnswer = "?????",                          //存放正确答案的字符串
//存放试题和用户提交的答案的字符串
testContent = "",
selection = "";
int score = 0;                                    //考试者的得分
boolean 批分结束 = false;
File f = null;
FileReader in = null;
BufferedReader buffer = null;
public void setFilename(String name){
    filename = name;
   //当选择了新的考题文件后,将用户的答案字符串清空
   //将分数设为 0
    score = 0;
    selection = "";
    批分结束 = false;
    //读取试题文件的第 1 行:标准答案
    try{ f = new File(filename);
         in = new FileReader(f);
         buffer = new BufferedReader(in);
         correctAnswer = (buffer.readLine()).trim();              //读取一行,去掉前后空格
    }
    catch(Exception e){
         testContent = "没有选择" + f.getAbsolutePath() + "试题";
         buffer = null;
    }
}
public String getFilename(){
    return filename;
}
public String getTestContent(){                   //获取试题的内容
   try { String s = null;
         StringBuffer temp = new StringBuffer();
         if(buffer! = null) {                     //如果用户选择了试题文件,buffer 就不是空对象
            while((s = buffer.readLine())! = null){                          //继续取某个试题
              if(s.startsWith("**"))              //试题结束标志
                 break;
              temp.append(s);
              if(s.startsWith("endend")){         //试题文件结束标志
                 in.close();                      //关闭和文件的连接
                 buffer.close();
              }
               testContent = new String(temp);
            }
         }
         else{
            testContent = new String("没有选择" + f.getAbsolutePath() + "试题");
         }
   }
   catch(Exception e){
         testContent = "试题无内容,考试结束了!";
```

```
            try{ in.close();
                   buffer.close();
            }
            catch(IOException exp){}
        }
        return testContent;
    }
    public void setSelection(String s){
        selection = selection + s;                    //将用户提交的答案依次尾加到 selection
    }
    public int getScore(){
        int length1 = selection.length();
        int length2 = correctAnswer.length();
        int i = length1 - 1;                          //用户提交的第 i 题答案在 selection 中的位置
        if((i! = - 1)&&(i <= length2 - 1)){           //判定分数
          if((selection.charAt(i) == correctAnswer.charAt(i))&&(批分结束 == false))
             score++;
          if(i == length2 - 1)
             批分结束 = true;
        }
      return score;
    }
}
```

2. JSP 页面

JSP 页面将试题的路径和名字传递给所使用的 bean。例如,将 D:/2000/A.txt 传递给 bean,然后 bean 将试题的内容显示给当前 JSP 页面,并负责判定用户的答案是否正确,效果如图 7-25 所示。

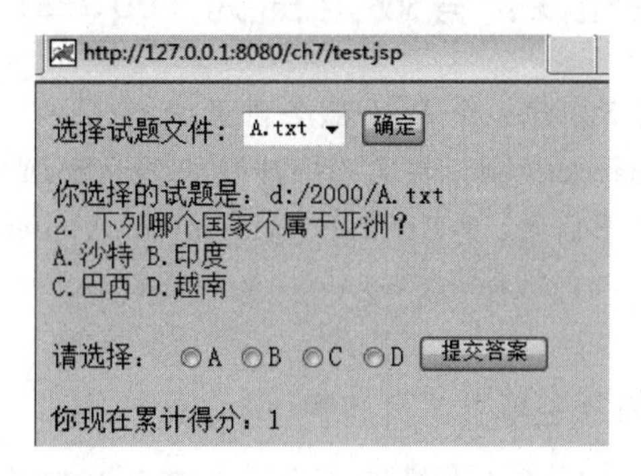

图 7-25 使用 bean 进行标准化考试

test.jsp(效果如图 7-25 所示)

```
<%@ page contentType = "text/html;charset = GB2312" %>
<%@ page import = "tom.jiafei.Test" %>
<jsp:useBean id = "test" class = "tom.jiafei.Test" scope = "session"/>
<HTML><BODY bgcolor = cyan>
<FORM action = "" method = "post">
选择试题文件:
<Select name = "filename" value = "A.txt">
```

```
            < Option value = "d:/2000/A.txt" > A.txt
            < Option value = "d:/1000/B.txt"> B.txt
            < Option value = "d:/1000/C.txt"> C.txt
</Select >
< Input type = "submit" name = "sub" value = "确定">
</FORM >
<% -- bean 设置文件的名字 ,但只有提交了下面相应的表单后 bean 才能到文件名字: -- %>
 < jsp:setProperty name = "test" property = "filename" param = "filename"/>
 你选择的试题是:< jsp:getProperty name =  "test" property = "filename"/>
<% -- 通过 bean 获取试题的内容 -- %>
< br >< Font size = 3 color = blue >
    < jsp:getProperty name = "test" property = "testContent"/>
    </font >
< BR >< FORM action = "" method = "post">
     请选择:
    < Input type = "radio" name = "selection" value = "A"> A
    < Input type = "radio" name = "selection" value = "B"> B
    < Input type = "radio" name = "selection" value = "C"> C
    < Input type = "radio" name = "selection" value = "D"> D
    < Input type = "submit" name = "tijiao" value = "提交答案">
</FORM >
<% -- 下面的标签只有提交了相应的表单才被执行: -- %>
 < jsp:setProperty name = "test" property = "selection"/>
 你现在累计得分:< jsp:getProperty name = "test" property = "score"/>
</BODY ></HTML >
```

7.8 实验 1:有效范围为 request 的 bean

要求在 webapps 目录下新建一个 Web 服务目录:practice7。除特别要求外,实验所涉及的 JSP 页面均保存在 practice7 中。在实验中涉及的 JavaBean 的包名均为 tom.jiafei,因此要求在 practice7 下建立子目录:WEB-INF\classes\tom\jiafei,JavaBean 的字节码文件保存在该子目录中。

1. 相关知识点

JSP 页面使用 useBean 标记调用一个 bean:

```
< jsp: useBean id = "名字" class = "创建 bean 的类" scope = "request"></jsp: useBean >
```

或

```
< jsp: useBean id =  "名字" class = "创建 bean 的类" scope =  "request"/>
```

JSP 引擎分配给每个用户的有效范围是 request 的 bean 是互不相同的,也就是说,尽管每个用户的 bean 的功能相同,但它们占有不同的内存空间。JSP 引擎对请求做出响应之后,取消分配给用户的这个 bean。

2. 实验目的

本实验的目的是让学生掌握使用有效范围是 request 的 bean 显示汽车的基本信息。

3. 实验要求

编写一个 JSP 页面：inputAndShow.jsp 和一个名字为 car 的 JavaBean，其中 car 由 Car.class 类负责创建。

1）inputAndShow.jsp 的具体要求

inputAndShow.jsp 页面提供一个表单。其中表单允许用户输入汽车的牌号、名称和生产日期，该表单将用户输入的信息提交给当前页面，当前页面调用名字为 car 的 bean，并使用表单提交的数据设置 car 的有关属性的值，然后显示 car 的各个属性的值。inputAndShow.jsp 页面的效果如图 7-26 所示。

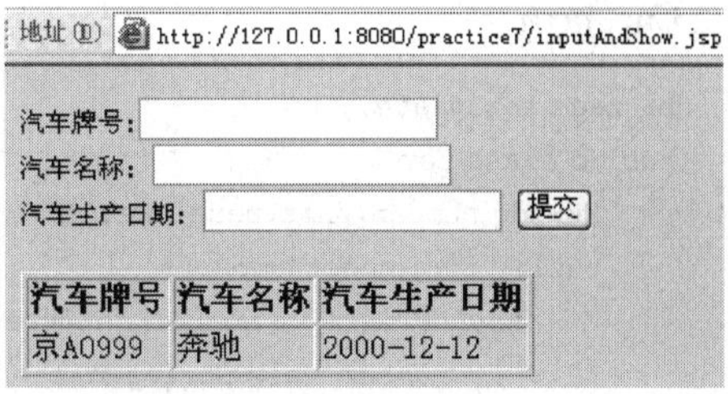

图 7-26　inputAndShow.jsp 页面的效果

2）Car.java 的具体要求

编写的 Car.java 应当有汽车号码、名称和生产日期的属性，并提供相应的 getXxx 和 setXxx 方法来获取和修改这些属性的值。Car.java 中使用 package 语句，起的包名是 tom.jiafei。将 Car.java 编译后的字节码文件 Car.class 保存到

```
practice7\WEB-INF\classes\tom\jiafei
```

目录中。

4. 参考代码

代码仅供参考，学生可按着实验要求，参考本代码编写代码。

JSP 页面参考代码如下：

inputAndShow.jsp

```
<%@ page contentType="text/html;charset=GB2312" %>
<%@ page import="tom.jiafei.Car" %>
<jsp:useBean id="car" class="tom.jiafei.Car" scope="request"/>
<HTML><BODY bgcolor=yellow>
<FONT size=2>
   <FORM action="" Method="post">
     汽车牌号:<Input type=text name="number">
     <br>汽车名称:<Input type=text name="name">
     <br>汽车生产日期:<Input type=text name="madeTime">
     <Input type=submit value="提交">
</FORM>
<jsp:setProperty name="car" property="*"/>
<table border=1>
    <tr><th>汽车牌号</th>
          <th>汽车名称</th>
          <th>汽车生产日期</th>
    </tr>
    <tr>
         <td><jsp:getProperty name="car" property="number"/></td>
         <td><jsp:getProperty name="car" property="name"/></td>
         <td><jsp:getProperty name="car" property="madeTime" />
    </tr>
```

```
</FONT>
</BODY></HTML>
```

JavaBean 源文件参考代码如下：

Car. java

```
package tom.jiafei;
public class Car{
    String number,name,madeTime;
    public String getNumber(){
        try{ byte b[] = number.getBytes("ISO-8859-1");
            number = new String(b);
        }
        catch(Exception e){}
        return number;
    }
    public void setNumber(String number){
        this.number = number;
    }
    public String getName(){
        try{ byte b[] = name.getBytes("ISO-8859-1");
            name = new String(b);
        }
        catch(Exception e){}
        return name;
    }
    public void setName(String name){
        this.name = name;
    }
    public String getMadeTime(){
        try{ byte b[] = madeTime.getBytes("ISO-8859-1");
            madeTime = new String(b);
        }
        catch(Exception e){}
        return madeTime;
    }
    public void setMadeTime(String time){
        madeTime = time;
    }
}
```

7.9 实验 2：有效范围为 session 的 bean

要求在 webapps 目录下新建一个 Web 服务目录：practice7。除特别要求外，实验所涉及的 JSP 页面均保存在 practice7 中。在实验中涉及的 JavaBean 的包名均为 tom.jiafei，因此要求在 practice7 下建立子目录：WEB-INF\classes\tom\jiafei，JavaBean 的字节码文件保存在该子目录中。

1. 相关知识点

JSP 页面使用 useBean 标记调用一个有效范围是 session 的 bean：

```
<jsp: useBean id = "名字" class = "创建 bean 的类" scope = "session"></jsp: useBean>
```

或

```
<jsp: useBean id = "名字" class = "创建 bean 的类" scope = "session"/>
```

如果用户在某个 Web 服务目录多个页面中相互连接，每个页面都含有一个 useBean 标记，而且各个页面的 useBean 标记中 ID 的值相同、scope 的值都是 session，那么，该用户在这些页面得到的 bean 是相同的一个(占有相同的内存空间)。如果用户在某个页面更改了这个 bean 的属性，其他页面的这个 bean 的属性也将发生同样的变化。当用户的会话(session)消失，例如用户关闭浏览器时，JSP 引擎取消分配的 bean，即释放 bean 所占有的内存空间。需要注意的是，不同用户的 scope 取值是 session 的 bean 是互不相同的(占有不同的内存空间)，也就是说，当两个用户同时访问一个 JSP 页面时，一个用户对自己 bean 的属性的改变，不会影响到另一个用户。

2. 实验目的

本实验的目的是让学生掌握使用有效范围是 session 的 bean 显示汽车的基本信息。

3. 实验要求

本实验 2 要求和实验 1 类似，但是和实验 1 不同的是，要求编写两个 JSP 页面：input.jsp 和 show.jsp。编写一个名字为 car 的 JavaBean，其中 car 由 Car.class 类负责创建。

1) input.jsp 的具体要求

input.jsp 页面提供一个表单。其中表单允许用户输入汽车的牌号、名称和生产日期，该表单将用户输入的信息提交给当前页面，当前页面调用名字为 car 的 bean，并使用表单提交的数据设置 car 的有关属性的值。要求在 input.jsp 提供一个超链接，以便用户单击这个超链接访问 show.jsp 页面。input.jsp 页面的效果如图 7-27 所示。

2) show.jsp 的具体要求

show.jsp 调用名字为 car 的 bean，并显示该 bean 的各个属性的值。show.jsp 页面的效果如图 7-28 所示。

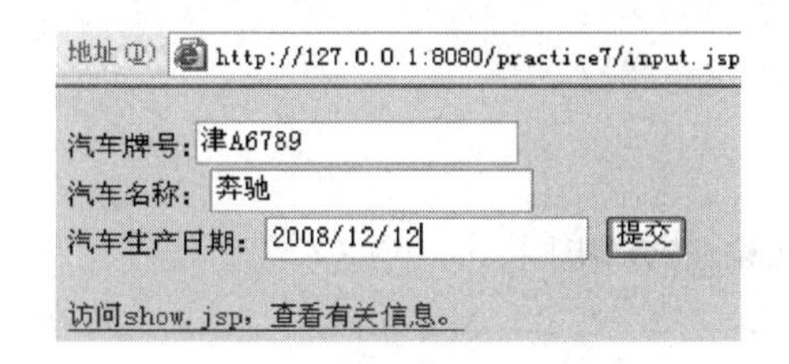

图 7-27 input.jsp 页面的效果

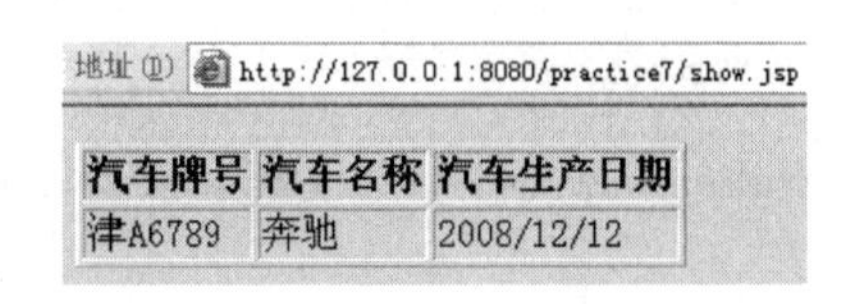

图 7-28 show.jsp 页面的效果

3) Car.java 的具体要求

编写的 Car.java 应当有汽车号码、名称和生产日期的属性，并提供相应的 getXxx 和 setXxx 方法，来获取和修改这些属性的值。Car.java 中使用 package 语句，起的包名是 tom.jiafei。将 Car.java 编译后的字节码文件 Car.class 保存到

```
practice7\WEB-INF\classes\tom\jiafei
```

目录中。

4. 参考代码

代码仅供参考,学生可按着实验要求,参考本代码编写代码。

JSP 页面参考代码如下：

input.jsp

```
<%@ page contentType="text/html; charset=GB2312" %>
<%@ page import="tom.jiafei.Car" %>
<jsp:useBean id="car" class="tom.jiafei.Car" scope="session"/>
<HTML><BODY bgcolor=yellow>
<FONT size=2>
  <FORM action="" Method="post">
    汽车牌号：<Input type=text name="number">
    <br>汽车名称：<Input type=text name="name">
    <br>汽车生产日期：<Input type=text name="madeTime">
    <Input type=submit value="提交">
</FORM>
<jsp:setProperty name="car" property="*"/>
<A href="show.jsp">访问 show.jsp,查看有关信息。</A>
</FONT></BODY></HTML>
```

show.jsp

```
<%@ page contentType="text/html; charset=GB2312" %>
<%@ page import="tom.jiafei.Car" %>
<jsp:useBean id="car" class="tom.jiafei.Car" scope="session"/>
<HTML><BODY bgcolor=yellow>
<table border=1>
    <tr><th>汽车牌号</th>
          <th>汽车名称</th>
          <th>汽车生产日期</th>
    </tr>
    <tr>
          <td><jsp:getProperty name="car" property="number"/></td>
          <td><jsp:getProperty name="car" property="name"/></td>
          <td><jsp:getProperty name="car" property="madeTime" />
    </tr>
</FONT></BODY></HTML>
```

JavaBean 源文件参考代码和实验 1 中的 Car.java 相同。

7.10 实验 3：有效范围为 application 的 bean

要求在 webapps 目录下新建一个 Web 服务目录：practice7。除特别要求外,实验所涉及的 JSP 页面均保存在 practice7 中。在实验中涉及的 JavaBean 的包名均为 tom.jiafei,因此要求在 practice7 下建立子目录：WEB-INF\classes\tom\jiafei,JavaBean 的字节码文件保存在该子目录中。

1. 相关知识点

JSP 页面使用 useBean 标记调用一个 bean：

```
<jsp:useBean id = "名字" class = "创建 bean 的类" scope = "application"></jsp:useBean>
```

或

```
<jsp:useBean id = "名字" class = "创建 bean 的类" scope = "application"/>
```

JSP 引擎为 Web 服务目录下所有的 JSP 页面分配一个共享的 bean，不同用户的 scope 取值是 application 的 bean 也都是相同的一个，也就是说，当多个用户同时访问一个 JSP 页面时，任何一个用户对自己 bean 的属性的改变，都会影响到其他的用户。

2. 实验目的

本实验的目的是让学生掌握使用有效范围是 application 的 bean 制作一个简单的留言板。

3. 实验要求

要求编写两个 JSP 页面：inputMess.jsp 和 showMess.jsp。编写一个名字为 board 的 JavaBean，其中 board 由 MessBoard.class 类负责创建。

1）inputMess.jsp 的具体要求

inputMess.jsp 页面提供一个表单。其中表单允许用户输入留言者的姓名、留言标题和留言内容，该表单将用户输入的信息提交给当前页面，当前页面调用名字为 board 的 bean，并使用表单提交的数据设置 board 的有关属性的值。要求在 inputMess.jsp 提供一个超链接，以便用户单击这个超链接访问 showMess.jsp 页面。inputMess.jsp 页面的效果如图 7-29 所示。

2）showMess.jsp 的具体要求

showMess.jsp 调用名字为 board 的 bean，并显示该 bean 的 allMessage 属性的值。showMess.jsp 页面的效果如图 7-30 所示。

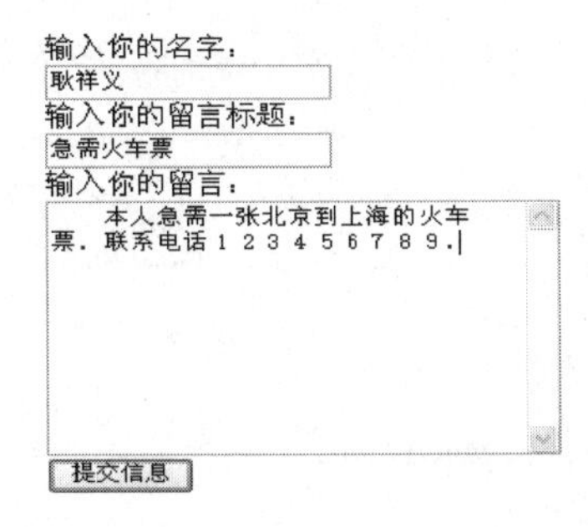

图 7-29　inputMess.jsp 页面的效果

留言者姓名	留言标题	留言内容	留言时间
耿祥义	急需火车票	本人急需一张北京到上海的火车票.	2009-02-05,11:29:28
张林	书籍	本人需要Java方面的书，联系：	2009-02-05,11:30:23

图 7-30　showMess.jsp 页面的效果

3）MessBoard.java 的具体要求

编写的 MessBoard.java 应当有留言者的姓名、留言标题和留言内容属性，并且有全部留言信息的属性：allMessage。将 MessBoard.java 编译后的字节码文件 MessBoard.class 保存到

```
practice7\WEB-INF\classes\tom\jiafei
```

目录中。

4. 参考代码

代码仅供参考,学生可按着实验要求,参考本代码编写代码。

JSP 页面参考代码如下:

inputMess.jsp

```
<%@ page contentType = "text/html; charset = GB2312" %>
<%@ page import = "tom.jiafei.MessBoard" %>
<jsp:useBean id = "board" class = "tom.jiafei.MessBoard" scope = "application"/>
<HTML><BODY>
  <FORM action = "" method = "post" name = "form">
      输入你的名字:<BR><INPUT type = "text" name = "name">
      <BR>输入你的留言标题:<BR><INPUT type = "text" name = "title">
      <BR>输入你的留言:<BR><TEXTAREA name = "content" ROWs = "10"
  COLS = 36 WRAP = "physical"></TEXTAREA>
      <BR><INPUT type = "submit" value = "提交信息" name = "submit">
  </FORM>
    <jsp:setProperty name = "board" property = "*"/>
  <A href = "showMess.jsp">查看留言板</A>
</BODY></HTML>
```

showMess.jsp

```
<%@ page contentType = "text/html; charset = GB2312" %>
<%@ page import = "tom.jiafei.MessBoard" %>
<jsp:useBean id = "board" class = "tom.jiafei.MessBoard" scope = "application"/>
<HTML><BODY bgcolor = yellow>
   <jsp:getProperty name = "board" property = "allMessage"/>
   <A href = "inputMess.jsp">我要留言</A>
</FONT>
</BODY></HTML>
```

JavaBean 源文件参考代码如下:

MessBoard.java

```
package tom.jiafei;
import java.util.*;
import java.text.SimpleDateFormat;
public class MessBoard{
    String name,title,content;
    StringBuffer allMessage;
    ArrayList<String> savedName,savedTitle,savedContent,savedTime;
    public MessBoard(){
        savedName = new ArrayList<String>();
        savedTitle = new ArrayList<String>();
        savedContent = new ArrayList<String>();
        savedTime = new ArrayList<String>();
    }
    public void setName(String s){
        try{
           byte bb[] = s.getBytes("iso-8859-1");
           s = new String(bb);
```

```
        }
        catch(Exception exp){}
        name = s;
        savedName.add(name);
        Date time = new Date();
        SimpleDateFormat matter = new SimpleDateFormat("yyyy - MM - dd,HH: mm: ss");
        String messTime = matter.format(time);
        savedTime.add(messTime);
    }
    public void setTitle(String t){
        try{
            byte bb[] = t.getBytes("iso - 8859 - 1");
            t = new String(bb);
        }
        catch(Exception exp){}
        title = t;
        savedTitle.add(title);
    }
    public void setContent(String c){
        try{
            byte bb[] = c.getBytes("iso - 8859 - 1");
            c = new String(bb);
        }
        catch(Exception exp){}
        content = c;
        savedContent.add(content);
    }
    public StringBuffer getAllMessage(){
        allMessage = new StringBuffer();
        allMessage.append("< table border = 1 >");
        allMessage.append("< tr >");
        allMessage.append("< th>留言者姓名</th>");
        allMessage.append("< th>留言标题</th>");
        allMessage.append("< th>留言内容</th>");
        allMessage.append("< th>留言时间</th>");
        allMessage.append("</tr >");
        for(int k = 0; k < savedName.size(); k++ ){
          allMessage.append("< tr >");
          allMessage.append("< td >");
          allMessage.append(savedName.get(k));
          allMessage.append("</td >");
          allMessage.append("< td >");
          allMessage.append(savedTitle.get(k));
          allMessage.append("</td >");
          allMessage.append("< td >");
          allMessage.append("< textarea >");
          allMessage.append(savedContent.get(k));
          allMessage.append("</textarea >");
          allMessage.append("</td >");
          allMessage.append("< td >");
          allMessage.append(savedTime.get(k));
```

```
            allMessage.append("</td>");
            allMessage.append("<tr>");
        }
        allMessage.append("</table>");
        return allMessage;
    }
}
```

习 题 7

1. 设 Web 服务目录 mymoon 中的 JSP 页面要使用一个 bean,该 bean 的包名为 blue.sky。请说明,应当怎样保存 bean 的字节码。

2. 创建了一个名字为 moon 的 bean,该 bean 有一个 String 类型、名字为 number 的属性。如果创建 moon 的 Java 类没有提供 public String getNumber()方法,在 JSP 页面中是否允许使用 getProperty 标记获取 moon 的 number 属性的值?

3. tom.jiafei.Circle 是创建 bean 的类,下列 A~D 中哪个标记是正确创建 session 周期 bean 的标记?(　　)

```
(A) <jsp:useBean id="circle" class="tom.jiafei.Circle" scope="page"/>
(B) <jsp:useBean id="circle" class="tom.jiafei.Circle " scope="request"/>
(C) <jsp:useBean id="circle" class="tom.jiafei.Circle " scope="session"/>
(D) <jsp:useBean id="circle" type="tom.jiafei.Circle " scope="session"/>
```

4. 假设创建 bean 的类有一个 int 型的属性 number,下列 A~D 中哪个方法是设置该属性值的正确方法?(　　)

```
(A) public void setNumber(int n){
      number = n;
    }
(B) void setNumber(int n){
      number = n;
    }
(C) public void SetNumber(int n){
      number = n;
    }
(D) public void Setnumber(int n){
      number = n;
    }
```

5. 编写两个 JSP 页面 a.jsp 和 b.jsp,a.jsp 页面提供一个表单,用户可以通过表单输入矩形的 2 个边长提交给 b.jsp 页面,b.jsp 调用一个 bean 完成计算矩形面积的任务。b.jsp 页面使用 getProperty 动作标记显示矩形的面积。

第8章 Java Servlet 基础

本章导读

主要内容

- Servlet 类与 servlet 对象
- 编写 web. xml
- servlet 对象的创建与运行
- servlet 对象的工作原理
- 通过 JSP 页面访问 servlet
- 共享变量
- doGet 和 doPost 方法
- 重定向与转发
- 使用 session

难点

- servlet 对象的工作原理
- 重定向与转发

关键实践

- 使用 servlet 读文件

在第 1 章学习了 JSP 页面的运行原理：当用户请求一个 JSP 页面时，Tomcat 服务器自动生成 Java 文件、编译 Java 文件，并用编译得到的字节码文件在服务器端创建一个对象来响应用户的请求。JSP 的根基是 Java Servlet 技术，该技术的核心就是在服务器端创建能响应用户请求的对象，被创建的对象习惯上称作一个 servlet 对象。在 JSP 技术出现之前，Web 应用开发人员自己编写创建 servlet 对象的类，并负责编译生成字节码文件、复制这个字节码文件到服务器的特定目录中，以便服务器使用这个字节码创建一个 servlet 对象来响应用户的请求。

JSP 技术不是 Java Servlet 技术的全部，它只是 Java Servlet 技术的一个成功应用。JSP 技术屏蔽了 servlet 对象创建的过程，使得 Web 程序设计者只需关心 JSP 页面本身的结构，设计好各种标记。例如，使用 HTML 标记设计页面的视图，使用 JavaBean 标记有效地分离页面的视图和数据处理等。有些 Web 应用可能只需要 JSP＋Tag 或 JSP＋JavaBean 就能设计得很好，但是有些 Web 应用，就可能需要 JSP＋JavaBean＋servlet 来完成，即需要服务器再创建一些 servlet 对象，配合 JSP 页面来完成整个 Web 应用程序的工作，关于这一点，将在第 9 章的 MVC 模式中讲述。

本章将使用 javax. servlet. http 包中的类，javax. servlet. http 包不在 JDK 的核心类库中，因此需要将 Tomcat 安装目录 lib 子目录中的 servlet-api. jar 文件复制到 Tomcat 服务器

所使用的 JDK 的扩展目录中,例如,复制到 D:\jdk1.6\jre\lib\ext 中。

在本章中,在 webapps 目录下新建一个 Web 服务目录 ch8,因此,除非特别约定,本章例子中涉及的 JSP 页面均保存在 ch8 中。

8.1 Servlet 类与 servlet 对象

Java Servlet 的核心思想是在服务器端创建能响应用户请求的对象,即创建 servlet 对象。因此,学习 Java Servlet 的首要任务是掌握怎样编写创建 servlet 对象的类,怎样在 Tomcat 服务器上保存编译这个类所得到的字节码,怎样编写部署文件,怎样请求 Tomcat 服务器创建一个 servlet 对象。有关 servlet 对象的运行原理以及使用细节将在后续节中讲述。

1. 编写 Servlet 类

写一个创建 servlet 对象的类就是编写一个特殊类的子类,这个特殊的类就是 javax.servlet.http 包中的 HttpServlet 类。HttpServlet 类实现了 Servlet 接口,实现了响应用户的方法(这些方法将在后续节中讲述)。在 Java Servlet 中,HttpServlet 类的子类被习惯地称作一个 Servlet 类,这样的类创建的对象习惯地被称作一个 servlet 对象。

注意:不要忘记将 Tomcat 安装目录 lib 子目录中的 servlet-api.jar 文件复制到 Tomcat 服务器所使用的 JDK 的扩展目录中,例如,复制到 D:\jdk1.6\jre\lib\ext 中。

下面的 Hello.java 是一个简单的 Servlet 类(如果使用 Tomcat5.x 后的 Tomcat 服务器,要求 Servlet 类必须有包名),该类创建的 servlet 对象可以响应用户的请求,即用户请求这个 servlet 对象时,会在浏览器看到"你好,欢迎学习 servlet"这样的响应信息。

Hello.java

```
package china.dalian;
import java.io.*;
import javax.servlet.*;
import javax.servlet.http.*;
public class Hello extends HttpServlet{
    public void init(ServletConfig config) throws ServletException{
        super.init(config);
    }
    public void service(HttpServletRequest request,HttpServletResponse response)
                throws IOException{
        response.setContentType("text/html; charset = GB2312");    //设置响应的 MIME 类型
        PrintWriter out = response.getWriter();          //获得一个向用户发送数据的输出流
        out.println("<html><body>");
        out.println("<h2>你好,欢迎学习 servlet。</h2>");
        out.println("</body></html>");
    }
}
```

将上述 Servlet 类的源文件:Hello.java 保存到某个目录中,例如 D:\5000 中,然后对其进行编译：

```
D:\5000\javac Hello.java
```

编译成功后，将得到字节码文件 Hello. class。

2. 字节码文件的保存

为了能让 Tomcat 服务器使用 Hello. class 字节码创建一个 servlet 对象，需要将 Hello. class 保存到某个 Web 服务目录中的特定子目录中。本章使用的 Web 服务目录是 ch8，那么需要在 ch8 下建立如下的目录结构：

```
ch8\WEB-INF\classes
```

为了让 Tomcat 服务器启用上述目录，必须重新启动 Tomcat 服务器。

上述目录被启用后，根据 Servlet 类的包名，在 classes 下再建立相应的子目录，例如 Servlet 类的包名为 china. dalian，那么在 classes 下建立子目录：\china\dalian。现在就可以将 Servlet 类的字节码，例如 Hello. class，复制到 ch8\WEB-INF\classes\china\dalian 中，如图 8-1 所示。

图 8-1　Servlet 类的字节码的存放目录

8.2　编写 web. xml

Servlet 类的字节码保存到指定的目录后，必须为 Tomcat 服务器编写一个部署文件，只有这样，Tomcat 服务器才会按用户的请求使用 Servlet 字节码文件创建对象。

该部署文件是一个 XML 文件，名字是 web. xml，该文件由 Tomcat 服务器负责管理。在这里，不需要深刻理解 XML 文件，只需要知道 XML 文件是由标记组成的文件，使用该 XML 文件的应用程序(例如 Tomcat 服务器)配有内置的解析器，可以解析 XML 标记中的数据。可以在 Tomcat 服务器的 webapps 目录中的 root 目录找到一个 web. xml 文件，参照它编写自己的 web. xml 文件。

编写的 web. xml 文件保存到 Web 服务目录的 WEB-INF 子目录中，例如 ch8\WEB-INF 中。编写的 web. xml 文件的内容如下(需要用纯文本编辑器编辑 web. xml)：

web. xml

```
<?xml version = "1.0" encoding = "ISO-8859-1"?>
<web-app>
    <servlet>
        <servlet-name>hello</servlet-name>
        <servlet-class>china.dalian.Hello</servlet-class>
    </servlet>
    <servlet-mapping>
        <servlet-name>hello</servlet-name>
        <url-pattern>/lookHello</url-pattern>
    </servlet-mapping>
</web-app>
```

一个XML文件应当以XML声明作为文件的第一行,在其前面不能有空白、其他的处理指令或注释。XML声明以“<?xml”标识开始,以“?>”标识结束。注意“<?”和“xml”之间以及“?”和“>”之间不要有空格。如果在XML声明中没有显式地指定encoding属性的值,那么该属性的默认值为UTF-8编码。如果在编写XML文件时只准备使用ASCII字符,也可以将encoding属性的值设置为“ISO-8859-1”。例如:

```
<?xml version = "1.0" encoding = " ISO - 8859 - 1" ?>
```

这时,XML文件必须使用“ANSI”编码保存(如图8-2所示),Tomcat服务器中的XML解析器根据encoding属性的值识别XML文件中的标记并正确解析标记中的内容。

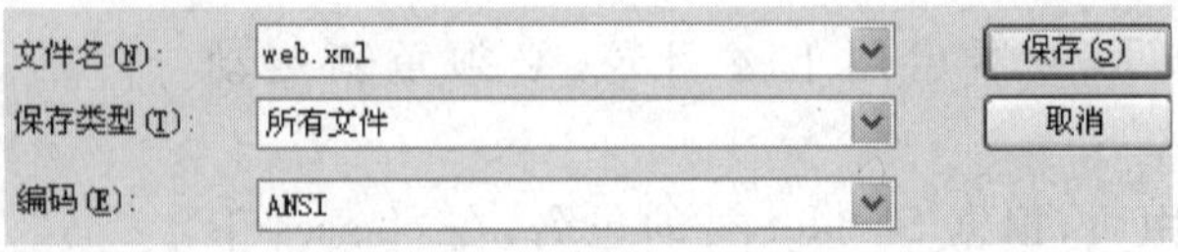

图8-2　encoding值是ISO-8859-1时XML文件的保存

如果encoding属性的值为UTF-8,XML文件必须按着UTF-8编码来保存(如图8-3所示)。如果XML使用UTF-8编码,那么标记以及标记的内容除了可以使用ASCII字符外,还可以使用汉字、日文中的片假名、平假名等字符,Tomcat服务器中的XML解析器就会识别这些标记并正确解析标记中的内容。

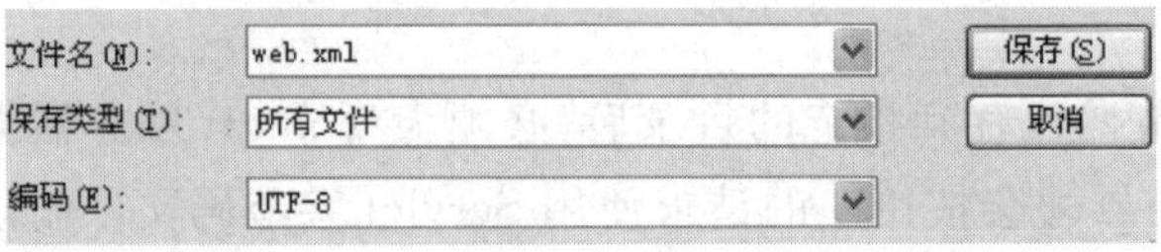

图8-3　encoding值是UTF-8时XML文件的保存

下面是web.xml文件中标记的具体内容及其作用。

1. 根标记

xml文件必须有一个根标记,web.xml文件的根标记是<web-app>。

2. <servlet>标记及子标记

web.xml文件中可以有若干个<servlet>标记,该标记的内容由Tomcat服务器负责处理。<servlet>标记需要有两个子标记:<servlet-name>和<servlet-class>,其中<servlet-name>标记的内容是Tomcat服务器创建的servlet对象的名字,<servlet-class>标记的内容指定Tomcat服务器用哪个Servlet类来创建servlet对象,在给出的web.xml文件中,让Tomcat服务器使用Hello类创建的名字是hello的servlet对象。web.xml文件可以有若干个<servlet>标记,但要求它们的<servlet-name>子标记的内容互不相同。

3. <servlet-mapping>标记及子标记

web.xml文件中出现一个<servlet>标记就会对应地出现一个<servlet-mapping>标记,该标记和<servlet>标记都是根标记的直接子标记。不同的是,<servlet-mapping>标记需要有两个子标记:<servlet-name>和<url-pattern>,其中<servlet-name>标记的内容是Tomcat服务器创建的servlet对象的名字(该名字必须和<servlet>标记的子标记

＜servlet-name＞标记的内容相同)；＜ url-pattern ＞标记用来指定用户用怎样的 URL 格式来请求 servlet 对象，例如，＜url-pattern ＞标记的内容是：/lookHello，那么用户必须在浏览器的地址栏中输入：

```
http：//127.0.0.1：8080/ch7/lookHello
```

来请求 servlet 对象 hello。

Web 服务目录的 WEB-INF 子目录下的 web. xml 文件负责管理该 Web 服务目录下的 servlet 对象，当该 web 服务目录需要提供更多的 servlet 对象时，只要在 web. xml 文件中增加＜servlet＞和＜servlet-mapping＞子标记即可。

如果修改并重新保存 web. xml 文件，Tomcat 服务器就会立刻重新读取 web. xml 文件，因此，修改 web. xml 文件不必重新启动 Tomcat 服务器。但是，如果修改导致 web. xml 文件出现错误，Tomcat 服务器就会关闭当前 Web 服务目录下的所有 servlet 的使用权限。所以必须保证 web. xml 文件正确无误，才能成功启动 Tomcat 服务器。

本章中涉及的 JSP 页面存放在 Web 服务目录 ch8 中，负责创建 servlet 的字节码文件存放在 ch8\WEB-INF\classes\china\dalian 中(本章 Servlet 类的包名均为 china. dalian)。每当 Web 服务目录增加新的 servlet 时，都需要为 web. xml 文件添加＜servlet＞标记和＜servlet-mapping＞标记。

8.3 servlet 对象的创建与运行

servlet 对象由 Tomcat 服务器负责创建，Web 设计者只需为 Tomcat 服务器预备好 Servlet 类的字节码文件、编写好相应的配置文件 web. xml 即可，用户就可以根据 web. xml 部署文件来请求服务器创建并运行一个 servlet 对象。如果服务器没有名字为 hello 的 servlet 对象，服务器就会根据 web. xml 文件中＜servlet＞标记的子标记＜servlet-class＞指定的 Servlet 类创建一个名字为 hello 的 servlet 对象。因此，当 servlet 对象 hello 被创建之后，如果修改了创建 servlet 的 Servlet 类，并希望服务器用新的字节码重新创建 servlet 对象，那么必须要重新启动 Tomcat 服务器。

Servlet 类可以使用 getServletName()方法返回配置文件中＜servlet-name＞标记给出的 servlet 的名字。

当用户请求服务器运行一个 servlet 对象时，必须根据 web. xml 文件中＜servlet-mapping＞标记的子标记＜url-pattern＞指定的格式输入请求。

根据 8. 2 节中的 web. xml 文件，在浏览器输入："http：//127. 0. 0. 1：8080/ch8/lookHello"，请求服务器运行 servlet 对象 hello，效果如图 8-4 所示。

地址(D) http://127.0.0.1:8080/ch8/lookHello

您好，欢迎学习servlet。

图 8-4　简单的 servlet 对象

8.4 servlet 对象的工作原理

servlet 对象由 Tomcat 服务器负责管理，Tomcat 服务器通过读取 web. xml 创建并运行 servlet 对象。本节将详细讲解 servlet 对象的运行原理。

8.4.1 servlet 对象的生命周期

servlet 对象是 javax. servlet 包中 HttpServlet 类的子类的一个实例,由服务器负责创建并完成初始化工作。当多个用户请求一个 servlet 时,服务器为每个用户启动一个线程而不是启动一个进程,这些线程由服务器来管理,与传统的 CGI 为每个用户启动一个进程相比较,效率要高得多。

一个 servlet 对象的生命周期主要有下列 3 个过程组成:

(1) 初始化 servlet 对象。servlet 对象第一次被请求加载时,服务器初始化这个 servlet 对象,即创建一个 servlet 对象,这个对象调用 init 方法完成必要的初始化工作。

(2) 产生的 servlet 对象再调用 service 方法响应用户的请求。

(3) 当服务器关闭时,调用 destroy 方法,消灭 servlet 对象。

init 方法只被调用一次,即在 servlet 第一次被请求加载时调用该方法。当后续的用户请求 servlet 服务时,Web 服务将启动一个新的线程,在该线程中,servlet 对象调用 service 方法响应用户的请求,也就是说,每个用户的每次请求都导致 service 方法被调用执行,其执行过程分别运行在不同的线程中。

8.4.2 init 方法

该方法是 HttpServlet 类中的方法,可以在子类中重写这个方法。init 方法的声明格式:

```
public void init(ServletConfig config) throws ServletException
```

servlet 对象第一次被请求加载时,服务器创建一个 servlet 对象,这个对象调用 init 方法完成必要的初始化工作。该方法在执行时,服务器会把一个 ServletConfig 类型的对象传递给 init()方法,这个对象就被保存在 servlet 对象中,直到 servlet 对象被消灭,这个 ServletConfig 对象负责向 servlet 传递服务设置信息,如果传递失败就会发生 ServletException,servlet 对象就不能正常工作。

8.4.3 service 方法

该方法是 HttpServlet 类中的方法,可以在子类中直接继承该方法或重写这个方法。service 方法的声明格式:

```
public void service(HttpServletRequest request HttpServletResponse response)
throw ServletException,IOException
```

当 servlet 对象成功创建和初始化之后,该对象就调用 service 方法来处理用户的请求并返回响应。服务器将两个参数传递给该方法,一个 HttpServletRequest 类型的对象,该对象封装了用户的请求信息;另外一个参数对象是 HttpServletResponse 类型的对象,该对象用来响应用户的请求。和 init 方法不同的是,init 方法只被调用一次,而 service 方法可能被多次调用,已经知道,当后续的用户请求该 servlet 对象服务时,服务器将启动一个新的线程,在该线程中,servlet 对象调用 service 方法响应用户的请求,也就是说,每个用户的每次请求都导致 service 方法被调用执行,调用过程运行在不同的线程中,互不干扰。因此,不同线程的 service 方法中的局部变量互不干扰,一个线程改变了自己的 service 方法中局部变量的值不会影响其他线程的 service 方法中的局部变量。

8.4.4 destroy 方法

该方法是 HttpServlet 类中的方法，子类可直接继承这个方法，一般不需要重写。destroy 方法的声明格式如下：

```
public destroy()
```

当服务器终止服务时，例如关闭服务器等，destroy()方法会被执行，消灭 servlet 对象。

8.5 通过 JSP 页面访问 servlet

用户除了可以在浏览器的地址栏中直接输入 servlet 对象的请求格式来请求运行一个 servlet 外，也可以通过 JSP 页面来请求一个 servlet。也就是说，可以让 JSP 页面负责数据的显示，而让一个 servlet 去做和处理数据有关的事情。

8.5.1 通过表单向 servlet 提交数据

Web 服务目录下的 JSP 页面都可以通过表单或超链接请求该 Web 服务目录下的某个 servlet。通过 JSP 页面访问 servlet 的好处是，JSP 页面可以负责页面的静态信息处理，动态信息处理交给 servlet 去完成。

需要特别注意的是，如果 web. xml 文件中＜servlet-mapping＞标记的子标记＜url-pattern＞指定的请求 servlet 的格式是“/lookHello”，那么 JSP 页面请求 servlet 时，必须要写成“lookHello”，不可以写成“/lookHello”，否则将变成请求 root 服务目录下的某个 servlet。

在例 8-1 中，JSP 页面 example8_1. jsp 通过表单向名字为 computer 的 servlet 对象提交一个正整数，computer 负责计算并显示该整数的全部因子(computer 由 Computer 类负责创建，访问它的 url-pattern 是 getNumber，见下面的 web. xml 中的配置)。

需要为 ch8\WEN-INF 目录下的 web. xml 文件添加如下的子标记：

```
<servlet>
      <servlet-name>computer</servlet-name>
      <servlet-class>china.dalian.Computer</servlet-class>
</servlet>
<servlet-mapping>
      <servlet-name>computer</servlet-name>
      <url-pattern>/getNumber</url-pattern>
</servlet-mapping>
```

例 8-1

1) JSP 页面

example8_1. jsp(效果如图 8-5 所示)

```
<%@ page contentType = "text/html; charset = GB2312" %>
<HTML><BODY bgcolor = cyan><Font size = 2>
<FORM action = "getNumber" method = post>
  输入一个正整数：<Input Type = text name = number>
```

```
    <br><Input Type = submit value = "提交">
</FORM>
</Font></BODY></HTML>
```

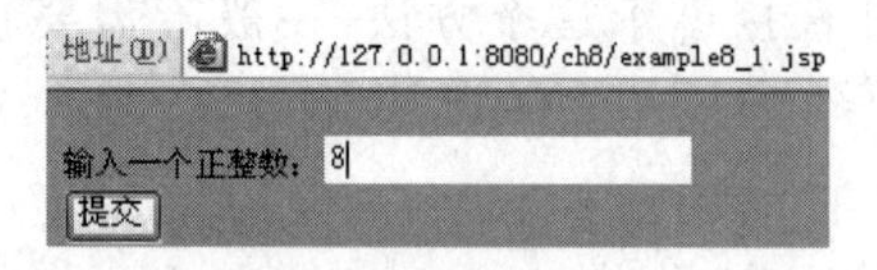

图 8-5 JSP 页面使用表单请求 servlet

地址(D) http://127.0.0.1:8080/ch8/getNumber

我是一个servlet对象，名字是:computer。
我负责计算并显示8的因子:
1 2 4 8

图 8-6 servlet 运行效果

2) Servlet 类

Computer.java(效果如图 8-6 所示)

```
package china.dalian;
import java.io.*;
import javax.servlet.*;
import javax.servlet.http.*;
public class Computer extends HttpServlet{
    String servletName;
    public void init(ServletConfig config) throws ServletException{
        super.init(config);
        servletName = getServletName();
    }
    public void service(HttpServletRequest request,HttpServletResponse response)
                      throws IOException{
        response.setContentType("text/html; charset = GB2312");
        PrintWriter out = response.getWriter();
        out.println("<html><body>");
        String str = request.getParameter("number");
        out.print("我是一个 servlet 对象,名字是: " + servletName + "。<br>");
        out.print("我负责计算并显示" + str + "的因子:<br>");
        int n = 0;
        try{  n = Integer.parseInt(str);
              for(int i = 1; i <= n; i++){
                  if(n % i == 0)
                   out.println(" " + i);
              }
        }
        catch(NumberFormatException e){
            out.print(" " + e);
        }
        out.println("</body></html>");
    }
}
```

8.5.2 通过超链接访问 servlet

在例 8-2 中,在 JSP 页面 example8_2.jsp 中单击一个超链接,请求一个 servlet 对象 show,该 servlet 对象负责输出俄文字母表(show 由 ShowLetter 类负责创建,访问它的 url-pattern 是 helpMeShow，见下面的 web.xml 中的配置）。

需要为 ch8\WEN-INF 目录下的 web.xml 文件添加如下的子标记：

```
<servlet>
    <servlet-name>show</servlet-name>
    <servlet-class>china.dalian.ShowLetter</servlet-class>
</servlet>
<servlet-mapping>
    <servlet-name>show</servlet-name>
    <url-pattern>/helpMeShow</url-pattern>
</servlet-mapping>
```

例 8-2

1）JSP 页面

example8_2.jsp（效果如图 8-7 所示）

```
<%@ page contentType="text/html;Charset=GB2312" %>
<HTML><BODY bgcolor=cyan><Font size=3>
    单击超链接查看俄文字母表：
    <BR><A href="helpMeShow">查看俄文字母表</A>
</Font></BODY></HTML>
```

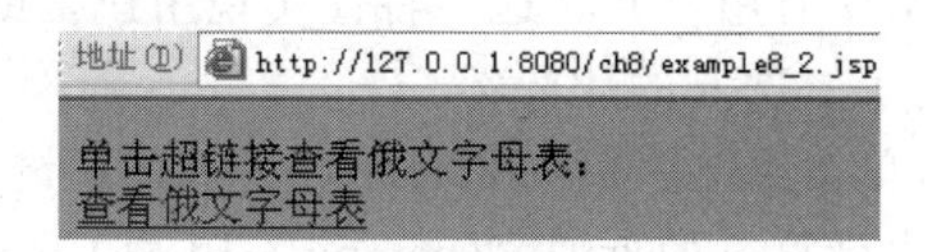

图 8-7　JSP 页面使用超链接请求 servlet

2）Servlet 类

ShowLetter.java（效果如图 8-8 所示）

```
package china.dalian;
import java.io.*;
import javax.servlet.*;
import javax.servlet.http.*;
public class ShowLetter extends HttpServlet{
    public void init(ServletConfig config) throws ServletException{
        super.init(config);
    }
    public void service(HttpServletRequest request,HttpServletResponse response)
                        throws IOException{
        response.setContentType("text/html;charset=GB2312");
        PrintWriter out=response.getWriter();
        out.println("<html><body>");
        out.print("<BR>小写俄文字母：");
        for(char c='а';c<='я';c++)
           out.print(" "+c);
        out.print("<BR>大写俄文字母：");
        for(char c='А';c<='Я';c++)
           out.print(" "+c);
```

```
        out.println("</body></html>");
    }
}
```


地址(D) http://127.0.0.1:8080/ch8/helpMeShow

小写俄文字母: а б в г д е ж з и й к л м н о п р с т у ф х ц ч ш щ ъ ы ь э ю я
大写俄文字母: А Б В Г Д Е Ж З И Й К Л М Н О П Р С Т У Ф Х Ц Ч Ш Щ Ъ Ы Ь Э Ю Я

图 8-8　servlet 运行效果

8.6　共 享 变 量

Servlet 类是 HttpServlet 的一个子类,那么在编写子类时就可以声明某些成员变量,那么,请求该 Servlet 类所创建的 servlet 的用户将共享该 servlet 的成员变量。

数学上有一个计算的公式:

$$\frac{\pi}{4}=1-\frac{1}{3}+\frac{1}{5}-\frac{1}{7}+\frac{1}{9}-\frac{1}{11}\cdots\cdots$$

例 8-3 利用成员变量被所有用户共享这一特性实现用户帮助计算 π 的值,即每当用户请求访问 servlet 时都参与了一次 π 的计算。

在例 8-3 中,用户通过单击 JSP 页面 example8_3.jsp 的超链接访问名字为 computerPI 的 servlet 对象,该对象负责计算 π 的近似值(computerPI 由 ComputerPI 类负责创建,访问它的 url-pattern 是 computerPI,见下面的 web.xml 中的配置)。

需要为 ch8\WEN-INF 目录下的 web.xml 文件添加如下的子标记:

```
<servlet>
    <servlet-name>computerPI</servlet-name>
    <servlet-class>china.dalian.ComputerPI</servlet-class>
</servlet>
<servlet-mapping>
    <servlet-name>computerPI</servlet-name>
    <url-pattern>/computerPI</url-pattern>
</servlet-mapping>
```

例 8-3

1) JSP 页面

example8_3.jsp(效果如图 8-9 所示)

```
<%@ page contentType = "text/html; charset = GB2312" %>
<HTML><BODY bgcolor = cyan><Font size = 3>
<A Href = "computerPI">参与计算 PI 的值<A>
</BODY></HTML>
```

2) Servlet 类

ComputerPI.java(效果如图 8-10 所示)

```
package china.dalian;
```

```
import java.io. * ;
import javax.servlet. * ;
import javax.servlet.http. * ;
public class ComputerPI extends HttpServlet{
    double sum = 0,i = 1,j = 1;
    int number = 0;
    public void init(ServletConfig config) throws ServletException{
        super.init(config);
    }
    public synchronized void service(HttpServletRequest request,
                HttpServletResponse response) throws IOException{
        response.setContentType("text/html; charset = GB2312");
        PrintWriter out = response.getWriter();
        out.println("< html >< body >");
        number ++ ;
        sum = sum + i/j;
        j = j + 2;
        i = - i;
        out.println("servlet: " + getServletName() + "已经被请求了" + number + "次");
        out.println("< BR >现在 PI 的值是: ");
        out.println(4 * sum);
        out.println("</body ></html >");
    }
}
```

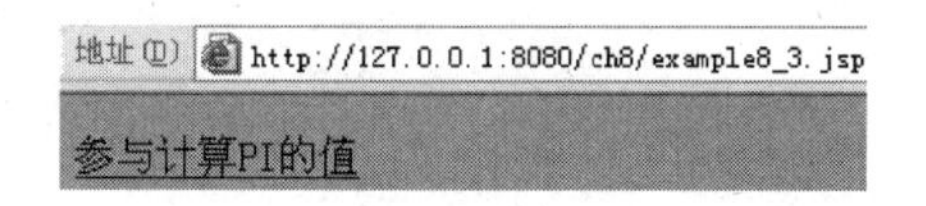

图 8-9　请求 servlet 计算圆周率

地址(D) http://127.0.0.1:8080/ch8/computerPI

servlet:computerPI已经被请求了121次
现在PI的值是: 3.149856975293274

图 8-10　servlet 负责计算圆周率

8.7　doGet 和 doPost 方法

HttpServlet 类除了 init、service、destroy 方法外，该类还有两个很重要的方法：doGet 和 doPost，用来处理用户的请求并作出响应。

当服务器创建 servlet 对象后，该对象会调用 init 方法初始化自己，以后每当服务器再接受到一个 servlet 请求时，就会产生一个新线程，并在这个线程中让 servlet 对象调用 service 方法，而 HttpServlet 类所给出的 service 方法的功能是检查 HTTP 请求类型(get、post 等)，并在 service 方法中根据用户的请求方式，在 service 方法中对应地再调用 doGet 或 doPost。因此，在编写 Servlet 类(HttpServlet 类的一个子类)时，不必重写 service 方法来响应用户，直接继承 service 方法即可。

可以在 Servlet 类中重写 doPost 或 doGet 方法来响应用户的请求，这样可以增加响应的灵活性，并降低服务器的负担。如果不论用户请求类型是 post 还是 get，服务器的处理过程完全相同，那么可以只在 doPost 方法中编写处理过程，而在 doGet 方法中再调用 doPost 方法即可，或只在 doGet 方法中编写处理过程，而在 doPost 方法中再调用 doGet 方法。如果根据请求的类型进行不同的处理,就需在两个方法中编写不同的处理过程。

在例 8-4 中,用户可以使用表单请求名字为“ok”的 servlet 对象,并提交字符串。其中一个表单的提交方式是 post;另一个表单的提交方式是 get。当表单的提交方式是 post 时,servlet 对象输出表单提交的字符串的长度;当表单的提交方式是 get 时,servlet 对象除了输出表单提交的字符串的长度外,还判断字符串中的前缀是否为 Hello(“ok”由 Show 类负责创建,访问它的 url-pattern 是 show,见下面的 web. xml 中的配置)。

需要为 ch8\WEN-INF 目录下的 web. xml 文件添加如下的子标记:

```
<servlet>
    <servlet-name>ok</servlet-name>
    <servlet-class>china.dalian.Show</servlet-class>
</servlet>
<servlet-mapping>
    <servlet-name>ok</servlet-name>
    <url-pattern>/show</url-pattern>
</servlet-mapping>
```

例 8-4

1) JSP 页面

example8_4. jsp(效果如图 8-11 所示)

```
<%@ page contentType = "text/html; charset = GB2312" %>
<HTML><BODY bgcolor = cyan><Font size = 2>
<FORM action = "show" method = post>
   输入字符串,提交给 servlet(Post 方式):
   <br><Input Type = text name = str>
   <Input Type = submit value = "提交">
</FORM>
<FORM action = "show" method = get>
   输入字符串,提交给 servlet(Get 方式):
   <br><Input Type = text name = str>
   <Input Type = submit value = "提交">
</FORM>
</FORM>
</BODY></HTML>
```

图 8-11　选择提交方式

2) Servlet 类

Show. java(效果如图 8-12 所示)

```
package china.dalian;
import java.io.*;
```

```
import javax.servlet.*;
import javax.servlet.http.*;
public class Show extends HttpServlet{
    public void init(ServletConfig config) throws ServletException{
        super.init(config);
    }
     public void doPost (HttpServletRequest request, HttpServletResponse response) throws
                                                    ServletException,IOException{
        response.setContentType("text/html; charset = GB2312");
        PrintWriter out = response.getWriter();
        out.println("<html><body>");
        String s = request.getParameter("str");
        byte bb[] = s.getBytes("iso-8859-1");
        s = new String(bb);
        int n = s.length();
        out.print("\"" + s + "\"" + "的长度: " + n + "<br>");
        out.println("</body></html>");
    }
    public void doGet(HttpServletRequest request,HttpServletResponse response)
                    throws ServletException,IOException{
        doPost(request,response);
        response.setContentType("text/html; charset = GB2312");
        PrintWriter out = response.getWriter();
        out.println("<html><body>");
        String s = request.getParameter("str");
        byte bb[] = s.getBytes("iso-8859-1");
        s = new String(bb);
        if(s.startsWith("Hello"))
            out.print("\"" + s + "\"" + "的前缀是: Hello");
        else
            out.print("\"" + s + "\"" + "的前缀不是: Hello");
        out.println("</body></html>");
    }
}
```

地址(D) http://127.0.0.1:8080/ch8/show?str=Hello%2C%C4%E3%BA%C3

"Hello,你好"的长度:8
"Hello,你好"的前缀是:Hello

图 8-12 servlet 调用 doPost 或 doGet 方法

8.8 重定向与转发

重定向的功能是将用户从当前页面或 servlet 定向到另一个 JSP 页面或 servlet；转发的功能是将用户对当前 JSP 页面或 servlet 对象的请求转发给另一个 JSP 页面或 servlet 对象。本节学习在 Servlet 类中使用 HttpServletResponse 类的重定向方法：sendRedirect 以及 RequestDispatcher 类的转发方法 forward 方法，并指出二者的区别。

8.8.1　sendRedirect 方法

重定向方法：void sendRedirect(String location)是 HttpServletResponse 类中的方法。当用户请求一个 servlet 时，该 servlet 在处理数据后，可以使用重定向方法将用户重新定向到另一个 JSP 页面或 servlet。重定向方法仅仅是将用户从当前页面或 servlet 定向到另一个 JSP 页面或 servlet，但不能将用户对当前页面或 servlet 的请求(HttpServletRequest 对象)转发给所定向的资源。也就是说，重定向的目标页面或 servlet 对象无法使用 request 获取用户提交的数据。

8.8.2　RequestDispatcher 对象

RequestDispatcher 对象可以把用户对当前 JSP 页面或 servlet 的请求转发给另一个 JSP 页面或 servlet，而且将用户对当前 JSP 页面或 servlet 的请求和响应(HttpServletRequest 对象和 HttpServletResponse 对象)传递给所转发的 JSP 页面或 servlet。也就是说，当前页面所要转发的目标页面或 servlet 对象可以使用 request 获取用户提交的数据。

实现转发的步骤如下：

1. 得到 RequestDispatcher 对象

用户所请求的当前 JSP 或 servlet 可以让 HttpServletRequest 对象 request 调用

```
public RequestDispatcher getRequestDispatcher(java.lang.String path)
```

方法返回一个 RequestDispatcher 对象，其中参数 path 是要转发的 JSP 页面或 servlet。例如：

```
RequestDispatcher dispatcher = request.getRequestDispatcher("a.jsp");
```

2. 转发

在步骤 1 中获取的 RequestDispatcher 对象调用

```
void forward(ServletRequest request,ServletResponse response)
            throws ServletException,ava.io.IOException
```

方法可以将用户对当前 JSP 页面或 servlet 的请求转发给 RequestDispatcher 对象所指定的 JSP 页面或 servlet。例如：

```
dispatcher.forward (request,response);
```

将用户对当前 JSP 页面或 servlet 的请求转变成对 a.jsp 页面的请求。

和重定向方法(sendRedirect)不同的是，用户在浏览器的地址栏中不能看到 forward 方法转发的页面的地址或 servlet 的地址，只能看到该页面或 servlet 的运行效果；用户在浏览器的地址栏中所看到的仍然是当前页面或 servlet 的地址(<url-pattern>标记的访问格式)。

在例 8-5 中，用户通过 example8_5.jsp 页面提供的表单输入实数，并提交给名字为 verify 的 servlet 对象(Verify 类负责创建)，如果用户的输入不符合要求或输入的实数大于 2000 或小于−2000，那么 verify 就将用户重新定向到 example8_5.jsp 页面；如果用户的输入符合要求，verify 就将用户对 example8_5.jsp 页面的请求转发给名字为 showMessage 的 servlet 对象，该 servlet 对象计算实数的平方(showMessage 由 ShowMessage 类负责创建，访问它的 url-pattern 是 forYouShowMessage，见下面的 web. xml 中的配置）。

需要为 ch8\WEN-INF 目录下的 web. xml 文件添加如下的子标记：

```
<servlet>
    <servlet-name>verify</servlet-name>
    <servlet-class>china.dalian.Verify</servlet-class>
</servlet>
<servlet>
    <servlet-name>showMessage</servlet-name>
    <servlet-class>china.dalian.ShowMessage</servlet-class>
</servlet>
<servlet-mapping>
    <servlet-name>verify</servlet-name>
    <url-pattern>/verifyYourMessage</url-pattern>
</servlet-mapping>
<servlet-mapping>
    <servlet-name>showMessage</servlet-name>
    <url-pattern>/forYouShowMessage</url-pattern>
</servlet-mapping>
```

例 8-5

1）JSP 页面

example8_5. jsp（效果如图 8-13 所示）

```
<%@ page contentType="text/html; charset=GB2312" %>
<HTML><BODY bgcolor=cyan><Font size=2>
<FORM action="verifyYourMessage" method=post>
   输入一个实数：<Input Type=text name=number>
<br><Input Type=submit value="提交">
</FORM></BODY></HTML>
```

2）Servlet 类

Verify. java

```
package china.dalian;
import java.io.*;
import javax.servlet.*;
import javax.servlet.http.*;
public class Verify extends HttpServlet{
    public void init(ServletConfig config) throws ServletException{
        super.init(config);
    }
    public void doPost(HttpServletRequest request,HttpServletResponse response)
                        throws ServletException,IOException{
        String number=request.getParameter("number");
        try{  double n=Double.parseDouble(number);
              if(n>2000||n<-2000)
                  response.sendRedirect("example8_5.jsp");      //重定向
              else{
                  RequestDispatcher dispatcher=
```

```
                    request.getRequestDispatcher("forYouShowMessage");
                    dispatcher.forward(request,response);        //转发到另一个 servlet
                }
            }
        catch(NumberFormatException e){
            response.sendRedirect("example8_5.jsp");            //重定向
        }
    }
    public void doGet(HttpServletRequest request,HttpServletResponse response)
                        throws ServletException,IOException{
        doPost(request,response);
    }
}
```

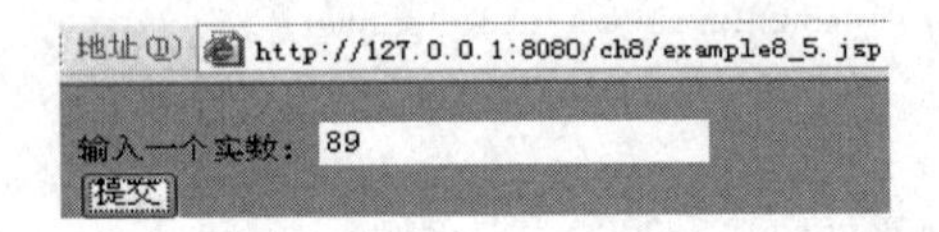

图 8-13 用户输入实数的页面

ShowMessage.java(效果如图 8-14 所示)

```
package china.dalian;
import java.io.*;
import javax.servlet.*;
import javax.servlet.http.*;
public class ShowMessage extends HttpServlet{
    public void init(ServletConfig config) throws ServletException{
        super.init(config);
    }
    public void doPost(HttpServletRequest request,HttpServletResponse response)
                        throws ServletException,IOException{
        response.setContentType("text/html; charset = GB2312");
        PrintWriter out = response.getWriter();
        String number = request.getParameter("number"); //获取用户提交的信息
        double n = Double.parseDouble(number);
        out.println(number + "的平方: " + (n * n));
    }
    public void doGet(HttpServletRequest request,HttpServletResponse response)
                        throws ServletException,IOException{
        doPost(request,response);
    }
}
```

地址(D) http://127.0.0.1:8080/ch8/verifyYourMessage

89的平方:7921.0

图 8-14 名字为 showMessage 的 servlet 运行效果

8.9　使用 session

HTTP 是一种无状态协议。一个用户向服务器发出请求(request),然后服务器返回响应(response),连接就被关闭了。在服务器端不保留连接的有关信息,因此当下一次连接时,服务器已经没有以前的连接信息,无法判断这一次连接和以前的连接是否属于同一用户。本节学习怎样在 Servlet 类中使用 session 对象记录有关连接的信息,session 对象的原理与第 4 章的 4.3 节讲解的完全相同。

HttpServletRequest 对象 request 调用 getSession 方法获取用户的 session 对象:

```
HttpSession session = request.getSession(true);
```

一个用户在访问服务器期间,可能请求多个 servlet 对象,那么服务器端为该用户获取的 session 对象是完全相同的一个,但是,服务器端为不同的用户获取的 session 对象互不相同。有关 session 对象常用方法可参见 4.3 节。

例 8-6 是一个猜数字游戏,当用户访问或刷新 example8_6.jsp 页面时,随机分配给用户一个 1～100 之间的整数,并将这个整数存在用户的 session 对象中。用户链接到 inputNumber.jsp 页面输入自己的猜测,并将该猜测提交给一个名字为 guess 的 servlet 对象(guess 由 HandleGuess 类负责创建,访问它的 url-pattern 是 handleGuess,见下面的 web.xml 中的配置),该 servlet 负责处理用户的猜测,具体处理方式是:

如果用户猜小了,就将用户重新定向到 inputNumber.jsp,并将“你猜小了”存放到用户的 session 中。

如果用户猜大了,就将用户重新定向到 inputNumber.jsp,并将“你猜大了”存放到用户的 session 中。

如果用户猜成功了,就将用户重新定向到 inputNumber.jsp,并将“你猜对了”存放到用户的 session 中。

需要为 ch8\WEN-INF 目录下的 web.xml 文件添加如下的子标记:

```
<servlet>
    <servlet-name>guess</servlet-name>
    <servlet-class>china.dalian.HandleGuess</servlet-class>
</servlet>
<servlet-mapping>
    <servlet-name>guess</servlet-name>
    <url-pattern>/handleGuess</url-pattern>
</servlet-mapping>
```

例 8-6

1) JSP 页面

example8_6.jsp(效果如图 8-15 所示)

```
<%@ page contentType = "text/html; charset = GB2312" %>
<HTML><BODY bgcolor = cyan><Font size = 2>
<% session.setAttribute("message","请你猜数字");
    int randomNumber = (int)(Math.random() * 100) + 1;          //获取一个随机数
```

```
        session.setAttribute("savedNumber",new Integer(randomNumber));
%>
访问或刷新该页面可以随机得到一个 1 至 100 之间的数。
<BR>单击超链接去猜出这个数:<A href = "inputNumber.jsp">去猜数字</A>
</Font></BODY></HTML>
```

地址(D) http://127.0.0.1:8080/ch8/example8_6.jsp

访问或刷新该页面可以随机得到一个1至100之间的数。
单击超链接去猜出这个数:去猜数字

图 8-15　访问该页面获得随机数

inputNumber.jsp(效果如图 8-16 所示)

```
<%@ page contentType = "text/html; charset = GB2312" %>
<HTML><BODY bgcolor = cyan><Font size = 2>
<% String message = (String)session.getAttribute("message"); //获取会话中的信息
%>
<Table border = 1>
<FORM action = "handleGuess" method = post>
  <tr><td> 输入你的猜测:</td>
  <td><Input Type = text name = clientGuess size = 4>
<Input Type = submit value = "提交"></td>
  </tr><td> 提示信息:</td>
        <td><% = message %></td>
</FORM>
  <FORM action = "example8_6.jsp" method = post>
    <tr><td>单击按钮重新开始:</td>
          <td><Input Type = submit value = "随机得到一个 1 至 100 之间的数字"></td>
    </tr>
</FORM>
</Font></BODY></HTML>
```

2) Servlet 类

HandleGuess.java

```
package china.dalian;
import java.io.*;
import javax.servlet.*;
import javax.servlet.http.*;
    public class HandleGuess extends HttpServlet{
        public void init(ServletConfig config) throws ServletException{
            super.init(config);
        }
        public void doPost(HttpServletRequest request,HttpServletResponse response)
                                throws ServletException,IOException{
          HttpSession session = request.getSession(true);          //获取用户的 session 对象
          String str = request.getParameter("clientGuess");       //获取用户猜测的数字
          int guessNumber = -1;
          try{ guessNumber = Integer.parseInt(str);
          }
```

```
            catch(Exception e){
                 response.sendRedirect("inputNumber.jsp");
            }
            int savedNumber =
            ((Integer)session.getAttribute("savedNumber")).intValue();
            if(guessNumber < savedNumber){
                 session.setAttribute("message","你猜小了");
                 response.sendRedirect("inputNumber.jsp");
            }
            if(guessNumber > savedNumber){
                 session.setAttribute("message","你猜大了");
                 response.sendRedirect("inputNumber.jsp");
          }
          if(guessNumber == savedNumber){
                 session.setAttribute("message","你猜对了");
                 response.sendRedirect("inputNumber.jsp");
          }
     }
     public void doGet(HttpServletRequest request,HttpServletResponse response)
                      throws ServletException,IOException{
         doPost(request,response);
     }
}
```

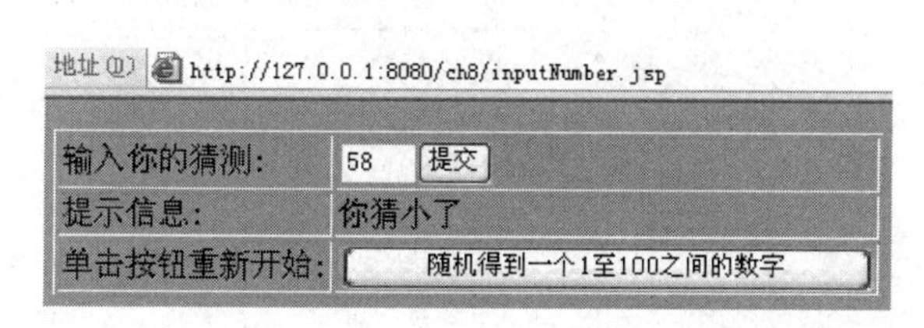

图 8-16　输入猜测

8.10　实验：使用 servlet 读取文件

要求在 webapps 目录下新建一个 Web 服务目录：practice8。除特别要求外，实验所涉及的 JSP 页面均保存在 practice8 中。实验中涉及的 servlet 的包名均为 my.servlet，因此要求在 practice8 下建立子目录：WEB-INF\classes\my\servlet，servlet 的字节码文件保存在该子目录中。

1. 相关知识点

servlet 对象是 javax.servlet.http 包中的 HttpServle 类的子类的实例，servlet 对象由服务器负责创建。需要编写一个 web.xml 文件保存到 Web 服务目录的 WEB-INF 子目录中，以便服务器根据用户的请求创建 servlet 对象。

HttpServlet 类除了 init、service、destroy 方法外，该类还有两个很重要的方法：doGet 和 doPost，用来处理用户的请求并作出响应。当服务器创建 servlet 对象后，该对象会调用 init 方法初始化自己，以后每当服务器再接受到一个 servlet 请求时，就会产生一个新线程，并在这个线程中让 servlet 对象调用 service 方法,而 HttpServlet 类所给出的 service 方法的功能是检查 HTTP 请求类型(get、post 等),并在 service 方法中根据用户的请求方式,对应地再调用 doGet 或 doPost 方法。因此,在编写 Servlet 类(HttpServlet 类的一个子类)时,不必重写 service 方法来响应用户,直接继承 service 方法即可。

2. 实验目的

本实验的目的是让学生掌握使用 servlet 读取文件的内容。

3. 实验要求

- 编写一个 JSP 页面:readFile.jsp,用户可以通过该页面选择服务器指定的某些文件。
- 编写一个 Servlet 类:ReadFile,该类创建的 servlet 可以读取 readFile.jsp 选择的文件。
- 配置 web.xml 文件,要求 Servlet 创建的 servlet 名字是 read,请求该 servlet 的 url-pattern 是 helpRead。

1) readFile.jsp 的具体要求

inputAndShow.jsp 页面提供一个表单。其中表单允许用户使用下拉列表选择服务器指定的文件,然后将文件名提交给 url-pattern 的是 helpRead 的 servlet。readFile.jsp 的页面效果如图 8-17 所示。

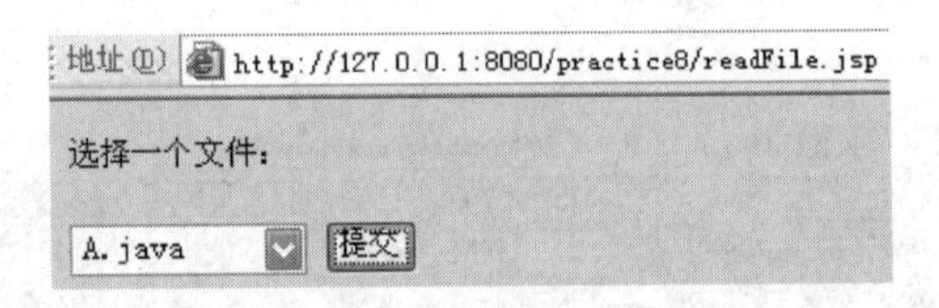

图 8-17　readFile.jsp 页面的效果

2) ReadFile.java 的具体要求

编写的 ReadFile.java 可以读取文件,并将所读内容用文本区组件显示给用户,效果如图 8-18 所示。ReadFile.java 中使用 package 语句,起的包名是 my.servlet。将 ReadFile.java 编译后的字节码文件 ReadFile.class 保存到

```
practice8\WEB-INF\classes\my\servlet
```

目录中。

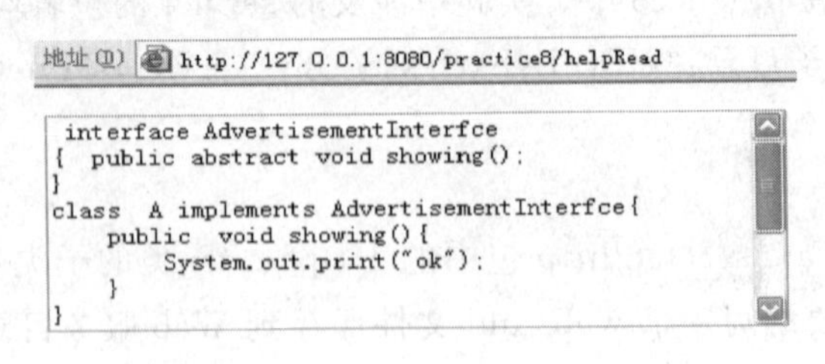

图 8-18　servlet 读取文件的效果

3) web.xml 的具体要求

需要为 practice8\WEN-INF 目录下的 web.xml 文件添加如下的子标记:

```
<servlet>
    <servlet-name>read</servlet-name>
    <servlet-class>my.servlet.ReadFile</servlet-class>
</servlet>
<servlet-mapping>
    <servlet-name>read</servlet-name>
    <url-pattern>/helpRead</url-pattern>
</servlet-mapping>
```

4. 参考代码

代码仅供参考,学生可按着实验要求,参考本代码编写代码。

JSP页面参考代码如下:

readFile.jsp

```
<%@ page contentType="text/html; charset=GB2312" %>
<HTML><BODY bgcolor=yellow><Font size=2>
<P>选择一个文件:
  <FORM action="helpRead" method=post>
      <Select name="fileName" >
          <Option value="D:/1000/A.java">A.java
          <Option value="D:/1000/Hello.java">Hello.java
          <Option value="D:/1000/E.java">E.java
      </Select>
<Input type=submit value="提交">
</FORM>
</FONT></BODY></HTML>
```

Servlet源文件参考代码如下:

ReadFile.java

```
package my.servlet;
import java.io.*;
import javax.servlet.*;
import javax.servlet.http.*;
public class ReadFile extends HttpServlet{
    public void init(ServletConfig config) throws ServletException{
        super.init(config);
    }
    public void doPost(HttpServletRequest request,HttpServletResponse response)
                throws ServletException,IOException{
        String fileName=request.getParameter("fileName");
        String content=getReadContent(fileName);
        response.setContentType("text/html; charset=GB2312");
        PrintWriter out=response.getWriter();
        out.println("<html><body>");
        out.println("<TextArea Rows=8 Cols=50>"+content+"</TextArea>");
        out.println("</body></html>");
    }
    public void doGet(HttpServletRequest request,HttpServletResponse response)
                      throws ServletException,IOException{
        doPost(request,response);
```

```
    }
    private String getReadContent(String fileName){                    //读取文件
        String readContent = "";
        try{  File file = new File(fileName);
              FileReader in = new FileReader(file);
              BufferedReader inTwo = new BufferedReader(in);
              StringBuffer stringbuffer = new StringBuffer();
              String s = null;
              while ((s = inTwo.readLine())!= null){
                  byte bb[] = s.getBytes();
                  s = new String(bb);
                  stringbuffer.append("\n" + s);
              }
              readContent = new String(stringbuffer);
        }
        catch(IOException e){
              readContent = "" + e;
        }
        return readContent;
    }
}
```

习　题　8

1. servlet 对象是在服务器端还是在用户端被创建?

2. servlet 对象被创建后将首先调用 init 方法还是 service 方法?

3. 下列说法是否正确:

servlet 第一次被请求加载时调用 init 方法。当后续的用户请求 servlet 对象时,servlet 对象不再调用 init 方法。

4. 假设创建 servlet 的类是 star.flower.Dalian,创建的 servlet 对象的名字是 myservlet,应当怎样配置 web.xml 文件?

5. 如果 Servlet 类不重写 service 方法,那么应当重写哪两个方法?

6. HttpServletResponse 类的 sendRedirect 方法和 RequestDispatcher 类的 forward 方法有何不同?

7. servlet 对象怎样获得用户的 session 对象?

第9章 MVC 模式

本章导读

主要内容

- MVC 模式介绍
- JSP 中的 MVC 模式
- 模型的生命周期与视图更新
- MVC 模式的简单实例
- MVC 模式与注册登录
- MVC 模式与数据库操作
- MVC 模式与文件操作

难点

- 模型的生命周期与视图更新
- MVC 模式与注册登录

关键实践

- 计算数列之和

MVC 模式的核心思想是有效的组合“视图”、“模型”和“控制器”。本章将介绍 MVC 模式，掌握该模式对于设计合理的 Web 应用以及学习使用某些流行的 Web 框架，如 Hibernate、Spring、Struts 等都有着十分重要的意义。

本章使用的 Web 服务目录是 ch9，ch9 是在 Tomcat 安装目录的 webapps 目录下建立的 Web 服务目录。

另外，需要在当前 web 服务目录下建立如下的目录结构：

```
ch9\WEB-INF\classes
```

为了让 Tomcat 服务器起用上述目录，必须重新启动 Tomcat 服务器。然后根据 servlet 的包名，在 classes 下再建立相应的子目录，比如本章的 Servlet 类的包名为 myservlet.control，那么在 classes 下建立子目录：\myservlet\control；JavaBean 类的包名为 mybean.data，那么在 classes 下建立子目录：\mybean\data，如图 9-1 所示。

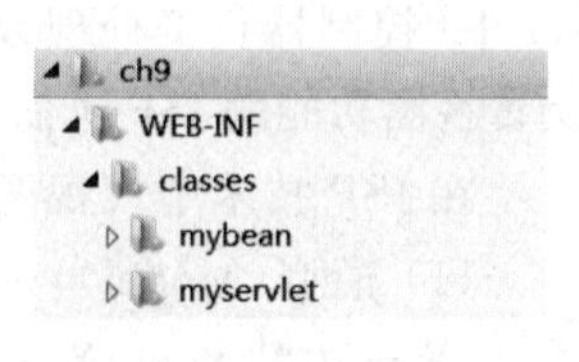

图 9-1 目录结构

9.1 MVC 模式介绍

模型-视图-控制器(Model-View-Controller),简称为 MVC。

MVC 是一种先进的设计模式,是 Trygve Reenskaug 教授于 1978 年最早开发的一个设计模板或基本结构,其目的是以会话形式提供方便的 GUI 支持。MVC 设计模式首先出现在 Smalltalk 编程语言中。

MVC 是一种通过三个不同部分构造一个软件或组件的理想办法:

(1) 模型(Model) 用于存储数据的对象。

(2) 视图(View) 向控制器提交所需数据、显示模型中的数据。

(3) 控制器(Controller) 负责具体的业务逻辑操作,即控制器根据视图提出的要求对数据做出处理,并将有关结果存储到模型中,并负责让模型和视图进行必要的交互,当模型中的数据变化时,让视图更新显示。

从面向对象的角度看,MVC 结构可以使程序更具有对象化特性,也更容易维护。在设计程序时,可以将某个对象看作"模型",然后为"模型"提供恰当的显示组件,即"视图"。在 MVC 模式中,"视图"、"模型"和"控制器"之间是松耦合结构,便于系统的维护和扩展。

9.2 JSP 中的 MVC 模式

目前,随着软件规模的扩大,MVC 模式正在被运用到各种应用程序的设计中。那么在 JSP 设计中,MVC 模式是怎样具体体现的呢?

JSP 页面擅长数据的显示,即适合作为用户的视图,应当尽量避免在 JSP 中使用大量的 Java 程序片来处理数据,即进行数据的逻辑操作,否则不利于代码的复用。servlet 擅长数据的处理,应当尽量避免在 servlet 中使用 out 流输出大量的 HTML 标记来显示数据,否则一旦需要修改显示外观就要重新编译 servlet。

通过前面的学习,特别是在学习了第 3 章和第 7 章后,已经体会到,一些小型的 Web 应用可以使用 JSP 页面调用 Tag 文件或 JavaBean 完成数据的处理,实现代码复用。在 JSP+JavaBean 模式中,JavaBean 不仅要提供修改和返回数据的方法,而且要经常参与数据的处理。当 Web 应用变得复杂时,希望 JavaBean 仅仅负责提供修改和返回数据的方法即可,不必参与数据的具体处理,而是把数据的处理交给称作控制器的 servlet 对象去完成,即 servlet 控制器负责处理数据,并将有关的结果存储到 JavaBean 中,实现存储与处理的分离。负责视图功能的 JSP 页面可以使用 Java 程序片或用 JavaBean 标记显示 JavaBean 中的数据。

在 JSP 技术中,"视图"、"模型"和"控制器"的具体实现如下。

(1) 模型(Model):一个或多个 JavaBean 对象,用于存储数据,JavaBean 主要提供简单的 setXxx 方法和 getXxx 方法,在这些方法中不涉及对数据的具体处理细节,以便增强模型的通用性。

(2) 视图(View):一个或多个 JSP 页面,其作用是向控制器提交必要的数据和显示数据。JSP 页面可以使用 HTML 标记、JavaBean 标记以及 Java 程序片或 Java 表达式来显示数据。视图的主要工作就是显示数据,对数据的逻辑操作由控制器负责。

(3) 控制器(Controller)：一个或多个 servlet 对象，根据视图提交的要求进行数据处理操作，并将有关的结果存储到 JavaBean 中，然后 servlet 使用转发或重定向的方式请求视图中的某个 JSP 页面显示数据，比如让某个 JSP 页面通过使用 JavaBean 标记显示控制器存储在 JavaBean 中的数据。

MVC 模式的结构如图 9-2 所示。

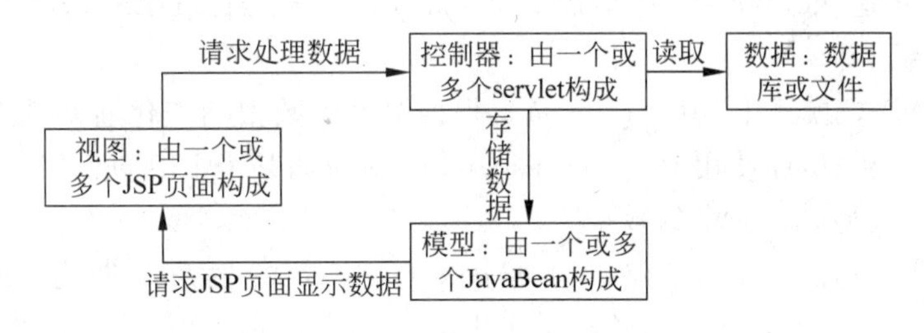

图 9-2 JSP 中的 MVC 模式

9.3 模型的生命周期与视图更新

JSP 中的 MVC 模式和前面学习的 JSP＋JavaBean 模式有很大的不同，在 JSP＋JavaBean 模式中，由 JSP 页面通过使用 useBean 标记：

```
<jsp:useBean id = "名字" class = "创建 bean 的类" scope = "生命周期"/>
```

创建 JavaBean。而在 JSP 中的 MVC 模式中，可以由控制器 servet 创建 JavaBean，并将有关数据存储到所创建的 JavaBean 中，然后 servlet 请求某个 JSP 页面使用 JavaBean 的 getProperty 动作标记：

```
<jsp:getProperty name =  "名字" property = "bean 的属性"/>
```

显示 JavaBean 的中的数据。

在 JSP 中的 MVC 模式中，当需要用控制器 servet 创建 JavaBean 时，就可以使用 JavaBean 类带参数的构造方法，类中的方法的命名继续保留"get"规则，但可以不遵守"set"规则，其理由是：不希望 JSP 页面修改 JavaBean 中的数据，只需要它显示 JavaBean 中的数据。

在 JSP 中的 MVC 模式中，servet 创建的 JavaBean 也涉及生命周期(有效期限)，生命周期分为 request、session 和 application。以下假设创建 JavaBean 类的名字是 BeanClass，该类的包名为 user. yourbean。以下分三种情形给予讨论。

9.3.1 request 周期的 JavaBean

1. JavaBean 的创建

servlet 负责创建 bean。那么创建生命周期为 request 的 bean 的步骤如下：

(1) 用 BeanClass 类的某个构造方法创建 bean 对象。例如：

```
BeanClass bean = new BeanClass();
```

(2) 将所创建的 bean 对象存放到 HttpSerletRequest 对象：request 中，并指定查找该 bean 的关键字，该步骤决定了 bean 的生命周期为 request。例如：

```
request.setAttribute("keyWord",bean);
```

执行上述操作，就会把 bean 存放到 Tomcat 引擎管理的内置对象 pageContext 中，该 bean 被指定的 id 是"keyWord"，生命周期是 PageContext.REQUEST_SCOPE(request)。

2. 视图更新

在 JSP 的 MVC 模式中，由 servlet 负责根据模型中数据的变化通知视图(JSP 页面)更新，其手段是使用转发，即使用 RequestDispatcher 对象向某个 JSP 页面发出请求，让所请求的 JSP 页面显示模型(bean)中的数据。

因为 servelt 创建 bean 的第(2)步骤决定了 bean 的生命周期为 request，因此，当 servlet 使用 RequestDispatcher 对象向某个 JSP 页面发出请求时(进行转发操作)，该 bean 只对 servlet 所请求的 JSP 页面有效。servlet 所请求的 JSP 页面可以使用相应的标记显示该 bean 中的数据，该 JSP 页面对请求做出响应之后，bean 所占有的内存被释放，结束自己的使命。

servlet 请求一个 JSP 页面，比如 show.jsp 的代码如下：

```
RequestDispatcher dispatcher = request.getRequestDispatcher("show.jsp");
dispatcher.forward(request,response);
```

servlet 所请求的 JSP 页面，比如 show.jsp 页面可以使用如下标记获得 servlet 所创建的 bean 的引用(type 属性使得该 JSP 页面不负责创建 bean)：

```
<jsp:useBean id = "keyWord" type = "user.yourbean.BeanClass" scope = "request"/>
```

该标记中的 id 是 servlet 所创建的 bean 索引关键字。

然后 JSP 页面使用：

```
<jsp:getProperty name = "keyWord" property = "bean 的变量"/>
```

标记显示 bean 中的数据。如果上述代码执行成功，用户就看到了 show.jsp 页面的执行效果。

需要特别注意的是，如果 servlet 所请求的 JSP 页面，比如 show.jsp 页面，使用如下标记获得 servlet 所创建的 bean 的引用(注意没有用 type 属性而是用 class 属性)：

```
<jsp:useBean id = "keyWord" class = "user.yourbean.BeanClass" scope = "request"/>
```

该标记中的 id 是 servlet 所创建的 bean 索引关键字。那么即使 servlet 所请求的 JSP 页面事先已经有了 id 是"keyWord"，scope 是"request"的 bean，那么这个 bean 也会被 servlet 所创建的 bean 替换。原因是 servlet 所请求的 JSP 页面会被刷新，就会根据当前页面使用的

```
<jsp:useBean id = "keyWord" class = "user.yourbean.BeanClass" scope = "request"/>
```

标记到 Tomcat 引擎管理的内置对象 PageContext 中寻找 id 是"keyWord"，生命周期是 request(有关 bean 的使用原理和有效期限见 7.1.2 节)，而该 bean 已经被 servlet 更新了(见本节前面的 JavaBean 的创建)。

9.3.2 session 周期的 JavaBean

1. JavaBean 的创建

servlet 创建生命周期为 session 的 bean 的步骤如下：

(1) 用 BeanClass 类的某个构造方法创建 bean 对象。例如：

```
BeanClass bean = new BeanClass();
```

(2) 将所创建的 bean 对象存放到 HttpSerletSession 对象：session 中，并指定查找该 bean 的关键字，该步骤决定了 bean 的生命周期为 session。例如：

```
HttpSession session = request.getSession(true);
session.setAttribute("keyWord",bean);
```

内置对象执行上述操作，就会把 bean 存放到 Tomcat 引擎管理的内置对象 pageContext 中，该 bean 被指定的 id 是"keyWord"，生命周期是 PageContext. SESSION_SCOPE(session)。

2. 视图更新

servelt 创建 bean 的第(2)步骤决定了 bean 的生命周期为 session，只要用户的 session 没有消失，该 bean 就一直存在，一个用户在访问 Web 服务目录的各个 JSP 中都可以使用

```
<jsp:useBean id = "keyWord" type = "usern.yourbean.BeanClass" scope = "session"/>
```

标记获得 servlet 所创建的 bean 的引用，然后使用

```
<jsp:getProperty name = "keyWord" property = "bean 的变量"/>
```

标记显示该 bean 中的数据，该标记中的 id 是 servlet 所创建的 bean 索引关键字。

对于生命周期为 session 的 bean，如果 servlet 希望某个 JSP 显示其中的数据，可以使用 RequestDispatcher 对象向该 JSP 页面发出请求，也可以使用 HttpServletResponse 类中的重定向方法(sendRedirect)。

需要注意的是，不同用户的 session 生命周期的 bean 是互不相同的，即占用不同的内存空间。另外需要特别注意的是，如果 servlet 所请求的 JSP 页面，比如 show.jsp 页面，使用如下标记获得 servlet 所创建的 bean 的引用(注意没有用 type 属性而是用 class 属性)：

```
<jsp:useBean id = "keyWord" class = "user.yourbean.BeanClass" scope = "session"/>
```

该标记中的 id 是 servlet 所创建的 bean 索引关键字。那么即使 servlet 所请求的 JSP 页面或其他页面事先已经有了 id 是"keyWord"，scope 是" session "的 bean，那么这个 bean 也会被 servlet 所创建的 bean 替换。原因是 servlet 所请求的 JSP 页面或其他页面被刷新时，就会根据当前页面使用的：

```
<jsp:useBean id = "keyWord" class = "user.yourbean.BeanClass" scope = " session "/>
```

标记到 Tomcat 引擎管理的内置对象 PageContext 中寻找 id 是"keyWord"，生命周期是 session(有关 bean 的使用原理和生命周期见 7.1.2 节)，而该 bean 已经被 servlet 更新(见本节前面的 JavaBean 的创建)。

9.3.3 application 周期的 JavaBean

1. JavaBean 的创建

servlet 创建生命周期为 application 的 bean 的步骤如下:

(1) 用 BeanClass 类的某个构造方法创建 bean 对象。例如:

```
BeanClass bean = new BeanClass();
```

(2) servlet 使用 getServletContext()方法返回服务器的 ServletContext 内置对象的引用,将所创建的 bean 对象存放到服务器这个 ServletContext 内置对象中,并指定查找该 bean 的关键字,该步骤决定了 bean 的生命周期为 application。例如:

```
getServletContext().setAttribute("keyWord",bean);
```

上述操作,就会把 bean 存放到 Tomcat 引擎管理的内置对象 pageContext 中,该 bean 被指定的 id 是"keyWord",有效期限(生命周期)是 PageContext. APPLICATION_SCOPE (application)。

2. 视图更新

servlet 创建 bean 的第(2)步骤决定了 bean 的生命周期为 application,当 servlet 创建创建生命周期为 application 的 bean 后,只要 Web 应用程序不结束,该 bean 就一直存在。一个用户在访问 Web 服务目录的各个 JSP 中都可以使用

```
<jsp:useBean id = "keyWord" type = "user.yourbean.BeanClass" scope = "application"/>
```

标记获得 servlet 所创建的 bean 的引用,然后使用

```
<jsp:getProperty name = "keyWord" property = "bean 的变量"/>
```

标记显示该 JavaBean 中的数据,该标记中的 id 是 servlet 所创建的 bean 索引关键字。

对于生命周期为 application 的 bean,如果 servlet 希望某个 JSP 显示其中的数据,可以使用 RequestDispatcher 对象向该 JSP 页面发出请求,也可以使用 HttpServletResponse 类中的重定向方法(sendRedirect)。

需要注意的是,所有用户在同一个 Web 服务目录中的 application 生命周期的 bean 是相同的,即占用相同的内存空间。另外需要特别注意的是,如果 servlet 所请求的 JSP 页面,比如 show.jsp 页面,使用如下标记获得 servlet 所创建的 bean 的引用(注意没有用 type 属性而是用 class 属性):

```
<jsp:useBean id = "keyWord" class = "user.yourbean.BeanClass" scope = "application"/>
```

该标记中的 id 是 servlet 所创建的 bean 索引关键字。那么即使 servlet 所请求的 JSP 页面或其他事先已经有了 id 是"keyWord",scope 是"application"的 bean,那么这个 bean 也会被 servlet 所创建的 bean 替换。原因是,servlet 所请求的 JSP 页面或其他页面被刷新时,就会根据当前页面使用的:

```
<jsp:useBean id = "keyWord" class = "user.yourbean.BeanClass" scope = "application"/>
```

标记到 Tomcat 引擎管理的内置对象 PageContext 中寻找 id 是"keyWord",生命周期是

application(有关 bean 的使用原理和生命周期见 7.1.2 节),而该 bean 已经被 servlet 更新(见本节前面的 JavaBean 的创建)。

9.4 MVC 模式的简单实例

本节结合一个简单的实例让学生知道 MVC 三个部分的设计与具体实现。

9.4.1 JavaBean 和 Servlet 与配置文件

按照本章的约定 JavaBean 类的包名均为 mybean.data;Servlet 类的包名均为 myservlet.control。由于 Servlet 类中要使用 JavaBean,所以为了能顺利地编译 Servlet 类,不要忘记将 Tomcat 安装目录 lib 子目录中的 servlet-api.jar 文件复制到 Tomcat 服务器所使用的 JDK 的扩展目录中,比如,复制到 D:\jdk1.7\jre\lib\ext 中。然后,按下列步骤进行编译和保存有关的字节码文件。

1. 保存 JavaBean 类和 Servlet 类的源文件

将 JavaBean 类和 Servlet 类源文件分别保存到:

```
D:\ mybean\data
```

和

```
D:\myservlet\control
```

目录中,注意保存时,Servlet 类的包名和 JavaBean 类的包名形成目录的父目录要相同。

2. 编译 JavaBean 类

用如下格式进行编译,即编译时带着包名形成的目录:

```
D:> javac mybean\data\JavaBean 的源文件
```

例如:

```
D:> javac mybean\data\Computer.java
```

3. 编译 Servlet 类

用如下格式进行编译,即编译时带着包名形成的目录:

```
D:> javac myservlet\control\servlet 的源文件
```

例如:

```
D:> javac myservlet\control\HandleComputer.java
```

4. 将字节码保存到服务器

将编译通过的 JavaBean 类和 Servlet 类的字节码分别复制到

```
ch9\WEB-INF\classes\mybean\data
```

和

```
ch9\WEB-INF\classes\myservlet\control
```

目录中。

注意:为了调试程序方便,可以将 JavaBean 和 Servlet 源文件分别保存到 Web 服务目录的下述目录中:

```
WEB-INF\classes\mybean\data
WEB-INF\classes\myservlet\control
```

这样就可以省略上面的最后步骤(步骤 4)。

5. web. xml 文件

编写 web. xml 文件,并保存到 Web 服务目录的 WEB-INF 子目录中,即 ch9\WEB-INF 中,web. xml 文件的内容如下(需要用纯文本编辑,使用"ANSI"编码保存):

```
<? xml version = "1.0" encoding = "ISO - 8859 - 1"? >
< web - app >
< servlet >
    < servlet - name > computerArea </servlet - name >
    < servlet - class > myservlet. control. HandleArea </servlet - class >
</servlet >
< servlet - mapping >
   < servlet - name > computerArea </servlet - name >
   < url - pattern >/lookArea </url - pattern >
</servlet - mapping >
</web - app >
```

如果 web. xml 文件已经存在,只需要将新内容,即新的<servlet>和<servlet = mapping>标记添加到已有的 web. xml 文件中即可(如果修改了 web. xml 文件,必须重新启动 Tomcat 服务器,有关 web. xml 文件的配置规定可参见 8.2 节)。

9.4.2 计算三角形和梯形的面积

设计一个 Web 应用,该 Web 应用提供两个 JSP 页面,一个页面使得用户可以输入三角形三边的值和梯形的上底、下底和高的值;另一个页面可以显示三角形和梯形的面积。Web 应用提供一个名字为 computerArea 的 servlet 对象,computerArea 负责计算三角形和梯形的面积(computerArea 由 HandleArea 类负责创建,访问它的 url-pattern 为 lookArea,见前面的 web. xml 中的配置),然后将有关数据存储到 JavaBean 中。Web 应用提供的 JavaBean 负责存储数据结果,该 JavaBean 提供简单的获取数据和修改数据的方法。

1. 模型(JavaBean)

本模型 Area. java 中的 getXxx 和 setXxx 方法不涉及对数据的具体处理细节,以便增强模型的通用性。比如,setArea(double s)仅仅将参数 s 的值赋给 area,因此,模型既可以存储三角形的面积也可以存储梯形的面积。如果 setArea(double s)参与具体的计算,比如,计算三角形的面积,然后将计算结果赋给 area,那么该模型就不能存储梯形的面积,从而减弱了模型的通用性(请读者比较 Area. java 和第 7 章中 7.4.1 中 Triange. java 的不同之处)。

Area. java

```
package mybean. data;
public class Area{
```

```
    double a,b,c,area;
    String mess;
    public void setMess(String mess){
        this.mess = mess;
    }
    public String getMess(){
        return mess;
    }
    public void setA(double a){
        this.a = a;
    }
    public void setB(double b){
        this.b = b;
    }
    public void setC(double c){
        this.c = c;
    }
    public void setArea(double s){
        area = s;
    }
    public double getArea(){
        return area;
    }
}
```

2. 控制器(servlet)

控制器是名字为 computerArea 的 servlet 对象(见 web.xml 中的配置),由下面的 Servlet 类负责创建。控制器使用 doPost 方法计算三角形的面积;使用 doGet 方法计算梯形的面积。

HandleArea.java

```
package myservlet.control;
import mybean.data.Area;
import java.io.*;
import javax.servlet.*;
import javax.servlet.http.*;
public class HandleArea extends HttpServlet{
    public void init(ServletConfig config) throws ServletException{
        super.init(config);
    }
    public void doPost(HttpServletRequest request,HttpServletResponse response)
                        throws ServletException,IOException{
        Area dataBean = new Area();                    //创建 JavaBean 对象
        request.setAttribute("data",dataBean);         //将 dataBean 存储到 request 对象中
        try{ double a = Double.parseDouble(request.getParameter("a"));
             double b = Double.parseDouble(request.getParameter("b"));
             double c = Double.parseDouble(request.getParameter("c"));
             dataBean.setA(a);                         //将数据存储在 dataBean 中
             dataBean.setB(b);
             dataBean.setC(c);
```

```
            double s = -1;
            double p = (a+b+c)/2.0;
            if(a+b>c&&a+c>b&&b+c>a)
              s = Math.sqrt(p*(p-a)*(p-b)*(p-c));
            dataBean.setArea(s);                    //将数据存储在 dataBean 中
            dataBean.setMess("三角形面积");
        }
        catch(Exception e){
            dataBean.setArea(-1);
            dataBean.setMess(""+e);
        }
        RequestDispatcher dispatcher = request.getRequestDispatcher
("showResult.jsp");
        //请求 showResult.jsp 显示 dataBean 中的数据
        dispatcher.forward(request,response);      //转发
    }
    public void doGet(HttpServletRequest request,HttpServletResponse response)
                              throws ServletException,IOException{
        Area dataBean = new Area();                //创建 JavaBean 对象
        request.setAttribute("data",dataBean);     //将 dataBean 存储到 request 对象中
        try{ double a = Double.parseDouble(request.getParameter("a"));
            double b = Double.parseDouble(request.getParameter("b"));
            double c = Double.parseDouble(request.getParameter("c"));
            dataBean.setA(a);                       //将数据存储在 dataBean 中
            dataBean.setB(b);
            dataBean.setC(c);
            double s = -1;
            s = (a+b)*c/2.0;
            dataBean.setArea(s);                    //将数据存储在 dataBean 中
            dataBean.setMess("梯形面积");
        }
        catch(Exception e){
            dataBean.setArea(-1);
            dataBean.setMess(""+e);
        }
        RequestDispatcher dispatcher = request.getRequestDispatcher
("showResult.jsp");
        //请求 showResult.jsp 显示 dataBean 中的数据
        dispatcher.forward(request,response);      //转发
    }
}
```

3. 视图(JSP 页面)

在 inputData.jsp 页面可以输入三角形三边的值或梯形的上、下底和高的值，并将所输入的三角形三边的值和梯形的上、下底和高的值分别用 post 方法和 get 方法提交给名字为 computerArea 的 servlet 对象。computerArea 使用 doPost 方法计算三角形的面积；使用 doGet 方法计算梯形的面积，并将结果存储到数据模型 bean 中，然后请求 showResult.jsp 页面显示模型中的数据。inputData.jsp 和 showResult.jsp 效果如图 9-3 和图 9-4 所示。

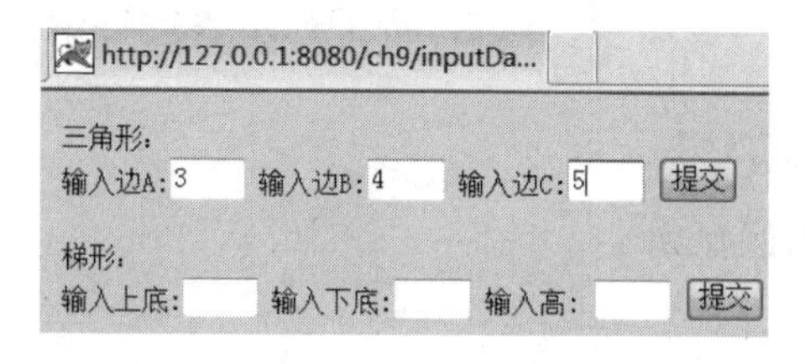

图 9-3　输入有关数据

http://127.0.0.1:8080/ch9/lookArea
三角形面积：6.0

图 9-4　显示结果

inputData.jsp（效果如图 9-3 所示）

```
<%@ page contentType = "text/html;charset = GB2312" %>
<HTML><BODY bgcolor = cyan><Font size = 2>
<FORM action = "lookArea" Method = "post" >
   三角形：
  <BR>输入边 A:<Input type = text name = "a" size = 4>
      输入边 B:<Input type = text name = "b" size = 4>
      输入边 C:<Input type = text name = "c" size = 4>
  <Input type = submit value = "提交">
</FORM>
<FORM action = "lookArea" Method = "get" >
   梯形：
  <BR>输入上底:<Input type = text name = "a" size = 4>
      输入下底:<Input type = text name = "b" size = 4>
      输入高: <Input type = text name = "c" size = 4>
  <Input type = submit value = "提交">
</FORM>
</Font></BODY></HTML>
```

showResult.jsp（效果如图 9-4 所示）

```
<%@ page contentType = "text/html;charset = GB2312" %>
<%@ page import = "mybean.data.Area" %>
<jsp:useBean id = "data" type = "mybean.data.Area" scope = "request"/>
<HTML><BODY bgcolor = yellow>
   <jsp:getProperty name = "data" property = "mess"/>:
   <jsp:getProperty name = "data" property = "area"/>
</BODY></HTML>
```

9.5　MVC 模式与注册登录

大部分 Web 应用都会涉及注册与登录模块。本节使用 MVC 模式讲述怎样设计注册、登录模块。

9.5.1　JavaBean 与 Servlet 管理

本节的 JavaBean 类的包名均为 mybean.data；Servlet 类的包名均为 myservlet.control。由于 Servlet 类中要使用 JavaBean，所以为了能顺利地编译 Servlet 类，不要忘记将 Tomcat 安装目录 lib 子目录中的 servlet-api.jar 文件复制到 Tomcat 服务器所使用的 JDK

的扩展目录中,比如,复制到 D:\jdk1.7\jre\lib\ext 中。然后,按下列步骤进行编译和保存有关的字节码文件。

1. 保存 JavaBean 类和 Servlet 类的源文件

将 JavaBean 类和 Servlet 类源文件分别保存到

```
D:\ mybean\data
```

和

```
D:\myservlet\control
```

目录中。保存时,让 Servlet 类的包名和 JavaBean 类的包名形成目录的父目录相同。

2. 编译 JavaBean 类

用如下格式进行编译,即带着包名形成的目录:

```
D:> javac mybean\data\JavaBean 的源文件
```

例如:

```
D:> javac mybean\data\Login.java
```

3. 编译 Servlet 类

用如下格式进行编译,即带着包名形成的目录:

```
D:> javac myservlet\control\servlet 的源文件
```

例如:

```
D:> javac myservlet\control\HandleLogin.java
```

4. 将字节码保存到服务器

将编译通过的 JavaBean 类和 Servlet 类的字节码分别复制到

```
ch9\WEB-INF\classes\mybean\data
```

和

```
ch9\WEB-INF\classes\myservlet\control
```

目录中。

9.5.2 配置文件管理

本节的 Servlet 类的包名均为 myservlet.control,需要配置 Web 服务目录的 web.xml 文件,即将下面的 web.xml 文件保存到 Tomcat 安装目录的 Web 服务目录 ch9 中。根据本书使用的 Tomcat 安装目录及 Web 服务目录,需要将 web.xml 文件保存到

```
D:\apache-tomcat-8.0.3\webapps\ch9\WEB-INF
```

目录中。如果 web.xml 文件已经存在,需要将下述内容添加到已有的 web.xml 文件中(有关 web.xml 文件的配置规定可参见 8.2 节)。web.xml 文件需要包含的内容如下:

```
<servlet>
    <servlet-name>register</servlet-name>
```

```
        <servlet-class>myservlet.control.HandleRegister</servlet-class>
    </servlet>
    <servlet-mapping>
        <servlet-name>register</servlet-name>
        <url-pattern>/helpRegister</url-pattern>
    </servlet-mapping>
    <servlet>
        <servlet-name>login</servlet-name>
        <servlet-class>myservlet.control.HandleLogin</servlet-class>
    </servlet>
    <servlet-mapping>
        <servlet-name>login</servlet-name>
        <url-pattern>/helpLogin</url-pattern>
    </servlet-mapping>
```

9.5.3 数据库设计与连接

1. 创建数据库和表

使用 MySQL 建立一个数据库 student,该库共有一个 user 表(如图 9-5 所示),有关建立数据库和表的操作细节见主教材的 6.1 节。

名	类型	长度	小数点	允许空值(	
logname	char			☐	🔑1
password	char			☑	
email	char			☑	

图 9-5　user 表

user 表的用途：存储用户的注册信息。即会员的注册信息存入 user 表中,user 表的主键是 logname,各个字段值的说明如下：

logname ：存储注册的用户名(属性是字符型,主键)。

password ：存储密码(属性是字符型)。

email ：存储 email(属性是字符型)。

2. 数据库连接

避免操作数据库出现中文乱码,需要使用

```
Connection getConnection(java.lang.String)
```

方法建立连接,连接中的代码是(用户是 root,其密码是空)：

```
String uri = "jdbc:mysql://127.0.0.1/student?" +
             "user=root&password=&characterEncoding=gb2312";
Connection con = DriverManager.getConnection(uri);
```

9.5.4 注册

当注册新会员时,该模块要求用户必须输入会员名、密码信息,否则不允许注册。用户的注册信息被存入数据库的 user 表中。

该模块视图部分由一个 JSP 页面构成,这个 JSP 页面 register.jsp 负责提交用户的注册信息到 servlet 控制器 register(见配置文件 web. xml),并负责显示注册是否成功的信息。该模块的 JavaBean 模型 userBean 存储用户的注册信息。servlet 控制器 register 负责将视图提交的信息写入数据库的 user 表中,并将有关反馈信息存储到 JavaBean 模型 userBean 中,然后将用户转发到 register. jsp,register. jsp 将显示 JavaBean 模型 userBean 中的数据(更新视图),效果如图 9-6 所示。

用户名由字母、数字、下划线构成,*注释的项必须填写。

*用户名称:		*用户密码:	
*重复密码:		email:	

提交

注册反馈:注册成功

会员名称:	gxygeng
email地址:	we@128.com

图 9-6 注册

1. 视图(JSP 页面)

register. jsp(效果如图 9-6 所示)

```
<%@ page contentType = "text/html;charset = GB2312" %>
<jsp:useBean id = "userBean" class = "mybean.data.Register" scope = "request"/>
<title>注册页面</title>
<HTML><BODY bgcolor = yellow><Font size = 2>
<div align = "center">
<FORM action = "helpRegister" method = "post" name = form>
<table>
    用户名由字母、数字、下划线构成,* 注释的项必须填写。
   <tr><td>* 用户名称:</td><td><Input type = text name = "logname" ></td>
       <td>* 用户密码:</td><td><Input type = password name = "password">
       </td></tr>
   <tr><td>* 重复密码:</td><td>
       <Input type = password name = "again_password"></td>
       <td>email:</td><td><Input type = text name = "email"></td></tr>
       <tr><td><Input type = submit name = "g" value = "提交"></td> </tr>
</table>
</Form>
</div>
<div align = "center">
<p> 注册反馈:
<jsp:getProperty name = "userBean" property = "backNews" />
<table border = 3>
     <tr><td>会员名称:</td>
     <td><jsp:getProperty name = "userBean" property = "logname"/></td>
     </tr>
     <tr><td>email 地址:</td>
     <td><jsp:getProperty name = "userBean" property = "email"/></td>
```

```
    </tr>
</table></div>
</Body></HTML>
```

2. 模型(JavaBean)

下列 JavaBean 用来描述用户注册的重要信息。在该模块中 JavaBean 的实例的 id 是 userBean,生命周期是 request。该 JavaBean 的实例由控制器负责创建或更新。

Register. java

```
package mybean.data;
public class Register{
    String logname = "" , email = "",
             backNews = "请填注册信息";
    public void setLogname(String logname){
        this.logname = logname;
    }
    public String getLogname(){
        return logname;
    }
    public void setEmail(String email){
        this.email = email;
    }
    public String getEmail(){
        return email;
    }
    public void setBackNews(String backNews){
        this.backNews = backNews;
    }
    public String getBackNews(){
        return backNews;
    }
}
```

3. 控制器(servlet)

控制器 servlet 对象的名字是 register(见 9.5.2 节的 web.xml 配置文件)。控制器 register 负责连接数据库,将用户提交的信息写入到 user 表中,并将用户转发到 inputRegisterMess.jsp 页面查看注册反馈信息。

HandleRegister. java

```
package myservlet.control;
import mybean.data.*;
import java.sql.*;
import java.io.*;
import javax.servlet.*;
import javax.servlet.http.*;
public class HandleRegister extends HttpServlet {
    public void init(ServletConfig config) throws ServletException {
        super.init(config);
```

```
        try { Class.forName("com.mysql.jdbc.Driver");
        }
        catch(Exception e){}
    }
    public String handleString(String s)
    { try{ byte bb[] = s.getBytes("iso - 8859 - 1");
              s = new String(bb);
         }
         catch(Exception ee){}
         return s;
    }
    public void doPost(HttpServletRequest request,HttpServletResponse response)
                              throws ServletException,IOException {
        String uri = "jdbc:mysql://127.0.0.1/student?" +
                                   "user = root&password = &characterEncoding = gb2312";
        Connection con;
        PreparedStatement sql;
        Register userBean = new Register();              //创建的 JavaBean 模型
        request.setAttribute("userBean",userBean);    //将会更新 id 是"userBean"的 bean
        String logname = request.getParameter("logname").trim();
        String password = request.getParameter("password").trim();
        String again_password = request.getParameter("again_password").trim();
        String email = request.getParameter("email").trim();
        if(logname == null)
             logname = "";
        if(password == null)
             password = "";
        if(! password.equals(again_password)) {
           userBean.setBackNews("两次密码不同,注册失败。");
           RequestDispatcher dispatcher =
           request.getRequestDispatcher("register.jsp");
           dispatcher.forward(request, response);     //转发
           return;
        }
        boolean isLD = true;
        for(int i = 0;i < logname.length();i++){
             char c = logname.charAt(i);
              if(! ((c <= 'z'&&c >= 'a')||(c <= 'Z'&&c >= 'A')||(c <= '9'&&c >= '0')))
                isLD = false;
        }
        boolean boo = logname.length()> 0&&password.length()> 0&&isLD;
        String backNews = "";
        try{ con = DriverManager.getConnection(uri);
               String insertCondition = "INSERT INTO user VALUES (?,?,?)";
               sql = con.prepareStatement(insertCondition);
               if(boo)
               { sql.setString(1,handleString(logname));
                 sql.setString(2,handleString(password));
```

```
                sql.setString(3,handleString(email));
                int m = sql.executeUpdate();
                if(m! = 0){
                    backNews = "注册成功";
                    userBean.setBackNews(backNews);
                    userBean.setLogname(logname);
                    userBean.setEmail(handleString(email));
                }
            }
            else {
                backNews = "信息填写不完整或名字中有非法字符";
                userBean.setBackNews(backNews);
            }
            con.close();
        }
        catch(SQLException exp){
            backNews = "该会员名已被使用,请你更换名字" + exp;
            userBean.setBackNews(backNews);
        }
        RequestDispatcher dispatcher =
        request.getRequestDispatcher("register.jsp");
        dispatcher.forward(request, response);        //转发
    }
    public void doGet(HttpServletRequest request,HttpServletResponse response)
                        throws ServletException,IOException {
        doPost(request,response);
    }
}
```

9.5.5 登录与验证

用户可在该模块中输入自己的会员名和密码,系统将对会员名和密码进行验证,如果输入的用户名或密码有错误,将提示用户输入的用户名或密码不正确。

该模块视图部分由两个 JSP 页面 login.jsp 和 lookPic.jsp 构成,login.jsp 页面负责提交用户的登录信息到控制器,并显示登录是否成功的信息,登录成功后 lookPic.jsp 页面负责显示一幅图像,如果没有登录,用户访问 lookPic.jsp 页面会被转发到 login.jsp 登录页面。该模块的 JavaBean 模型 loginBean 存储用户登录的信息。该模块的 servlet 控制器 login 负责验证会员名和密码是否正确,并负责让视图显示更新后的数据。

1. 视图(JSP 页面)

视图部分由一个 JSP 页面 login.jsp 构成。login.jsp 页面负责提供输入登录信息界面,并负责显示登录反馈信息,比如登录是否成功,是否已经登录等。效果如图 9-7 所示。

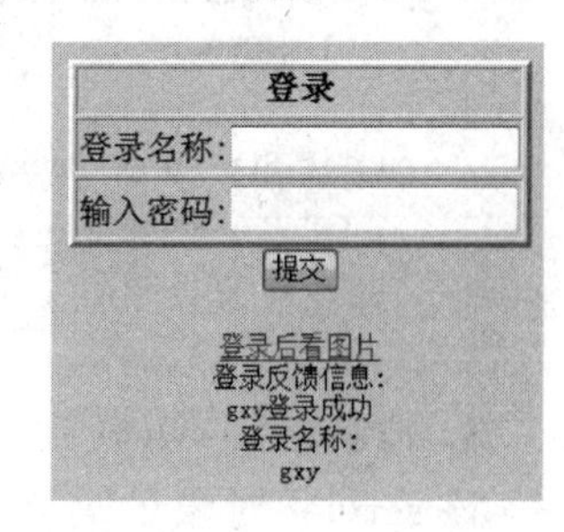

图 9-7 输入登录信息,并得到反馈信息

login.jsp(效果如图 9-7 所示)

```
<%@ page contentType = "text/html;charset = GB2312" %>
<jsp:useBean id = "loginBean" class = "mybean.data.Login" scope = "session"/>
<HTML><HEAD><%@ include file = "head.txt" %></HEAD>
<BODY bgcolor = pink><font size = 2>
<div align = "center">
<table border = 2>
<tr><th>登录</th></tr>
<FORM action = "loginServlet" Method = "post">
<tr><td>登录名称:<Input type = text name = "logname"></td></tr>
<tr><td>输入密码:<Input type = password name = "password"></td></tr>
</table>
<Input type = submit name = "g" value = "提交">
</form>
</div>
<div align = "center" >
登录反馈信息:<br>
<jsp:getProperty name = "loginBean" property = "backNews"/>
<br>登录名称:<br><jsp:getProperty name = "loginBean" property = "logname"/>
<div></font>
</BODY></HTML>
```

2. 模型(JavaBean)

下列 JavaBean 的实例用来存储用户登录信息,在该模块中 JavaBean 的实例的 id 是 loginBean,生命周期是 session。该 JavaBean 的实例由控制器负责创建或更新。

Login.java

```
package mybean.data;
public class Login {
   String logname = "",
          backNews = "未登录";
   public void setLogname(String logname){
      this.logname = logname;
   }
   public String getLogname(){
       return logname;
   }
   public void setBackNews(String s) {
       backNews = s;
   }
   public String getBackNews() {
      return backNews;
   }
}
```

3. 控制器(servlet)

该 servlet 对象的名字是 loginServlet(见 9.5.2 节的 web.xml 配置文件)。控制器 loginServlet 负责连接数据库,查询 user 表,验证用户输入的会员名和密码是否在该表中,即验证是否是已注册的用户,如果用户是已注册的用户,就将用户设置成登录状态,即将用户的名称存放到JavaBean模型loginBean中(见前面的JavaBean模型),并将用户转发到login.jsp页面查看登录反馈信息(见本模块前面的视图介绍)。如果用户不是注册用户,控制器将提示登录失败。

HandleLogin.java

```
package myservlet.control;
import mybean.data.*;
import java.sql.*;
import java.io.*;
import javax.servlet.*;
import javax.servlet.http.*;
public class HandleLogin extends HttpServlet{
   public void init(ServletConfig config) throws ServletException{
      super.init(config);
      try{
          Class.forName("com.mysql.jdbc.Driver");
      }
      catch(Exception e){}
   }
   public String handleString(String s){
      try{ byte bb[] = s.getBytes("iso-8859-1");
            s = new String(bb);
      }
      catch(Exception ee){}
      return s;
   }
   public void doPost(HttpServletRequest request,HttpServletResponse response)
                        throws ServletException,IOException{
      Connection con;
      Statement sql;
      String logname = request.getParameter("logname").trim(),
      password = request.getParameter("password").trim();
      logname = handleString(logname);
      password = handleString(password);
      String uri = "jdbc:mysql://127.0.0.1/student?" +
                              "user = root&password = &characterEncoding = gb2312";
      boolean boo = (logname.length()> 0)&&(password.length()> 0);
      try{
            con = DriverManager.getConnection(uri);
            String condition = "select * from user where logname = '" + logname +
             "' and password = '" + password + "'";
            sql = con.createStatement();
            if(boo){
               ResultSet rs = sql.executeQuery(condition);
               boolean m = rs.next();
               if(m == true){
                   //调用登录成功的方法
                   success(request,response,logname,password);
                   RequestDispatcher dispatcher =
                   request.getRequestDispatcher("login.jsp");    //转发
```

```
                    dispatcher.forward(request,response);
                }
                else{
                    String backNews = "你输入的用户名不存在,或密码不般配";
                    //调用登录失败的方法
                    fail(request,response,logname,backNews);
                }
            }
            else{
                    String backNews = "请输入用户名和密码";
                    fail(request,response,logname,backNews);
            }
            con.close();
        }
        catch(SQLException exp){
            String backNews = "" + exp;
            fail(request,response,logname,backNews);
        }
    }
    public void doGet(HttpServletRequest request,HttpServletResponse response)
                        throws ServletException,IOException{
        doPost(request,response);
    }
    public void success(HttpServletRequest request,HttpServletResponse response,
                        String logname,String password) {
        Login loginBean = null;
        HttpSession session = request.getSession(true);
        try{ loginBean = (Login)session.getAttribute("loginBean");
                if(loginBean == null){
                    loginBean = new Login();            //创建新的数据模型
                    session.setAttribute("loginBean",loginBean);
                    loginBean = (Login)session.getAttribute("loginBean");
                }
                String name  = loginBean.getLogname();
                if(name.equals(logname)) {
                    loginBean.setBackNews(logname + "已经登录了");
                    loginBean.setLogname(logname);
                }
                else { //数据模型存储新的登录用户
                     loginBean.setBackNews(logname + "登录成功");
                     loginBean.setLogname(logname);
                }
        }
        catch(Exception ee){
                loginBean = new Login();
                session.setAttribute("loginBean",loginBean);
                loginBean.setBackNews(logname + "登录成功");
                loginBean.setLogname(logname);
        }
    }
     public void fail(HttpServletRequest request,HttpServletResponse response,
```

```
                    String logname,String backNews) {
      response.setContentType("text/html;charset = GB2312");
      try {
       PrintWriter out = response.getWriter();
       out.println("< html >< body >");
       out.println("< h2 >" + logname + "登录反馈结果< br >" + backNews + "</h2 >") ;
       out.println("返回登录页面或主页< br >");
       out.println("< a href  = login.jsp >登录页面</a >");
       out.println("</body ></html >");
      }
      catch(IOException exp){}
   }
}
```

4. 验证

登录的用户可以有权利访问某些 JSP 页面。比如登录的用户可以通过超链接访问 lookPic.jsp(效果如图 9-8 所示)；lookPic.jsp 将验证用户是否为登录用户，如果用户没有登录，就单击超链接进入该页面，将被重定向到 login.jsp 登录页面。

图 9-8　登录用户访问效果

lookPic.jsp(效果如图 9-8 所示)

```
<%@ page contentType = "text/html;charset = GB2312" %>
<%@ page import = "mybean.data.Login" %>
<jsp:useBean id = "loginBean" class = "mybean.data.Login" scope = "session"/>
<%@ page import = "java.util.* " %>
<HTML><BODY bgcolor = cyan><Font size = 2><CENTER>
<table>
<td><A href = "register.jsp"><font size = 2>用户注册</font></A></td>
<td><A href = "login.jsp"><font size = 2>用户登录</font></A></td>
</table>
<% if(loginBean == null){
        response.sendRedirect("login.jsp");          //重定向到登录页面
    }
    else {
      boolean b = loginBean.getLogname() == null||
                  loginBean.getLogname().length() == 0;
      if(b)
```

```
            response.sendRedirect("login.jsp");          //重定向到登录页面
        }
    %>
    <image src = "image.jpg" width = 220 height = 200 ></image >
    </CENTER ></BODY ></HTML >
```


9.6 MVC 模式与数据库操作

本节的主要目的是学习使用 MVC 模式分页显示数据库表中的记录。

9.6.1 JavaBean 与 Servlet 管理

本节的 JavaBean 类的包名均为 mybean.data；Servlet 类的包名均为 myservlet.control。由于 Servlet 类中要使用 JavaBean，所以为了能顺利地编译 Servlet 类，不要忘记将 Tomcat 安装目录 lib 子目录中的 servlet-api.jar 文件复制到 Tomcat 服务器所使用的 JDK 的扩展目录中，比如，复制到 D:\jdk1.7\jre\lib\ext 中。然后，按下列步骤进行编译和保存有关的字节码文件。

1. 保存 JavaBean 类和 Servlet 类的源文件

将 JavaBean 类和 Servlet 类源文件分别保存到：

```
D:\ mybean\data
```

和

```
D:\myservlet\control
```

目录中。保存时，让 Servlet 类的包名和 JavaBean 类的包名形成目录的父目录相同。

2. 编译 JavaBean 类

用如下格式进行编译，即带着包名形成的目录：

```
D:> javac mybean\data\JavaBean 的源文件
```

3. 编译 Servlet 类

用如下格式进行编译，即带着包名形成的目录：

```
D:> javac myservlet\control\servlet 的源文件
```

4. 将字节码保存到服务器

将编译通过的 JavaBean 类和 Servlet 类的字节码分别复制到

```
ch9\WEB - INF\classes\mybean\data
```

和

```
ch9\WEB - INF\classes\myservlet\control
```

目录中。

9.6.2 配置文件与数据库连接

本节的 Servlet 类的包名均为 myservlet.control，需要配置 Web 服务目录的 web.xml

文件，即将下面的 web.xml 文件保存到 Tomcat 安装目录的 Web 服务目录 ch9 中。根据本书使用的 Tomcat 安装目录及 Web 服务目录，需要将 web.xml 文件保存到

```
D:\apache-tomcat-8.0.3\webapps\ch9\WEB-INF
```

目录中。如果 web.xml 文件已经存在，需要将下述内容添加到已有的 web.xml 文件中（有关 web.xml 文件的配置规定可参见 8.2 节）。web.xml 文件需要包含的内容如下：

```
<servlet>
    <servlet-name>database</servlet-name>
    <servlet-class>myservlet.control.HandleDatabase</servlet-class>
</servlet>
<servlet-mapping>
    <servlet-name>database</servlet-name>
    <url-pattern>/helpReadRecord</url-pattern>
</servlet-mapping>
```

避免操作数据库出现中文乱码，连接中的代码是（用户是 root，其密码是空）：

```
String uri = "jdbc:mysql://127.0.0.1/数据库名?" +
             "user=root&password=&characterEncoding=gb2312";
Connection con = DriverManager.getConnection(uri);
```

9.6.3 MVC 设计细节

第 7 章曾使用 bean 读取数据库的记录（见 7.6 节），该 bean 不仅要负责查询记录，而且要负责存储所查询到的记录。在 MVC 模式中，查询记录的任务由 servlet 对象负责，bean 仅仅负责存储 servlet 对象所查询到的记录。请读者比较本节内容与 7.6 节的不同之处。

1. 视图（JSP 页面）

本节设计的 Web 应用有两个 Jsp 页面：choiceDatabase.jsp 和 showRecord.jsp。在 choiceDatabase.jsp 页面可以输入文件 MySQL 数据库的名字以及相应的表名（用户名是默认的 root 和密码是空），并提交给名字为 database 的 servlet 对象。database 负责读取数据库表中的记录，并将读取的内容以及相关的数据存储到数据模型 bean 中，然后请求 showRecord.jsp 页面显示模型中的数据。showRecord.jsp 页面提供了“上一页”和“下一页”提交按钮，用户在该页面可以继续请求控制器 servelt，以便继续查询记录。效果如图 9-9 和图 9-10 所示。

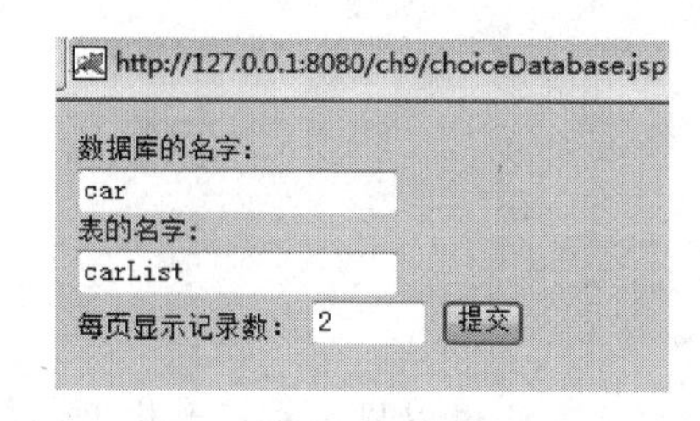

图 9-9　输入数据库和表名

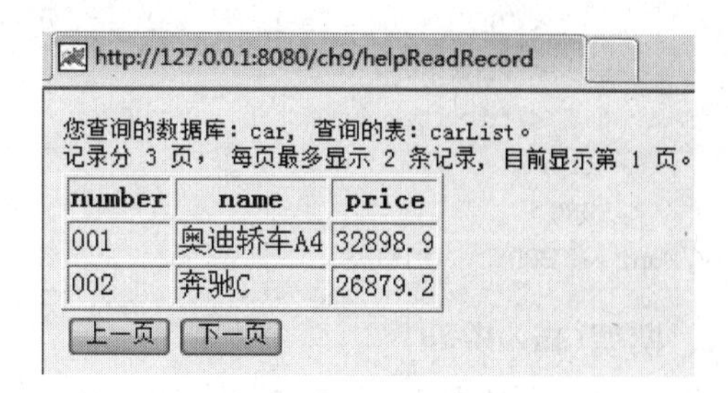

图 9-10　显示记录

choiceDatabase.jsp(效果如图 9-9 所示)

```
<%@ page contentType = "text/html;charset = GB2312" %>
<HTML><BODY bgcolor = cyan><Font size = 2>
    <FORM action = "helpReadRecord" method = "post" name = "form">
      数据库的名字:<BR><INPUT type = "text" name = "databaseName">
     <BR>表的名字:<BR><INPUT type = "text" name = "tableName">
     <BR>每页显示记录数:
        <INPUT type = "text" value = "2" name = "pageSize" size = 6>
        <INPUT type = "submit" value = "提交" name = "submit">
    </FORM>
</Font></BODY></HTML>
```

showRecord.jsp(效果如图 9-10 所示)

```
<%@ page contentType = "text/html;charset = GB2312" %>
<%@ page import = "mybean.data.ShowRecordByPage" %>
<HTML><BODY bgcolor = yellow><Font size = 2>
<jsp:useBean id = "database"
      type = "mybean.data.ShowRecordByPage" scope = "session"/>
   你查询的数据库:<jsp:getProperty name = "database" property = "databaseName"/>,
   查询的表:<jsp:getProperty name = "database" property = "tableName"/>。
   <BR>记录分 <jsp:getProperty name = "database" property = "pageAllCount"/> 页,
   每页最多显示 <jsp:getProperty name = "database" property = "pageSize"/> 条记录,
   目前显示第 <jsp:getProperty name = "database" property = "showPage"/> 页。
   <table border = 1>
     <jsp:getProperty name = "database" property = "formTitle"/>
     <jsp:getProperty name = "database" property = "presentPageResult"/>
   </table>
    <table>
      <tr><td>
          <FORM action = "helpReadRecord" method = "post" name = "form">
          <INPUT type = "hidden" value = "previousPage" name = "whichPage">
          <INPUT type = "submit" value = "上一页" name = "submit">
          </FORM>
          </td>
         <td>
          <FORM action = "helpReadRecord" method = "post" name = "form">
          <INPUT type = "hidden" value = "nextPage" name = "whichPage">
          <INPUT type = "submit" value = "下一页" name = "submit">
          </FORM>
          </td>
     </tr>
    </FORM>
</Font></BODY></HTML>
```

2. 模型(JavaBean)

ShowRecordByPage.java 模型中的 getXxx 和 setXxx 方法可以显示和修改模型中的数据,但不参与数据的处理。请读者比较本节中的 ShowRecordByPage.java 和 7.6.2 节中的 ShowRecordByPage.java 之间的不同之处。

ShowRecordByPage. java

```
package mybean.data;
import com.sun.rowset.*;
public class ShowRecordByPage{
    CachedRowSetImpl rowSet = null;              //存储表中全部记录的行集对象
    int pageSize = 10;                           //每页显示的记录数
    int pageAllCount = 0;                        //分页后的总页数
    int showPage = 1 ;                           //当前显示页
    StringBuffer presentPageResult;              //显示当前页内容
    String databaseName = "";                    //数据库名称
    String tableName = "";                       //表的名字
    StringBuffer formTitle = null;               //表头
    public void setRowSet(CachedRowSetImpl set){
        rowSet = set;
    }
    public CachedRowSetImpl getRowSet(){
        return rowSet;
    }
    public void setPageSize(int size){
        pageSize = size;
    }
    public int getPageSize(){
        return pageSize;
    }
    public int getPageAllCount(){
        return pageAllCount;
    }
    public void setPageAllCount(int n){
        pageAllCount = n;
    }
    public void setShowPage(int n){
        showPage = n;
    }
    public int getShowPage(){
        return showPage;
    }
    public void setPresentPageResult(StringBuffer p){
        presentPageResult = p;
    }
    public StringBuffer getPresentPageResult(){
        return presentPageResult;
    }
    public void setDatabaseName(String s){
        databaseName = s.trim();
    }
    public String getDatabaseName(){
        return databaseName;
    }
    public void setTableName(String s){
        tableName = s.trim();
```

```
    }
    public String getTableName(){
        return tableName;
    }
    public void setFormTitle(StringBuffer s){
        formTitle = s;
    }
    public StringBuffer getFormTitle(){
        return formTitle;
    }
}
```

3. 控制器(servlet)

控制器是名字为 database 的 servlet 对象(见 web. xml 中的配置),由 HandleDatabase 类负责创建。

HandleDatabase. java

```
package myservlet. control;
import mybean. data. ShowRecordByPage;
import com. sun. rowset. * ;
import java. sql. * ;
import java. io. * ;
import javax. servlet. * ;
import javax. servlet. http. * ;
public class HandleDatabase extends HttpServlet{
   int 字段个数;
   CachedRowSetImpl rowSet = null;
   public void init(ServletConfig config) throws ServletException{
      super. init(config);
      try { Class. forName("com. mysql. jdbc. Driver");
      }
      catch(Exception e){}
   }
   public void doPost(HttpServletRequest request,HttpServletResponse response)
  throws ServletException,IOException{
      Connection con;
      StringBuffer presentPageResult = new StringBuffer();
      ShowRecordByPage databaseBean = null;
      HttpSession session = request. getSession(true);
      try{ databaseBean = (ShowRecordByPage)session. getAttribute("database");
              if(databaseBean == null){
                  databaseBean = new ShowRecordByPage();   //创建 JavaBean 对象
                  session. setAttribute("database",databaseBean);
              }
      }
      catch(Exception exp){
              databaseBean = new ShowRecordByPage();
              session. setAttribute("database",databaseBean);
      }
      String databaseName = request. getParameter("databaseName");
```

```
String tableName = request.getParameter("tableName");
String ps = request.getParameter("pageSize");
if(ps! = null){
   try{ int mm = Integer.parseInt(ps);
         databaseBean.setPageSize(mm);
   }
   catch(NumberFormatException exp){
         databaseBean.setPageSize(1);
   }
}
int showPage = databaseBean.getShowPage();
int pageSize = databaseBean.getPageSize();
boolean boo = databaseName! = null&&tableName! = null&&
            databaseName.length()> 0&&tableName.length()> 0;
if(boo){
   databaseBean.setDatabaseName(databaseName);     //数据存储在 databaseBean 中
   databaseBean.setTableName(tableName);           //数据存储在 databaseBean 中
   String uri = "jdbc:mysql://127.0.0.1/" + databaseName;
   try{ 字段个数 = 0;
          con = DriverManager.getConnection(uri,"root","");
          DatabaseMetaData metadata = con.getMetaData();
          ResultSet rs1 = metadata.getColumns(null,null,tableName,null);
          int k = 0;
          String 字段[] = new String[100] ;
          while(rs1.next()){
            字段个数++;
            字段[k] = rs1.getString(4);              //获取字段的名字
            k++;
          }
          StringBuffer str = new StringBuffer();
          str.append("<tr>");
          for(int i = 0;i<字段个数;i++)
             str.append("<th>" + 字段[i] + "</th>");
          str.append("</tr>");
          databaseBean.setFormTitle(str);            //数据存储在 databaseBean 中
          Statement sql =
          con.createStatement(ResultSet.TYPE_SCROLL_SENSITIVE,
                                         ResultSet.CONCUR_READ_ONLY);
          ResultSet rs = sql.executeQuery("SELECT * FROM " + tableName);
          rowSet = new CachedRowSetImpl();           //创建行集对象
          rowSet.populate(rs);
          con.close();                               //关闭连接
          databaseBean.setRowSet(rowSet);            //数据存储在 databaseBean 中
          rowSet.last();
          int m = rowSet.getRow();                   //总行数
          int n = pageSize;
          int pageAllCount = ((m % n) == 0)? (m/n):(m/n + 1);
          databaseBean.setPageAllCount(pageAllCount);
   }
   catch(SQLException exp){}
}
```

```
        String whichPage = request.getParameter("whichPage");
        if(whichPage == null||whichPage.length() == 0){
              showPage = 1;
              databaseBean.setShowPage(showPage);
              CachedRowSetImpl rowSet = databaseBean.getRowSet();
              if(rowSet! = null){
                 presentPageResult = show(showPage,pageSize,rowSet);
                 databaseBean.setPresentPageResult(presentPageResult);
              }
        }
        else if(whichPage.equals("nextPage")){
              showPage++;
              if(showPage > databaseBean.getPageAllCount())
                 showPage = 1;
              databaseBean.setShowPage(showPage);
              CachedRowSetImpl rowSet = databaseBean.getRowSet();
              if(rowSet! = null){
                 presentPageResult = show(showPage,pageSize,rowSet);
                 databaseBean.setPresentPageResult(presentPageResult);
              }
        }
        else if(whichPage.equals("previousPage")){
              showPage -- ;
              if(showPage <= 0)
                 showPage = databaseBean.getPageAllCount();
              databaseBean.setShowPage(showPage);
              CachedRowSetImpl rowSet = databaseBean.getRowSet();
              if(rowSet! = null){
                 presentPageResult = show(showPage,pageSize,rowSet);
                 databaseBean.setPresentPageResult(presentPageResult);
              }
        }
        databaseBean.setPresentPageResult(presentPageResult);
        RequestDispatcher dispatcher = request.getRequestDispatcher
("showRecord.jsp");
        dispatcher.forward(request,response);                  //请求 showRecord.jsp 显示数据
    }
    public StringBuffer show(int page,int pageSize,CachedRowSetImpl rowSet){
        StringBuffer str = new StringBuffer();
        try{ rowSet.absolute((page - 1) * pageSize + 1);
                for(int i = 1;i <= pageSize;i++){
                    str.append("<tr>");
                    for(int k = 1;k <= 字段个数;k++)
                       str.append("<td>" + rowSet.getString(k) + "</td>");
                    str.append("</tr>");
                    rowSet.next();
                }
         }
         catch(SQLException exp){}
         return str;
     }
```

```
    public void doGet(HttpServletRequest request,
HttpServletResponse response)
                        throws ServletException,IOException{
        doPost(request,response);                    //转发
    }
}
```

9.7 MVC 模式与文件操作

在第 7 章,曾使用 bean 读取文件的内容(见 7.5 节),该 bean 不仅要负责读取文件,而且要负责存储所读取的内容。在 MVC 模式中,读取文件的工作由 servlet 对象负责,bean 仅仅负责存储 servlet 对象所读取的文件内容。

本节设计一个 Web 应用,在该 Web 应用中有两个 JSP 页面: choiceFile. jsp 和 showFile. jsp,使用一个 JavaBean 和一个 servlet。用户在 JSP 页面 choiceFile. jsp 选择一个文件,提交给 servlet,该 servlet 负责读取文件的有关信息存放到 JavaBean 中,并请求 JSP 页面 showFile. jsp 显示 Javanean 中的数据。

需要为 ch9\WEB-INF 中的 web. xml 文件添加如下子标记(本节 servlet 的包名是 myservlet. control):

```
<servlet>
    <servlet-name>helpReadFile</servlet-name>
    <servlet-class>myservlet.control.HandleFile</servlet-class>
</servlet>
<servlet-mapping>
    <servlet-name>helpReadFile</servlet-name>
    <url-pattern>/helpReadFile</url-pattern>
</servlet-mapping>
```

9.7.1 模型(JavaBean)

本节的 JavaBean 类的包名为 mybean. data。FileMessage. java 模型中的 getXxx 和 setXxx 方法可以显示和修改模型中的数据,但不参与数据的处理。请读者比较本节中的 FileMessage. java 和 7.5 节中 ReadFile. java 的不同之处。

JavaBean 类的源文件 FileMessage. java 保存到: D:\mybean\data,如下列格式编译源文件,即带着包名形成的目录:

```
D:> javac mybean\data\FileMessage.java
```

将编译得到的字节码文件 FileMessage. class 复制到

```
ch9\WEB-INF\classes\mybean\data
```

目录中。

FileMessage. java

```
package mybean.data;
public class FileMessage {
```

```
    String filePath,fileName,fileContent;
    long fileLength;
    public void setFilePath(String str){
       filePath = str;
    }
    public String getFilePath(){
       return filePath;
    }
    public void setFileName(String str){
       fileName = str;
    }
    public String getFileName(){
       return fileName;
    }
    public void setFileContent(String str){
       fileContent = str;
    }
    public String getFileContent(){
       return fileContent;
    }
    public void setFileLength(long len){
       fileLength = len;
    }
    public long getFileLength(){
       return fileLength;
    }
}
```

9.7.2 控制器(servlet)

控制器是名字为 helpReadFile 的 servlet 对象(见 web. xml 中的配置),由下面的 HandleFile 类负责创建。由于 Servlet 类中要使用 JavaBean,所以为了能顺利地编译 Servlet 类,不要忘记将 Tomcat 安装目录 lib 子目录中的 servlet-api. jar 文件复制到 Tomcat 服务器所使用的 JDK 的扩展目录中,比如,复制到 D:\jdk1. 7\jre\lib\ext 中。Servlet 类的包名为 myservlet. control,将下面的 Servlet 类的源文件 HandeFile. java 保存到

```
D:\myservlet\control
```

目录中,即保存时,让 Servlet 类的包名和 JavaBean 类的包名形成的目录的父目录相同。用如下格式进行编译,即带着包名形成的目录:

```
D:> javac myservlet\control\HandleFile. java
```

将编译得到的字节码文件 HandeFile. class 复制到

```
ch9\WEB - INF\classes\myservlet\control
```

目录中。

HandeFile. java

```
package myservlet. control;
import mybean. data. FileMessage;
```

```
import java.io.*;
import javax.servlet.*;
import javax.servlet.http.*;
public class HandleFile extends HttpServlet{
   public void init(ServletConfig config) throws ServletException{
      super.init(config);
   }
   public void doPost(HttpServletRequest request,HttpServletResponse response)
                        throws ServletException,IOException{
      FileMessage file = new FileMessage();                    //创建 JavaBean 对象
      request.setAttribute("file",file);
      String filePath = request.getParameter("filePath");
      String fileName = request.getParameter("fileName");
      file.setFilePath(filePath);                              //将数据存储在 file 中
      file.setFileName(fileName);
      try{ File f = new File(filePath,fileName);
             long length = f.length();
             file.setFileLength(length);
             FileReader in = new FileReader(f) ;
             BufferedReader inTwo = new BufferedReader(in);
             StringBuffer stringbuffer = new StringBuffer();
             String s = null;
             while ((s = inTwo.readLine())! = null)
                stringbuffer.append("\n" + s);
             String content = new String(stringbuffer);
             file.setFileContent(content);
      }
      catch(IOException exp){}
      RequestDispatcher dispatcher = request.getRequestDispatcher("showFile.jsp");
      dispatcher.forward(request, response);
   }
   public void doGet(HttpServletRequest request,
HttpServletResponse response)
                        throws ServletException,IOException{
      doPost(request,response);
   }
}
```

9.7.3 视图(JSP 页面)

在 choiceFile.jsp 页面可以输入文件的路径和名字,并提交给名字为 handleFile 的 servlet 对象。servlet 对象负责读取文件,并将读取的内容以及相关的数据存储到数据模型 bean 中,然后请求 showFile.jsp 页面显示模型中的数据。choiceFile.jsp 和 showFile.jsp 的效果如图 9-11 和图 9-12 所示。

choiceFile.jsp(效果如图 9-11 所示)

```
<%@ page contentType = "text/html;Charset = GB2312" %>
<HTML><BODY bgcolor = cyan><Font size = 2>
<FORM action = "helpReadFile" method = "post" name = "form">
```

```
输入文件的路径(如:d:/2000):
<INPUT type="text" name="filePath" size=12>
<BR>输入文件的名字(如:Hello.java):
<INPUT type="text" name="fileName" size=9>
<BR><INPUT type="submit" value="读取" name="submit">
</FORM>
</Font></BODY></HTML>
```

showFile.jsp(效果如图 9-12 所示)

```
<%@ page import="mybean.data.FileMessage" %>
<%@ page contentType="text/html;charset=GB2312" %>
<jsp:useBean id="file" type="mybean.data.FileMessage" scope="request"/>
<HTML><BODY bgcolor=yellow><Font size=2>
  文件的位置:<jsp:getProperty name="file" property="filePath"/>,
  文件的名字:<jsp:getProperty name="file" property="fileName"/>,
  文件的长度:<jsp:getProperty name="file" property="fileLength"/> 字节。
  <BR>文件的内容:
  <BR><TextArea rows="6" cols="60">
       <jsp:getProperty name="file" property="fileContent"/>
     </TextArea>
</Font></BODY></HTML>
```

地址(D) http://127.0.0.1:8080/ch9/choiceFile.jsp
输入文件的路径(如:d:/2000): d:/2000
输入文件的名字(如:Hello.java): Hello.java
读取

图 9-11　输入目录和文件名

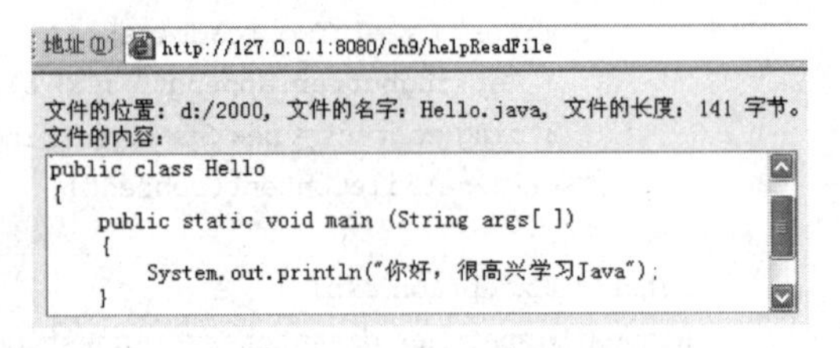

图 9-12　显示文件内容和相关数据

9.8　实验:计算等差、等比数列的和

要求在 webapps 目录下新建一个 Web 服务目录: practice9。除特别要求外,实验所涉及的 JSP 页面均保存在 practice9 中。实验中涉及的 servlet 的包名均为 user.yourservlet,因此要求在 practice9 下建立子目录: WEB-INF\classes\user\yourservlet,servlet 的字节码文件保存在该子目录中;实验中涉及的 JavaBean 的包名均为 user.yourbean,因此要求在 practice9 下建立子目录: WEB-INF\classes\user\yourbean,JavaBean 的字节码文件保存在该子目录中。

1. 相关知识点

(1) 模型(Model):一个或多个 JavaBean 对象,用于存储数据,JavaBean 主要提供简单的 setXxx 方法和 getXxx 方法,在这些方法中不涉及对数据的具体处理细节,以便增强模型的通用性。

(2) 视图(View):一个或多个 JSP 页面,其作用是向控制器提交必要的数据和为模型提供数据显示,JSP 页面使用 HTML 标记和 JavaBean 标记来显示数据。

(3) 控制器(Controller)：一个或多个 servlet 对象,根据视图提交的要求进行数据处理操作,并将有关的结果存储到 JavaBean 中,然后 servlet 使用重定向方式请求视图中的某个 JSP 页面更新显示,即让该 JSP 页面通过使用 JavaBean 标记显示控制器存储在 JavaBean 中的数据。

2. 实验目的

本实验的目的是让学生掌握 MVC 模式。

3. 实验要求

1) 视图

视图由两个 JSP 页面组成：inputData. jsp 和 showResult. jsp。inputData. jsp 页面提供一个表单,用户可以输入等差数列的首项、公差、求和项数,也可以输入等比数列的首项、公比和求和项数。inputData. jsp 页面将用户输入的有关数据提交给一个名字为 computerSum 的 servlet 对象,computerSum 负责计算等差数列的和以及等比数列的和。showResult. jsp 页面可以显示等差数列和等比数列的求和结果。

2) 数据模型

模型即可以存储等差数列的和也可以存储等比数列的和。数据模型 JavaBean(Series. java 类的实例)中的 getXxx 和 setXxx 方法不涉及对数据的具体处理细节,以便增强模型的通用性。

3) 控制器

提供一个名字为 computerSum 的 servlet 对象(HandleSum. java 类的实例),computerSum 负责计算等差数列和等比数列的和,将有关数据存储到数据模型 JavaBean 中,然后请求 showResult. jsp 显示。

4) 配置文件

读者需要根据 servlet 包名在 web. xml 文件增加适当的内容(见主教材的 8. 2 节)。如果根据后面给出的参考代码,web. xml 配置文件中需要增加如下内容：

```
<servlet>
    <servlet-name>computerSum</servlet-name>
<servlet-class>myservlet.control.HandleSum</servlet-class>
</servlet>
<servlet-mapping>
   <servlet-name>computerSum</servlet-name>
   <url-pattern>/lookSum</url-pattern>
</servlet-mapping>
```

4. 参考代码

代码仅供参考,学生可按着实验要求,参考本代码编写代码。

视图参考代码参考如下：

inputData. jsp

```
<%@ page contentType="text/html;charset=GB2312" %>
<HTML><BODY bgcolor=cyan>
<Font size=2>
<FORM action="lookSum" Method="post">
   等差数列求和：
```

```
  <BR>输入首项:<Input type=text name="firstItem" size=4>
      输入公差:<Input type=text name="var" size=4>
      求和项数:<Input type=text name="number" size=4>
  <Input type=submit value="提交">
</FORM>
<FORM action="lookSum" Method="get">
    等比数列求和:
  <BR>输入首项:<Input type=text name="firstItem" size=4>
      输入公比:<Input type=text name="var" size=4>
      求和项数:<Input type=text name="number" size=4>
  <Input type=submit value="提交">
</FORM>
</Font></BODY></HTML>
```

showResult.jsp

```
<%@ page contentType="text/html;charset=GB2312" %>
<%@ page import="mybean.data.Series" %>
<jsp:useBean id="lie" type="mybean.data.Series" scope="request"/>
<HTML><BODY bgcolor=yellow>
  <Table border=1>
    <tr>
     <th> 数列的首项</th>
     <th><jsp:getProperty name="lie" property="name"/></th>
     <th> 所求项数</th>
     <th> 求和结果</th>
    </tr>
     <td><jsp:getProperty name="lie" property="firstItem"/></td>
     <td><jsp:getProperty name="lie" property="var"/></td>
     <td><jsp:getProperty name="lie" property="number"/></td>
     <td><jsp:getProperty name="lie" property="sum"/></td>
    </tr>
    </Table>
</BODY></HTML>
```

数据模型(JavaBean)参考代码如下:

Series.java

```
package mybean.data;
public class Series{
    double firstItem;                    //数列首项
    double var;                          //公差或公比
    int number;                          //求和项数
    double sum;                          //求和结果
    String name="";                      //数列类别
    public void setFirstItem(double a){
        firstItem=a;
    }
    public double getFirstItem(){
        return firstItem;
    }
    public void setVar(double b){
```

```
        var = b;
    }
    public double getVar(){
        return var;
    }
    public void setNumber(int n){
        number = n;
    }
    public double getNumber(){
        return number;
    }
    public void setSum(double s){
        sum = s;
    }
    public double getSum(){
        return sum;
    }
    public void setName(String na){
        name = na;
    }
    public String getName(){
       return name;
    }
}
```

控制器(servlet)参考代码如下：

HandleSum.java

```
package myservlet.control;
import mybean.data.*;
import java.io.*;
import javax.servlet.*;
import javax.servlet.http.*;
public class HandleSum extends HttpServlet{
    public void init(ServletConfig config) throws ServletException{
        super.init(config);
    }
    public void doPost(HttpServletRequest request,HttpServletResponse response)
                              throws ServletException,IOException{
        Series shulie = new Series();                //创建 JavaBean 对象
        request.setAttribute("lie",shulie);    //将 shulie 存储到 request 对象中
        double a = Double.parseDouble(request.getParameter("firstItem"));
        double d = Double.parseDouble(request.getParameter("var"));
        int n = Integer.parseInt(request.getParameter("number"));
        shulie.setFirstItem(a);                      //将数据存储在 shulie 中
        shulie.setVar(d);
        shulie.setNumber(n);
        double sum = 0,item = a;
        int i = 1;
        shulie.setName("等差数列的公差");
        while(i <= n){                               //计算等差数列的和
```

```
            sum = sum + item;
            i++;
            item = item + d;
        }
        shulie.setSum(sum);
        RequestDispatcher dispatcher = request.getRequestDispatcher("showResult.jsp");
        dispatcher.forward(request,response); //请求 showResult.jsp 显示 shulie 中的数据
    }
    public void doGet(HttpServletRequest request,HttpServletResponse response)
                          throws ServletException,IOException{
        Series shulie = new Series();
        request.setAttribute("lie",shulie);
        double a = Double.parseDouble(request.getParameter("firstItem"));
        double d = Double.parseDouble(request.getParameter("var"));
        int n = Integer.parseInt(request.getParameter("number"));
        shulie.setFirstItem(a);
        shulie.setVar(d);
        shulie.setNumber(n);
        double sum = 0,item = a;
        int i = 1;
        shulie.setName("等比数列的公比");
        while(i<=n){                                //计算等比数列的和
             sum = sum + item;
             i++;
             item = item * d;
        }
        shulie.setSum(sum);
        RequestDispatcher dispatcher = request.getRequestDispatcher("showResult.jsp");
        dispatcher.forward(request,response);
    }
}
```

习 题 9

1. 在JSP中,MVC模式中的数据模型的角色由谁担当?
2. 在JSP中,MVC模式中的控制器的角色由谁担当?
3. 在JSP中,MVC模式中的视图的角色由谁担当?
4. MVC的好处是什么?
5. MVC模式中用到的JavaBean是由JSP页面还是servlet负责创建?
6. 参照9.4.1节设计一个Web应用。用户可以通过JSP页面输入一元二次方程的系数给一个servlet控制器,控制器负责计算方程的根,并将结果存储到数据模型中,然后请求JSP页面显示数据模型中的数据。

第 10 章　手机销售网

这一章讲述如何用 JSP 技术建立一个简单的手机销售网(货到付款),其目的是掌握一般 Web 应用中常用基本模块的开发方法。JSP 引擎为 Tomcat 8.0,系统采用 MVC 模式实现各个模块,数据库使用的是 MySQL 数据库。

10.1　系统模块构成

系统主要模块如图 10-1 所示。

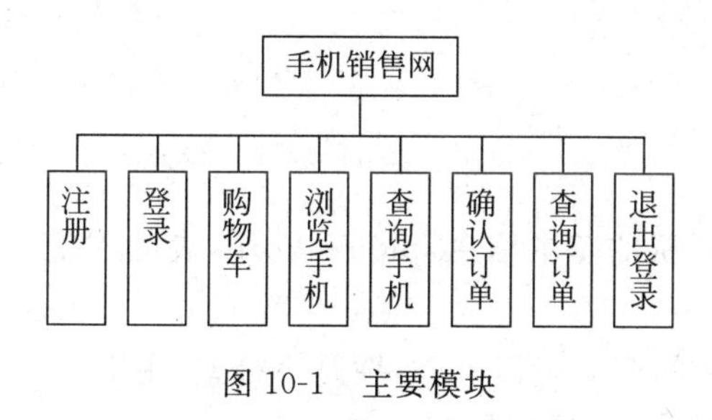

图 10-1　主要模块

10.2　数据库设计与连接

10.2.1　数据库设计

使用 MySQL 建立一个数据库 mobileshop,该库共有 4 个表,以下是这些表的名称、结构和用途(有关建立数据库和表的操作细节见主教材的 6.1 节)。

1. user 表

user 表用于存储用户的注册信息,user 表的主键是 logname,各个字段值的说明如下:

logname ：存储注册的用户名(属性是字符型,主键)。

password ：存储密码(属性是字符型)。

phone ：存储电话(属性是字符型)。

addess：存储地址(属性是字符型)。

realname ：存储姓名(属性是字符型)。

user 表如图 10-2 所示。

2. mobileClassify 表

mobileClassify 表存储手机的类别,各个字段值的说明如下:

栏位 索引 外键 触发器 选项 注释 SQL 预览

名	类型	长度	小数点	允许空值(	
logname	char	255	0	☐	
password	char	255	0	☑	
phone	char	255	0	☑	
address	char	255	0	☑	
realname	char	255	0	☑	

图 10-2　user 表

id：手机的分类号(属性是整型,号码自动增加,主键)。

name：手机的分类名称(属性是字符型)。

mobileClassify 表如图 10-3 所示。

使用 MySQL 数据库客户端管理程序或 MySQL 监视器在 maobileCalssify 表中事先输入了几条记录,如图 10-4 所示。

栏位 索引 外键 触发器 选项 注释 SQL 预览

名	类型	长度	小数点	允许空值(	
id	int	11	0	☐	1
name	char	200	0	☐	

图 10-3　mobileClassify 表

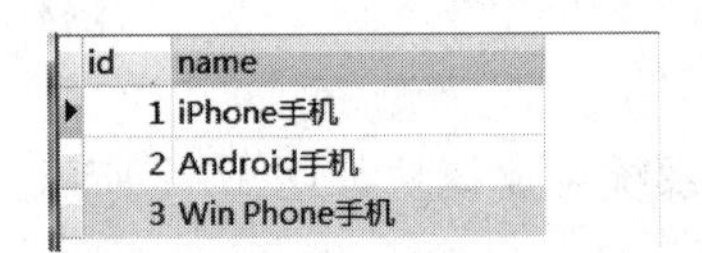

id	name
1	iPhone手机
2	Android手机
3	Win Phone手机

图 10-4　mobileClassify 表中的记录

3. mobileForm 表

mobileForm 表存储手机的基本信息,mobileForm 表的主键是 mobile_version,各个字段值的说明如下：

mobile_version：手机的产品标识号(属性是字符型,主键)。

mobile_name：手机的名称(属性是字符型)。

mMobile_made：手机的制造商(属性是字符型)。

mobile_price：手机的价格(属性是单精度浮点型)。

mobile _mess：手机产品介绍(属性是字符型)

mobile_pic：存储和手机相关的一幅图像文件的名字(属性是字符型,主键)。

id:作为 mobileClassify 表中 id 的外键。

mobileForm 表如图 10-5 所示。mobileForm 表使用 mobileClassify 表的 id 作外键与 mobileClassify 相关联。mobileClassify 表和 mobileForm 表是一对多关系。在表中事先输入了几条记录,如图 10-6 所示。

栏位 索引 外键 触发器 选项 注释 SQL 预览

名	类型	长度	小数点	允许空值	
mobile_version	char	50	0	☐	1
mobile_name	char	60	0	☑	
mobile_made	char	200	0	☑	
mobile_price	float	4	0	☑	
mobile_mess	char	255	0	☑	
mobile_pic	char	200	0	☐	2
id	int	11	0	☐	

图 10-5　mobileForm 表

mobile_versic	mobile_name	mobile_made	mobile_	mobile_mess	mobile_pic	id
A89S6	苹果S5手机	苹果公司	9876	高清大屏幕	apple.jpg	1
B8978	三星A98手机	三星公司	8976	支持图形非常好	aa.jpg	2
C555	中兴 N986	中兴 公司	3567	双卡双待	cc.jpg	2

图 10-6　mobileForm 表中的记录

4. orderForm 表

orderForm 表存储订单信息，orderForm 表的主键是 orderNumber，字段值的说明如下：

id：存储订单序号(属性是整型，自动增加)。

logname ：存储注册的用户名(属性是字符型)。

mess：订单信息(属性是字符型)。

sum：所定图书的价格总和(属性是单精度浮点型)。

orderForm 表如图 10-7 所示。

栏位 | 索引 | 外键 | 触发器 | 选项 | 注释 | SQL 预览

名	类型	长度	小数点	允许空值	
id	int	10	0	☐	🔑1
logname	char	255	0	☑	
mess	char	255	0	☑	
sum	float	0	0	☑	

图 10-7　orderform 表

读者可以参见主教材的 6.1 节，选择自己喜欢的方式，比如，用某种 MySQL 客户端管理工具创建数据库和上述相关的表。

10.2.2　数据库连接

为了避免操作数据库时出现中文乱码，需要使用

```
Connection getConnection(java.lang.String)
```

方法建立连接，连接中的代码是(用户是 root，密码为空)：

```
String uri = "jdbc:mysql://127.0.0.1/mobileshop?" +
             "user = root&password = &characterEncoding = gb2312";
Connection con = DriverManager.getConnection(uri);
```

10.3　系 统 管 理

本系统使用的 Web 服务目录是 ch10，ch10 是在 Tomcat 安装目录的 webapps 目录下建立的 Web 服务目录。

现在需要在当前 web 服务目录下建立如下的目录结构：

```
ch10\WEB - INF\classes
```

然后根据 servlet 的包名，在 classes 下建立相应的子目录，比如 Servlet 类的包名为

myservlet.control,那么在 classes 下建立子目录:\myservlet\control;如果 JavaBean 类的包名为 mybean.data,那么在 classes 下建立子目录:\mybean\data。为了让 Tomcat 服务器起用上述目录,必须重新启动 Tomcat 服务器。

10.3.1 页面管理

本系统用的 JSP 页面全部保存在 Web 服务目录 ch10 中。所有的页面将包括一个导航条,该导航条由注册、登录、浏览手机、查看订单等组成。为了便于维护,其他页面通过使用 JSP 的<%@ include …… %>标记将 head.txt(导航条文件)嵌入到自己的页面中。head.txt 保存在 Web 服务目录 ch10 中。head.txt 的内容如下:

head.txt

```
<%@ page contentType = "text/html;charset = GB2312" %>
<div align = "center">
  <H2>"智多星"智能手机销售网</H2>
  <table cellSpacing = "1" cellPadding = "1" width = "660" align = "center" border = "0" >
   <tr valign = "bottom">
   <td><A href = "inputRegisterMess.jsp"><font size = 2>注册</font></A></td>
   <td><A href = "login.jsp"><font size = 2>登录</font></A></td>
   <td><A href = "lookMobile.jsp"><font size = 2>浏览手机</font></A></td>
   <td><A href = "searchMobile.jsp"><font size = 2>查询手机</font></A></td>
   <td><A href = "lookShoppingCar.jsp"><font size = 2>查看购物车</font></A></td>
   <td><A href = "lookOrderForm.jsp"><font size = 2>查看订单</font></A></td>
   <td><A href = "exitServlet"><font size = 2>退出</font></A></td>
   <td><A href = "index.jsp"><font size = 2>主页</font></A></td>
  </tr>
  </Font>
</table>
</div>
```

主页 index.jsp 由导航条、一个欢迎语和一幅图片 welcome.jpg 组成,welcome.jpg 保存在 ch10 的 image 子目录中(网站用的图像文件均存在 image 目录中,见 10.3.4 节的图像管理)。

用户可以通过在浏览器的地址栏中键入"http://服务器 IP:8080/index.jsp"或"http://服务器 IP:8080/"访问该主页,主页运行效果如图 10-8 所示。

index.jsp(效果如图 10-8 所示)

```
<%@ page contentType = "text/html;charset = GB2312" %>
<HTML><BODY>
<HEAD><%@ include file = "head.txt" %></HEAD>
<title>首页</title>
<CENTER><h1><font Size = 4 color = blue>
          欢迎光临"智多星"智能手机销售网
          </font></h1>
<img src = "image/welcome.jpg" width = 500 height = 400 ></img>
</CENTER>
</BODY></HTML>
```

图 10-8 主页 index.jsp

10.3.2 JavaBean 与 Servlet 管理

本系统的 JavaBean 类的包名均为 mybean.data；Servlet 类的包名均为 myservlet.control。由于 Servlet 类中要使用 JavaBean，所以为了能顺利地编译 Servlet 类，不要忘记将 Tomcat 安装目录 lib 子目录中的 servlet-api.jar 文件复制到 Tomcat 服务器所使用的 JDK 的扩展目录中，比如，复制到 D:\jdk1.7\jre\lib\ext 中。然后，按下列步骤编译和保存有关的字节码文件。

1. 保存 JavaBean 类和 Servlet 类的源文件

将 JavaBean 类和 Servlet 类源文件分别保存到：

```
D:\ mybean\data
```

和

```
D:\myservlet\control
```

目录中。保存时，让 Servlet 类的包名和 JavaBean 类的包名形成的目录的父目录相同。

2. 编译 JavaBean 类

用如下格式进行编译，即带着包名形成的目录：

```
D:> javac mybean\data\JavaBean 的源文件
```

例如：

```
D:> javac mybean\data\Login.java
```

3. 编译 Servlet 类

用如下格式进行编译，即带着包名形成的目录：

```
D:> javac myservlet\control\servlet 的源文件
```

例如：

```
D:> javac myservlet\control\HandleLogin.java
```

4. 将字节码保存到服务器

将编译通过的 JavaBean 类和 Servlet 类的字节码文件分别复制到

```
ch10\WEB-INF\classes\mybean\data
```

和

```
ch10\WEB-INF\classes\myservlet\control
```

目录中。

注意：为了调试程序方便，可以将 JavaBean 和 Servlet 源文件分别保存到 Web 服务目录的下述目录中：

```
WEB-INF\classes\mybean\data
WEB-INF\classes\myservlet\control
```

这样就可以省略上面的最后步骤。

10.3.3 配置文件管理

本系统的 Servlet 类的包名均为 myservlet.control，需要配置 Web 服务目录的 web.xml 文件，即将下面的 web.xml 文件保存到 Tomcat 安装目录的 Web 服务目录的 WEB-INF 子目录中。根据本书使用的 Tomcat 安装目录及 Web 服务目录，需要将下面的 web.xml 文件保存到

```
D:\apache-tomcat-8.0.3\webapps\ch10\WEB-INF
```

目录中(有关 web.xml 文件的配置规定可参见 8.2 节)。

web.xml

```
<?xml version="1.0" encoding="ISO-8859-1"?>
<web-app>
<servlet>
    <servlet-name>registerServlet</servlet-name>
    <servlet-class>myservlet.control.HandleRegister</servlet-class>
</servlet>
<servlet-mapping>
    <servlet-name>registerServlet</servlet-name>
    <url-pattern>/registerServlet</url-pattern>
</servlet-mapping>
<servlet>
    <servlet-name>loginServlet</servlet-name>
    <servlet-class>myservlet.control.HandleLogin</servlet-class>
</servlet>
<servlet-mapping>
    <servlet-name>loginServlet</servlet-name>
    <url-pattern>/loginServlet</url-pattern>
</servlet-mapping>
<servlet>
    <servlet-name>deleteServlet</servlet-name>
    <servlet-class>myservlet.control.HandleDelete</servlet-class>
```

```
</servlet>
<servlet-mapping>
    <servlet-name>deleteServlet</servlet-name>
    <url-pattern>/deleteServlet</url-pattern>
</servlet-mapping>
<servlet>
    <servlet-name>buyServlet</servlet-name>
    <servlet-class>myservlet.control.HandleBuyGoods</servlet-class>
</servlet>
<servlet-mapping>
    <servlet-name>buyServlet</servlet-name>
    <url-pattern>/buyServlet</url-pattern>
</servlet-mapping>
<servlet>
    <servlet-name>queryServlet</servlet-name>
    <servlet-class>myservlet.control.QueryAllRecord</servlet-class>
</servlet>
<servlet-mapping>
    <servlet-name>queryServlet</servlet-name>
    <url-pattern>/queryServlet</url-pattern>
</servlet-mapping>
<servlet>
    <servlet-name>putGoodsServlet</servlet-name>
    <servlet-class>myservlet.control.PutGoodsToCar</servlet-class>
</servlet>
<servlet-mapping>
    <servlet-name>putGoodsServlet</servlet-name>
    <url-pattern>/putGoodsServlet</url-pattern>
</servlet-mapping>
<servlet>
    <servlet-name>searchByConditionServlet</servlet-name>
    <servlet-class>myservlet.control.SearchByCondition</servlet-class>
</servlet>
<servlet-mapping>
    <servlet-name>searchByConditionServlet</servlet-name>
    <url-pattern>/searchByConditionServlet</url-pattern>
</servlet-mapping>
<servlet>
    <servlet-name>exitServlet</servlet-name>
    <servlet-class>myservlet.control.HandleExit</servlet-class>
</servlet>
<servlet-mapping>
    <servlet-name>exitServlet</servlet-name>
    <url-pattern>/exitServlet</url-pattern>
</servlet-mapping>
</web-app>
```

10.3.4 图像管理

本系统用到的图片所对应的图像文件，比如手机图片用的图像文件等，均保存在 Web

服务目录 ch10 的子目录 image 中。

10.4 会员注册

当新会员注册时,该模块要求用户必须输入会员名、密码信息,否则不允许注册。用户的注册信息被存入数据库的 user 表中。

10.4.1 视图(JSP 页面)

该模块视图部分由一个 JSP 页面构成,这个 JSP 页面 inputRegisterMess.jsp 负责提交用户的注册信息到 servlet 控制器 registerServlet(见配置文件 web.xml),并负责显示注册是否成功的信息。inputRegisterMess.jsp 效果如图 10-9 所示。

图 10-9 注册

inputRegisterMess.jsp(效果如图 10-9 所示)

```
<%@ page contentType = "text/html;charset = GB2312" %>
<jsp:useBean id = "userBean" class = "mybean.data.Register" scope = "request"/>
<HEAD><%@ include file = "head.txt" %></HEAD>
<title>注册页面</title>
<HTML><BODY bgcolor = pink><Font size = 2>
<div align = "center">
<FORM action = "registerServlet" method = "post" name = form>
<table>
    用户名由字母、数字、下划线构成,* 注释的项必须填写。
   <tr><td>* 用户名称:</td><td><Input type = text name = "logname" ></td>
       <td>* 用户密码:</td><td><Input type = password name = "password">
       </td></tr>
   <tr><td>* 重复密码:</td><td>
       <Input type = password name = "again_password"></td>
       <td>联系电话:</td><td><Input type = text name = "phone"></td></tr>
   <tr><td>邮寄地址:</td><td><Input type = text name = "address"></td>
       <td>真实姓名:</td><td><Input type = text name = "realname"></td>
       <td><Input type = submit name = "g" value = "提交"></td> </tr>
</table>
</Form>
</div>
<div align = "center">
```

```
<p> 注册反馈:
<jsp:getProperty name = "userBean" property = "backNews" />
<table border = 3 >
    <tr><td>会员名称:</td>
    <td><jsp:getProperty name = "userBean" property = "logname"/></td>
    </tr>
    <tr><td>姓名:</td>
    <td><jsp:getProperty name = "userBean" property = "realname"/></td>
    </tr>
    <tr><td>地址:</td>
    <td><jsp:getProperty name = "userBean" property = "address"/></td>
    </tr>
    <tr><td>电话:</td>
    <td><jsp:getProperty name = "userBean" property = "phone"/></td>
    </tr>
</table></div>
</Body></HTML>
```

10.4.2 模型(JavaBean)

下列 JavaBean 用来描述用户注册的重要信息。在该模块中 JavaBean 的实例的 id 是 userBean,生命周期是 request。该 JavaBean 的实例由控制器负责创建或更新。

Register.java

```
package mybean.data;
public class Register{
   String logname = "" , phone = "",
          address = "",realname = "",backNews = "请输入信息";
   public void setLogname(String logname){
      this.logname = logname;
   }
   public String getLogname(){
      return logname;
   }
   public void setPhone(String phone){
      this.phone = phone;
   }
   public String getPhone(){
      return phone;
   }
   public void setAddress(String address){
      this.address = address;
   }
   public String getAddress(){
      return address;
   }
   public void setRealname(String realname){
      this.realname = realname;
   }
   public String getRealname(){
```

```
        return realname;
    }
    public void setBackNews(String backNews){
        this.backNews = backNews;
    }
    public String getBackNews(){
        return backNews;
    }
}
```

10.4.3 控制器(servlet)

servlet 控制器是 registerServlet(见 10.3.3 节给出的 web.xml 配置文件),负责连接数据库,将用户提交的信息写入到 user 表中,并将用户转发到 inputRegisterMess.jsp 页面查看注册反馈信息。

HandleRegister.java

```
package myservlet.control;
import mybean.data.*;
import java.sql.*;
import java.io.*;
import javax.servlet.*;
import javax.servlet.http.*;
public class HandleRegister extends HttpServlet {
    public void init(ServletConfig config) throws ServletException {
        super.init(config);
        try { Class.forName("com.mysql.jdbc.Driver");
        }
        catch(Exception e){}
    }
    public String handleString(String s)
    { try{ byte bb[] = s.getBytes("iso-8859-1");
             s = new String(bb);
       }
       catch(Exception ee){}
       return s;
    }
    public void doPost(HttpServletRequest request,HttpServletResponse response)
                          throws ServletException,IOException {
        String uri = "jdbc:mysql://127.0.0.1/mobileshop?" +
                               "user=root&password=&characterEncoding=gb2312";
        Connection con;
        PreparedStatement sql;
        Register userBean = new Register(); //创建的 JavaBean 模型
        request.setAttribute("userBean",userBean);
        String logname = request.getParameter("logname").trim();
        String password = request.getParameter("password").trim();
        String again_password = request.getParameter("again_password").trim();
        String phone = request.getParameter("phone").trim();
        String address = request.getParameter("address").trim();
```

```
String realname = request.getParameter("realname").trim();
if(logname == null)
     logname = "";
if(password == null)
     password = "";
if(! password.equals(again_password)) {
   userBean.setBackNews("两次密码不同,注册失败。");
   RequestDispatcher dispatcher =
   request.getRequestDispatcher("inputRegisterMess.jsp");
   dispatcher.forward(request, response);     //转发
   return;
}
boolean isLD = true;
for(int i = 0;i < logname.length();i++){
    char c = logname.charAt(i);
     if(! ((c <= 'z'&&c >= 'a')||(c <= 'Z'&&c >= 'A')||(c <= '9'&&c >= '0')))
       isLD = false;
}
boolean boo = logname.length()> 0&&password.length()> 0&&isLD;
String backNews = "";
try{ con = DriverManager.getConnection(uri);
       String insertCondition = "INSERT INTO user VALUES (?,?,?,?,?)";
       sql = con.prepareStatement(insertCondition);
       if(boo)
       { sql.setString(1,handleString(logname));
         sql.setString(2,handleString(password));
         sql.setString(3,handleString(phone));
         sql.setString(4,handleString(address));
         sql.setString(5,handleString(realname));
         int m = sql.executeUpdate();
         if(m! = 0){
             backNews = "注册成功";
             userBean.setBackNews(backNews);
             userBean.setLogname(logname);
             userBean.setPhone(handleString(phone));
             userBean.setAddress(handleString(address));
             userBean.setRealname(handleString(realname));
         }
       }
       else {
            backNews = "信息填写不完整或名字中有非法字符";
            userBean.setBackNews(backNews);
       }
       con.close();
}
catch(SQLException exp){
       backNews = "该会员名已被使用,请你更换名字" + exp;
       userBean.setBackNews(backNews);
}
RequestDispatcher dispatcher =
request.getRequestDispatcher("inputRegisterMess.jsp");
```

```
        dispatcher.forward(request, response);   //转发
    }
    public void doGet(HttpServletRequest request,HttpServletResponse response)
                     throws ServletException,IOException {
        doPost(request,response);
    }
}
```

10.5 会员登录

用户可在该模块输入自己的会员名和密码,系统将对会员名和密码进行验证,如果输入的用户名或密码有错误,将提示用户名或密码不正确。

该模块视图部分由一个JSP页面login.jsp构成,该JSP页面负责提交用户的登录信息到控制器,并显示登录是否成功。该模块的JavaBean模型loginBean存储用户登录的信息。该模块的servlet控制器loginServlet负责验证会员名和密码是否正确,并负责让视图显示更新后的数据。

10.5.1 视图(JSP页面)

视图部分由一个JSP页面login.jsp构成。login.jsp页面负责提供输入登录信息界面,并负责显示登录反馈信息,比如登录是否成功,是否已经登录等。效果如图10-10所示。

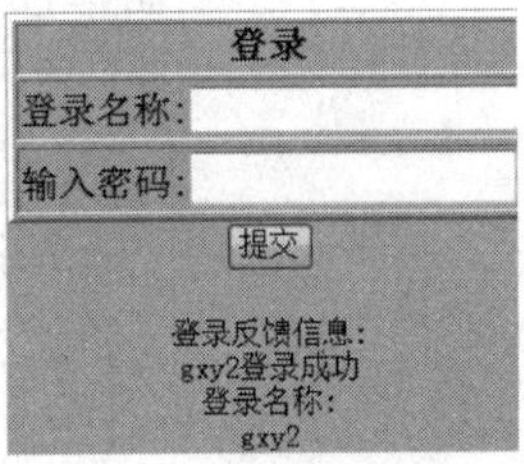

登录
登录名称:
输入密码:
提交
登录反馈信息:
gxy2登录成功
登录名称:
gxy2

图10-10 输入登录信息,并得到反馈信息

login.jsp(效果如图10-10所示)

```
<%@ page contentType = "text/html;charset = GB2312" %>
<jsp:useBean id = "loginBean" class = "mybean.data.Login"
scope = "session"/>
<HTML><HEAD><%@ include file = "head.txt" %></HEAD>
<BODY bgcolor = pink><font size = 2>
<div align = "center">
<table border = 2>
<tr><th>登录</th></tr>
<FORM action = "loginServlet" Method = "post">
<tr><td>登录名称:<Input type = text name = "logname"></td></tr>
<tr><td>输入密码:<Input type = password name = "password"></td></tr>
</table>
<Input type = submit name = "g" value = "提交">
</form></div>
<div align = "center" >
登录反馈信息:<br>
<jsp:getProperty name = "loginBean" property = "backNews"/>
<br>登录名称:<br><jsp:getProperty name = "loginBean" property = "logname"/>
<div></font>
</BODY></HTML>
```

10.5.2 模型(JavaBean)

下列 JavaBean 的实例用来存储用户登录信息,在该模块中 JavaBean 的实例的 id 是 loginBean,生命周期是 session。该 JavaBean 的实例由控制器负责创建或更新。

Login.java

```
package mybean.data;
import java.util.*;
public class Login {
   String logname = "",
          backNews = "未登录";
   LinkedList<String> car;                    //用户的购物车
   public Login() {
      car = new LinkedList<String>();
   }
   public void setLogname(String logname){
      this.logname = logname;
   }
   public String getLogname(){
      return logname;
   }
   public void setBackNews(String s) {
      backNews = s;
   }
   public String getBackNews(){
      return backNews;
   }
   public LinkedList<String> getCar() {
      return car;
   }
}
```

10.5.3 控制器(servlet)

Servlet 控制器 loginServlet(见 10.3 节给出的 web.xml 配置文件)负责连接数据库,查询 user 表,验证用户输入的会员名和密码是否在该表中,即验证是否是已注册的用户,如果用户是已注册的用户,就将用户设置成登录状态,即将用户的名称存放到 JavaBean 模型 loginBean 中(见前面的 JavaBean 模型),并将用户转发到 login.jsp 页面查看登录反馈信息(见本模块前面的视图介绍)。如果用户不是注册用户,控制器将提示登录失败。

HandleLogin.java

```
package myservlet.control;
import mybean.data.*;
import java.sql.*;
import java.io.*;
import javax.servlet.*;
import javax.servlet.http.*;
import java.util.*;
```

```
public class HandleLogin extends HttpServlet{
    public void init(ServletConfig config) throws ServletException{
        super.init(config);
        try{
            Class.forName("com.mysql.jdbc.Driver");
        }
        catch(Exception e){}
    }
    public String handleString(String s){
        try{ byte bb[] = s.getBytes("iso-8859-1");
              s = new String(bb);
        }
        catch(Exception ee){}
        return s;
    }
    public void doPost(HttpServletRequest request,HttpServletResponse response)
                            throws ServletException,IOException{
        Connection con;
        Statement sql;
        String logname = request.getParameter("logname").trim(),
        password = request.getParameter("password").trim();
        logname = handleString(logname);
        password = handleString(password);
        String uri = "jdbc:mysql://127.0.0.1/mobileshop?" +
                                "user = root&password = &characterEncoding = gb2312";
        boolean boo = (logname.length()> 0)&&(password.length()> 0);
        try{
            con = DriverManager.getConnection(uri);
            String condition = "select * from user where logname = '" + logname +
             "' and password = '" + password + "'";
            sql = con.createStatement();
            if(boo){
                ResultSet rs = sql.executeQuery(condition);
                boolean m = rs.next();
                if(m == true){
                    //调用登录成功的方法
                    success(request,response,logname,password);
                    RequestDispatcher dispatcher =
                    request.getRequestDispatcher("login.jsp");   //转发
                    dispatcher.forward(request,response);
                }
                else{
                    String backNews = "你输入的用户名不存在,或密码不匹配";
                    //调用登录失败的方法
                    fail(request,response,logname,backNews);
                }
            }
            else{
                    String backNews = "请输入用户名和密码";
                    fail(request,response,logname,backNews);
            }
```

```
            con.close();
        }
        catch(SQLException exp){
            String backNews = "" + exp;
            fail(request,response,logname,backNews);
        }
    }
    public void doGet(HttpServletRequest request,HttpServletResponse response)
                            throws ServletException,IOException{
        doPost(request,response);
    }
    public void success(HttpServletRequest request,HttpServletResponse response,
                        String logname,String password) {
        Login loginBean = null;
        HttpSession session = request.getSession(true);
        try{ loginBean = (Login)session.getAttribute("loginBean");
             if(loginBean == null){
                 loginBean = new Login();      //创建新的数据模型
                 session.setAttribute("loginBean",loginBean);
                 loginBean = (Login)session.getAttribute("loginBean");
             }
             String name  = loginBean.getLogname();
             if(name.equals(logname)) {
                 loginBean.setBackNews(logname + "已经登录了");
                 loginBean.setLogname(logname);
             }
             else { //数据模型存储新的登录用户
                  loginBean.setBackNews(logname + "登录成功");
                  loginBean.setLogname(logname);
             }
        }
        catch(Exception ee){
             loginBean = new Login();
             session.setAttribute("loginBean",loginBean);
             loginBean.setBackNews(logname + "登录成功");
             loginBean.setLogname(logname);
        }
    }
     public void fail(HttpServletRequest request,
HttpServletResponse response,String logname,String backNews) {
         response.setContentType("text/html;charset = GB2312");
         try {
          PrintWriter out = response.getWriter();
          out.println("<html><body>");
          out.println("<h2>" + logname + "登录反馈结果<br>" + backNews + "</h2>") ;
          out.println("返回登录页面或主页<br>");
          out.println("<a href  = login.jsp>登录页面</a>");
          out.println("<br><a href  = index.jsp>主页</a>");
          out.println("</body></html>");
         }
         catch(IOException exp){}
     }
}
```

10.6 浏览手机

用户选择手机分类后(见 10.2.1 节给出的 mobileClassfies 表),该模块可以分页显示 mobileForm 表中的记录。

10.6.1 视图(JSP 页面)

视图部分由 3 个 JSP 页面 lookMobile.jsp、byPageShow.jsp 和 showDetail.jsp 构成。在 lookMobile.jsp 页面选择某个分类,比如 iPhone 手机,然后提交给 servlet 控制器 queryServlet,该控制器将查询结果放到 JavaBean 模型 dataBean 中,然后将显示 dataBean 中的数据的任务交给 byPageShow.jsp 页面。byPageShow.jsp 可以分页显示 dataBean 中的数据,即分页显示记录。用户在 byPageShow.jsp 页面看到产品后,可以选择"查看细节",到 showDetail.jsp 页面查看该产品的细节。lookMobile.jsp、byPageShow.jsp 和 showDetail.jsp 页面的效果如图 10-11、图 10-12 和图 10-13 所示。

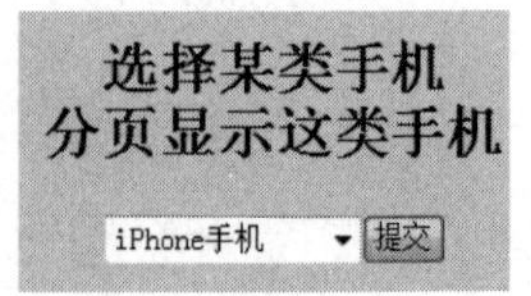

图 10-11 选择分类

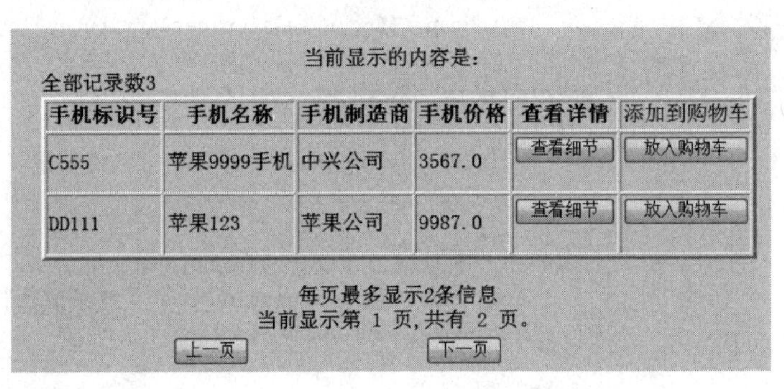

图 10-12 分页浏览

图 10-13 查看产品细节

lookMobile.jsp(效果如图 10-11 所示)

```
<%@ page contentType = "text/html;charset = GB2312" %>
<%@ page import = "java.sql.*" %>
<HTML><HEAD><%@ include file = "head.txt" %></HEAD>
<BODY bgcolor = cyan><font size = 2>
```

```
<div align="center">
<% try { Class.forName("com.mysql.jdbc.Driver");
      }
      catch(Exception e){}
      String uri="jdbc:mysql://127.0.0.1/mobileshop?"+
                          "user=root&password=&characterEncoding=gb2312";
      Connection con;
      Statement sql;
      ResultSet rs;
      try {
        con=DriverManager.getConnection(uri);
        sql=con.createStatement();
        //读取 mobileClassify 表,获得分类
        rs=sql.executeQuery("SELECT * FROM mobileClassify ");
        out.print("<form action='queryServlet' method ='post'>") ;
        out.print("<select name='fenleiNumber'>") ;
        while(rs.next()){
           int id = rs.getInt(1);
           String mobileCategory = rs.getString(2);
           out.print("<option value ="+id+">"+mobileCategory+"</option>");
        }
        out.print("</select>");
        out.print("<input type ='submit' value ='提交'>");
        out.print("</form>");
        con.close();
      }
      catch(SQLException e){
        out.print(e);
      }
%>
</div></font>
</BODY></HTML>
```

byPageShow.jsp(效果如图 10-12 所示)

```
<%@ page contentType="text/html;charset=GB2312" %>
<%@ page import="mybean.data.DataByPage" %>
<%@ page import="com.sun.rowset.*" %>
<jsp:useBean id="dataBean" class="mybean.data.DataByPage" scope="session"/>
<%@ include file="head.txt" %></HEAD>
<HTML><Body bgcolor=#66FFAA><center>
<BR>当前显示的内容是:
  <table border=2>
  <tr>
    <th>手机标识号</th>
    <th>手机名称</th>
    <th>手机制造商</th>
    <th>手机价格</th>
    <th>查看详情</th>
    <td><font color=blue>添加到购物车</font></td>
  </tr>
```

```
<jsp:setProperty name = "dataBean" property = "pageSize" param = "pageSize"/>
<jsp:setProperty name = "dataBean" property = "currentPage" param = "currentPage"/>
<%
       CachedRowSetImpl rowSet = dataBean.getRowSet();
       if(rowSet == null) {
          out.print("没有任何查询信息,无法浏览");
          return;
       }
       rowSet.last();
       int totalRecord = rowSet.getRow();
       out.println("全部记录数" + totalRecord);        //全部记录数
       int pageSize = dataBean.getPageSize();          //每页显示的记录数
       int totalPages  =  dataBean.getTotalPages();
       if(totalRecord % pageSize == 0)
            totalPages  =  totalRecord/pageSize;       //总页数
       else
            totalPages  =  totalRecord/pageSize + 1;
       dataBean.setPageSize(pageSize);
       dataBean.setTotalPages(totalPages);
       if(totalPages >= 1) {
          if(dataBean.getCurrentPage()< 1)
               dataBean.setCurrentPage(dataBean.getTotalPages());
          if(dataBean.getCurrentPage()> dataBean.getTotalPages())
               dataBean.setCurrentPage(1);
          int index = (dataBean.getCurrentPage() - 1) * pageSize + 1;
          rowSet.absolute(index);                       //查询位置移动到 currentPage 页起始位置
          boolean boo = true;
          for(int i = 1;i <= pageSize&&boo;i++) {
             String number = rowSet.getString(1);
             String name = rowSet.getString(2);
             String maker = rowSet.getString(3);
             String price = rowSet.getString(4);
             String goods  =
             "(" + number + "," + name + "," + maker +
              "," + price + ")#" + price;               //便于购物车计算价格,尾缀上"#价格值"
             goods  =  goods.replaceAll("\\p{Blank}","");
             String button = "<form action = 'putGoodsServlet' method  =  'post'>" +
                      "<input type  =  'hidden' name = 'java' value =  " + goods + ">" +
                      "<input type  =  'submit' value = '放入购物车'></form>";
             String detail = "<form action = 'showDetail.jsp' method  =  'post'>" +
                      "<input type  =  'hidden' name = 'xijie' value =  " + number + ">" +
                      "<input type  =  'submit' value = '查看细节'></form>";
             out.print("<tr>");
             out.print("<td>" + number + "</td>");
             out.print("<td>" + name + "</td>");
             out.print("<td>" + maker + "</td>");
             out.print("<td>" + price + "</td>");
             out.print("<td>" + detail + "</td>");
             out.print("<td>" + button + "</td>");
             out.print("</tr>");
             boo = rowSet.next();
```

```
            }
        }
%>
</table>
<BR>每页最多显示<jsp:getProperty name = "dataBean" property = "pageSize"/>条信息
<BR>当前显示第<Font color = blue>
      <jsp:getProperty name = "dataBean" property = "currentPage"/>
   </Font>页,共有
   <Font color = blue><jsp:getProperty name = "dataBean" property = "totalPages"/>
   </Font>页。
<Table>
  <tr><td><FORM action = "" method = post>
            <Input type = hidden name = "currentPage" value =
          "<% = dataBean.getCurrentPage() - 1 %>">
             <Input type = submit name = "g" value = "上一页"></FORM></td>
      <td><FORM action = "" method = post>
            <Input type = hidden name = "currentPage"
             value = "<% = dataBean.getCurrentPage() + 1 %>">
            <Input type = submit name = "g" value = "下一页"></FORM></td></tr>
<tr><td><FORM action = "" method = post>
            每页显示<Input type = text name = "pageSize" value = 2 size = 3>
            条记录<Input type = submit name = "g" value = "确定"></FORM></td>
      <td><FORM action = "" method = post>
             输入页码:<Input type = text name = "currentPage" size = 2>
             <Input type = submit name = "g" value = "提交"></FORM></td></tr>
</Table>
</Center>
</BODY></HTML>
```

showDetail.jsp(效果如图 10-13 所示)

```
<%@ page contentType = "text/html;charset = GB2312" %>
<%@ page import = "mybean.data.Login" %>
<%@ page import = "java.sql.*" %>
<jsp:useBean id = "loginBean" class = "mybean.data.Login" scope = "session"/>
<%@ include file = "head.txt" %></HEAD>
<HTML><Body bgcolor = #99FFCC><center>
<% if(loginBean == null){
        response.sendRedirect("login.jsp");          //重定向到登录页面
    }
    else {
        boolean b = loginBean.getLogname() == null||
                    loginBean.getLogname().length() == 0;
        if(b)
          response.sendRedirect("login.jsp");        //重定向到登录页面
    }
    String mobileID = request.getParameter("xijie");
    out.print("<th>产品号" + mobileID);
    if(mobileID == null) {
        out.print("没有产品号,无法查看细节");
```

```
        return;
    }
    Connection con;
    Statement sql;
    ResultSet rs;
    try { Class.forName("com.mysql.jdbc.Driver");
    }
    catch(Exception e){}
    String uri = "jdbc:mysql://127.0.0.1/mobileshop";
    try{
      con = DriverManager.getConnection(uri,"root","");
      sql = con.createStatement();
      String cdn = "SELECT * FROM mobileForm where mobile_version = '" + mobileID + "'";
      rs = sql.executeQuery(cdn);
      out.print("<table border = 2>");
      out.print("<tr>");
      out.print("<th>产品号");
      out.print("<th>名称");
      out.print("<th>制造商");
      out.print("<th>价格");
      out.print("<th><font color = blue>放入购物车</font>");
      out.print("</TR>");
      String picture = "welcome.jpg";
      String detailMess = "";
      while(rs.next()){
        String number = rs.getString(1);
        String name = rs.getString(2);
        String maker = rs.getString(3);
        String price = rs.getString(4);
        detailMess = rs.getString(5);
        picture = rs.getString(6);
        String goods =
        "(" + number + "," + name + "," + maker +
              "," + price + ")#" + price;              //便于购物车计算价格,尾缀上"#价格值"
         goods = goods.replaceAll("\\p{Blank}","");
         String button = "<form action = 'putGoodsServlet' method = 'post'>" +
                      "<input type = 'hidden' name = 'java' value = " + goods + ">" +
                      "<input type = 'submit' value = '放入购物车'></form>";
         out.print("<tr>");
         out.print("<td>" + number + "</td>");
         out.print("<td>" + name + "</td>");
         out.print("<td>" + maker + "</td>");
         out.print("<td>" + price + "</td>");
         out.print("<td>" + button + "</td>");
         out.print("</tr>");
      }
      out.print("</table>");
      out.print("产品详情:<br>");
      out.println("<div align = center>" + detailMess + "<div>");
      String pic = "<img src = 'image/" + picture + "' width = 260 height = 200 ></img>";
      out.print(pic);                                  //产品图片
```

```
        con.close();
    }
    catch(SQLException exp){}
  %>
</Center></BODY></HTML>
```

10.6.2 模型(JavaBean)

本模块的模型是 dataBean,生命周期是 session,用于存储数据库中的记录。servlet 控制器 queryServlet 把从数据库查询到的记录存到 dataBean 中。

DataByPage.java

```
package mybean.data;
import com.sun.rowset.*;
public class DataByPage{
    CachedRowSetImpl rowSet = null;                        //存储表中全部记录的行集对象
    int pageSize = 1;                                      //每页显示的记录数
    int totalPages = 1;                                    //分页后的总页数
    int currentPage  = 1;                                  //当前显示页
    public void setRowSet(CachedRowSetImpl set){
       rowSet = set;
    }
    public CachedRowSetImpl getRowSet(){
       return rowSet;
    }
    public void setPageSize(int size){
       pageSize = size;
    }
    public int getPageSize(){
       return pageSize;
    }
    public int getTotalPages(){
       return totalPages;
    }
    public void setTotalPages(int n){
       totalPages = n;
    }
    public void setCurrentPage(int n){
       currentPage  = n;
    }
    public int getCurrentPage(){
       return currentPage ;
    }
}
```

10.6.3 控制器(servlet)

本模块有两个控制器 queryServlet 和 putGoodsServlet(见 10.3 节给出的 web.xml 配置文件)。servlet 控制器 queryServlet 把从数据库 mobileForm 表中查询到的记录存到

dataBean 中,然后将用户重定向到 byPageShow.jsp 页面。当用户在 byPageShow.jsp 页面或 showDetail.jsp 看到产品时,每个产品都后缀了一个“添加到购物车”按钮,用户单击该按钮后,putGoodsServlet 控制器将该产品放入用户的购物车。

QueryAllRecord.java

```
package myservlet.control;
import mybean.data.DataByPage;
import com.sun.rowset.*;
import java.sql.*;
import java.io.*;
import javax.servlet.*;
import javax.servlet.http.*;
public class QueryAllRecord extends HttpServlet{
   CachedRowSetImpl rowSet = null;
   public void init(ServletConfig config) throws ServletException{
      super.init(config);
      try { Class.forName("com.mysql.jdbc.Driver");
      }
      catch(Exception e){}
   }
   public void doPost(HttpServletRequest request,HttpServletResponse response)
                     throws ServletException,IOException{
      request.setCharacterEncoding("gb2312");
      String idNumber = request.getParameter("fenleiNumber");
      if(idNumber == null)
           idNumber = "0";
      int id = Integer.parseInt(idNumber);
      HttpSession session = request.getSession(true);
      Connection con = null;
      DataByPage dataBean = null;
      try{
           dataBean = (DataByPage)session.getAttribute("dataBean");
           if(dataBean == null){
              dataBean = new DataByPage();          //创建 JavaBean 对象
              session.setAttribute("dataBean",dataBean);
           }
      }
      catch(Exception exp){
           dataBean = new DataByPage();
           session.setAttribute("dataBean",dataBean);
      }
      String uri = "jdbc:mysql://127.0.0.1/mobileshop";
      try{
           con = DriverManager.getConnection(uri,"root","");
           Statement sql = con.createStatement(ResultSet.TYPE_SCROLL_SENSITIVE,
                                                    ResultSet.CONCUR_READ_ONLY);
           ResultSet rs = sql.executeQuery("SELECT * FROM mobileForm where id = " + id);
           rowSet = new CachedRowSetImpl();          //创建行集对象
           rowSet.populate(rs);
           dataBean.setRowSet(rowSet);              //行集数据存储在 dataBean 中
```

```
            con.close();                                    //关闭连接
        }
        catch(SQLException exp){}
        response.sendRedirect("byPageShow.jsp");        //重定向到 byPageShow.jsp
    }
    public void doGet(HttpServletRequest request,HttpServletResponse response)
                            throws ServletException,IOException{
         doPost(request,response);
    }
}
```

PutGoodsToCar.java

```
package myservlet.control;
import mybean.data.Login;
import java.util.*;
import java.io.*;
import javax.servlet.*;
import javax.servlet.http.*;
public class PutGoodsToCar extends HttpServlet {
    public void init(ServletConfig config) throws ServletException {
        super.init(config);
    }
    public void doPost(HttpServletRequest request,HttpServletResponse response)
                            throws ServletException,IOException {
        request.setCharacterEncoding("gb2312");
        String goods = request.getParameter("java");
        System.out.println(goods);
        Login loginBean = null;
        HttpSession session = request.getSession(true);
        try{ loginBean = (Login)session.getAttribute("loginBean");
                boolean b = loginBean.getLogname() == null||
                loginBean.getLogname().length() == 0;
                if(b)
                  response.sendRedirect("login.jsp"); //重定向到登录页面
                LinkedList<String> car = loginBean.getCar();
                car.add(goods);                             //产品放入购物车
                speakSomeMess(request,response,goods);
        }
        catch(Exception exp){
              response.sendRedirect("login.jsp");      //重定向到登录页面
        }
    }
    public void doGet(HttpServletRequest request,HttpServletResponse response)
                            throws ServletException,IOException {
        doPost(request,response);
    }
    public void speakSomeMess(HttpServletRequest request,
                        HttpServletResponse response,String goods) {
         response.setContentType("text/html;charset = GB2312");
```

```
        try {
         PrintWriter out = response.getWriter();
         out.print("<%@ include file = 'head.txt' %></HEAD>");
         out.println("<html><body>");
         out.println("<h2>" + goods + "放入购物车</h2>") ;
         out.println("查看购物车或返回<br>");
         out.println("<a href = lookShoppingCar.jsp>查看购物车</a>");
         out.println("<br><a href = byPageShow.jsp>主页</a>");
         out.println("</body></html>");
        }
        catch(IOException exp){}
    }
}
```

10.7 查看购物车

登录的用户可以通过该模块视图部分 lookShoppingCar.jsp 查看购物车中的物品,并选择是否删除某个货物。该模块有 deleteServlet 和 buyServlet 两个 servlet 控制器,deleteServlet 负责删除购物车中的物品,buyServlet 负责将用户购物车中的物品存放到数据库的 oderForm 表中,即生成订单。

10.7.1 视图(JSP 页面)

视图部分由一个 JSP 页面 lookShoppingCar.jsp 构成,负责显示和选择要删除的物品,即显示 JavaBean 模型 carBean 中的数据和删除购物车中的物品,并允许用户确定订单。效果如图 10-14 所示。

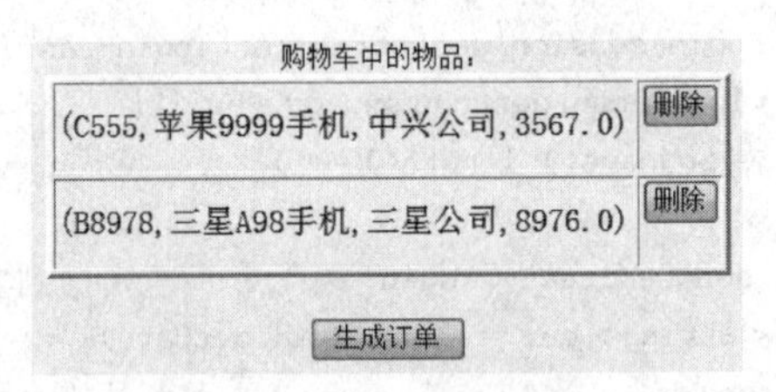

图 10-14 管理购物车

lookShoppingCar.jsp(效果如图 10-14 所示)

```
<%@ page contentType = "text/html;charset = GB2312" %>
<%@ page import = "mybean.data.Login" %>
<%@ page import = "java.util.*" %>
<jsp:useBean id = "loginBean" class = "mybean.data.Login" scope = "session"/>
<HTML><HEAD><%@ include file = "head.txt" %></HEAD>
<BODY bgcolor = yellow><font size = 2>
<div align = "center">
<% if(loginBean == null){
      response.sendRedirect("login.jsp");          //重定向到登录页面
   }
```

```
    else {
        boolean b = loginBean.getLogname() == null||
                    loginBean.getLogname().length() == 0;
        if(b)
          response.sendRedirect("login.jsp");        //重定向到登录页面
    }
    LinkedList car = loginBean.getCar();
    if(car == null)
      out.print("<h2>购物车没有物品.</h2>");
    else {
        Iterator<String> iterator = car.iterator();
        StringBuffer buyGoods = new StringBuffer();
        int n = 0;
        double priceSum = 0;
        out.print("购物车中的物品：<table border = 2>");
        while(iterator.hasNext()) {
            String goods = iterator.next();
            String showGoods = "";
            n++;
            //购车车物品的后缀是"#价格数字",比如"iPhone手机价格 3989 #3989"
            int index = goods.lastIndexOf("#");
            if(index! = -1){
               priceSum + = Double.parseDouble(goods.substring(index + 1));
               showGoods = goods.substring(0,index);
            }
            buyGoods.append(n + ":" + showGoods);
            String del = "<form action = 'deleteServlet' method = 'post'>" +
                         "<input type = 'hidden' name = 'delete' value = " + goods + ">" +
                         "<input type = 'submit' value = '删除'></form>";
            out.print("<tr><td>" + showGoods + "</td>");
            out.print("<td>" + del + "</td></tr>");
        }
        out.print("</table>");
        String orderForm = "<form action = 'buyServlet' method = 'post'>" +
                " <input type = 'hidden' name = 'buy' value = " + buyGoods + " >" +
                " <input type = 'hidden' name = 'price' value = " + priceSum + " >" +
                "<input type = 'submit' value = '生成订单'></form>";
        out.print(orderForm);
    }
%>
</div></font>
</BODY></HTML>
```

10.7.2 模型(JavaBean)

本模块用的模型是 id 为 loginBean,生命周期是 session 的 JavaBean,主要是需要 loginBean 中的 car,即需要用户的购物车(见 10.5.2 节)。

10.7.3 控制器(servlet)

本模块有两个 servlet 控制器 deleteServlet 和 buyServlet(见 10.3 节给出的 web.xml 配置文件)。用户查看购物车后,会看到购物车中的物品,并且每个物品都后缀一个删除按钮,用户单击这个删除按钮,deleteServlet 控制器从购物车中删除这件物品。用户在视图部分单击"生成订单",buyServlet 控制器负责把用户购物车中的物品存放到数据库的 oderform 表中,即生成订单,然后清空用户的购物车。

HandleDelete.java

```
package myservlet.control;
import mybean.data.Login;
import java.util.*;
import java.io.*;
import javax.servlet.*;
import javax.servlet.http.*;
public class HandleDelete extends HttpServlet {
   public void init(ServletConfig config) throws ServletException {
      super.init(config);
   }
   public void doPost(HttpServletRequest request,HttpServletResponse response)
                        throws ServletException,IOException {
      request.setCharacterEncoding("gb2312");
      String delete = request.getParameter("delete");
      Login loginBean = null;
      HttpSession session = request.getSession(true);
      try{ loginBean = (Login)session.getAttribute("loginBean");
            boolean b = loginBean.getLogname() == null||
            loginBean.getLogname().length() == 0;
            if(b)
              response.sendRedirect("login.jsp"); //重定向到登录页面
            LinkedList<String> car = loginBean.getCar();
            car.remove(delete);                          //从购物车中删除产品
      }
      catch(Exception exp){
           response.sendRedirect("login.jsp");     //重定向到登录页面
      }
      RequestDispatcher dispatcher =
      request.getRequestDispatcher("lookShoppingCar.jsp");
      dispatcher.forward(request, response);          //转发
   }
   public void doGet(HttpServletRequest request,HttpServletResponse response)
                        throws ServletException,IOException {
      doPost(request,response);
   }
}
```

HandleBuyGoods.java

```
package myservlet.control;
import mybean.data.Login;
import java.sql.*;
import java.util.*;
```

```
import java.io.*;
import javax.servlet.*;
import javax.servlet.http.*;
public class HandleBuyGoods extends HttpServlet {
   public void init(ServletConfig config) throws ServletException {
      super.init(config);
      try{
           Class.forName("com.mysql.jdbc.Driver");
      }
      catch(Exception e){}
   }
   public void doPost(HttpServletRequest request,HttpServletResponse response)
                        throws ServletException,IOException {
      request.setCharacterEncoding("gb2312");
      String buyGoodsMess = request.getParameter("buy");
      if(buyGoodsMess == null||buyGoodsMess.length() == 0) {
         fail(request,response,"购物车没有物品,无法生成订单");
         return;
      }
      String price = request.getParameter("price");
      if(price == null||price.length() == 0) {
         fail(request,response,"没有计算价格和,无法生成订单");
         return;
      }
      float sum = Float.parseFloat(price);
      Login loginBean = null;
      HttpSession session = request.getSession(true);
      try{ loginBean = (Login)session.getAttribute("loginBean");
            boolean b = loginBean.getLogname() == null||
            loginBean.getLogname().length() == 0;
            if(b)
              response.sendRedirect("login.jsp"); //重定向到登录页面
      }
      catch(Exception exp){
           response.sendRedirect("login.jsp");     //重定向到登录页面
      }
      String uri = "jdbc:mysql://127.0.0.1/mobileshop?" +
                          "user = root&password = &characterEncoding = gb2312";
      Connection con;
      PreparedStatement sql;
      try{ con = DriverManager.getConnection(uri);
           String insertCondition = "INSERT INTO orderform VALUES (?,?,?,?)";
           sql = con.prepareStatement(insertCondition);
           sql.setInt(1,0);                        //订单序号会自定增加
           sql.setString(2,loginBean.getLogname());
           sql.setString(3,buyGoodsMess);
           sql.setFloat(4,sum);
           sql.executeUpdate();
           LinkedList car = loginBean.getCar();
           car.clear();                            //清空购物车
           success(request,response,"生成订单成功");
```

```
    }
    catch(SQLException exp){
        fail(request,response,"生成订单失败" + exp);
    }
  }
  public void doGet(HttpServletRequest request,HttpServletResponse response)
                    throws ServletException,IOException {
    doPost(request,response);
  }
  public void success(HttpServletRequest request,HttpServletResponse response,
                    String backNews) {
      response.setContentType("text/html;charset = GB2312");
      try {
       PrintWriter out = response.getWriter();
       out.println("<html><body>");
       out.println("<h2>" + backNews + "</h2>") ;
       out.println("返回主页");
       out.println("<a href = index.jsp>主页</a>");
       out.println("<br>查看订单");
       out.println("<a href = lookOrderForm.jsp>查看订单</a>");
       out.println("</body></html>");
      }
      catch(IOException exp){}
   }
  public void fail(HttpServletRequest request,HttpServletResponse response,
                    String backNews) {
      response.setContentType("text/html;charset = GB2312");
      try {
       PrintWriter out = response.getWriter();
       out.println("<html><body>");
       out.println("<h2>" + backNews + "</h2>") ;
       out.println("返回主页:");
       out.println("<a href = index.jsp>主页</a>");
       out.println("</body></html>");
      }
      catch(IOException exp){}
   }
}
```

10.8 查询手机

本模块用到了 10.6 节给出的模型 dataBean 和视图 byPageShow.jsp。

在视图部分 searchMobile.jsp 输入查询条件给 servlet 控制器 searchByConditionServlet，searchByConditionServlet 控制器负责查询数据库，并将查询结果存放到数据模型 dataBean 中，然后将用户重定向到 byPageShow.jsp 页面查看 dataBean 中的数据。

10.8.1 视图(JSP 页面)

视图部分由两个 JSP 页面 searchMobile.jsp 和 byPageShow.jsp 构成，其中

byPageShow.jsp 见 10.6 节的视图部分。用户在 searchMobile.jsp 页面输入查询信息，提交给 searchByConditionServlet 控制器，该控制器将查询结果存放到 dataBean 中。byPageShow.jsp 页面负责显示 dataBean 中的数据。searchMobile.jsp 和 byPageShow.jsp 的效果如图 10-15 和图 10-16 所示。

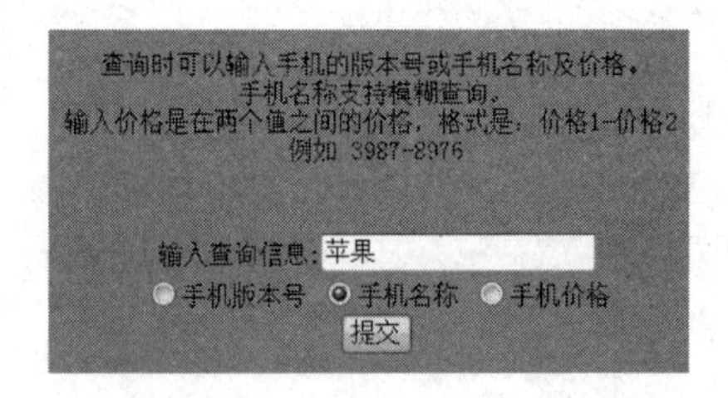

图 10-15　输入查询信息

全部记录数3

手机标识号	手机名称	手机制造商	手机价格	查看详情	添加到购物车
C555	苹果9999手机	中兴公司	3567.0	查看细节	放入购物车
DD111	2016苹果新款	苹果公司	9987.0	查看细节	放入购物车
DD897	苹果98手机	苹果公司	1999.0	查看细节	放入购物车

每页最多显示3条信息
当前显示第 1 页，共有 1 页。
上一页　下一页

图 10-16　显示查询结果

searMobile.jsp（效果如图 10-15 所示）

```
<%@ page contentType = "text/html;charset = GB2312" %>
<HTML><HEAD><%@ include file = "head.txt" %></HEAD>
<BODY bgcolor = #55BBDD><font size = 2>
<div align = "center">
<br>查询时可以输入手机的版本号或手机名称及价格。<br>
手机名称支持模糊查询。
<br>输入价格是在两个值之间的价格，格式是：价格 1 - 价格 2<br>
例如 3987 - 8976
<FORM action = "searchByConditionServlet" Method = "post" >
    <br>输入查询信息：<Input type = text name = "searchMess"><br>
    <Input type  = radio name = "radio" value = "mobile_version">手机版本号
    <Input type  = radio name = "radio" value = "mobile_name" checked = "ok">手机名称
    <Input type  = radio name = "radio" value = "mobile_price">手机价格
    <br><Input type = submit name = "g" value = "提交">
</Form>
</div>
</Font></BODY></HTML>
```

10.8.2　模型（JavaBean）

本模块用的模型是 id 为 dataBean，生命周期是 session 的 JavaBean（见 10.6.2 节）。

10.8.3 控制器(servlet)

servlet 控制器 searchByConditionServlet(见 10.3 节给出的 web.xml 配置文件)把从数据库 mobileForm 表中查询到的记录存到 dataBean 中,然后将用户重定向到 byPageShow.jsp 页面。

SearchByCondition.java

```
package myservlet.control;
import mybean.data.DataByPage;
import com.sun.rowset.*;
import java.sql.*;
import java.io.*;
import javax.servlet.*;
import javax.servlet.http.*;
public class SearchByCondition extends HttpServlet{
   CachedRowSetImpl rowSet = null;
   public void init(ServletConfig config) throws ServletException{
      super.init(config);
      try { Class.forName("com.mysql.jdbc.Driver");
      }
      catch(Exception e){}
   }
   public void doPost(HttpServletRequest request,HttpServletResponse response)
                        throws ServletException,IOException{
      request.setCharacterEncoding("gb2312");
      String searchMess = request.getParameter("searchMess");
      String radioMess = request.getParameter("radio");
      if(searchMess == null||searchMess.length() == 0) {
         fail(request,response,"没有查询信息,无法查询");
         return;
      }
      String condition = "";
      if(radioMess.equals("mobile_version")) {
          condition =
         "SELECT * FROM mobileForm where mobile_version = '" + searchMess + "'";
      }
      else if(radioMess.equals("mobile_name")) {
          condition =
         "SELECT * FROM mobileForm where mobile_name LIKE '%" + searchMess + "%'";
      }
      else if(radioMess.equals("mobile_price")) {
          double max = 0,min = 0;
          String regex = "[^0123456789.]";
          String [] priceMess = searchMess.split(regex);
          if(priceMess.length == 1) {
             max = min = Double.parseDouble(priceMess[0]);
          }
          else if(priceMess.length == 2) {
             min = Double.parseDouble(priceMess[0]);
```

```
            max = Double.parseDouble(priceMess[1]);
            if(max < min) {
               double t = max;
               max = min;
               min = t;
            }
         }
         else { fail(request,response,"输入的价格格式有错误");
                  return;
         }
         condition = "SELECT * FROM mobileForm where " +
         "mobile_price <= " + max + " AND mobile_price >= " + min ;
      }
      HttpSession session = request.getSession(true);
      Connection con = null;
      DataByPage dataBean = null;
      try{ dataBean = (DataByPage)session.getAttribute("dataBean");
           if(dataBean == null){
              dataBean = new DataByPage();             //创建 JavaBean 对象
              session.setAttribute("dataBean",dataBean);
           }
      }
      catch(Exception exp){
           dataBean = new DataByPage();
           session.setAttribute("dataBean",dataBean);
      }
      String uri = "jdbc:mysql://127.0.0.1/mobileshop?" +
                   "user = root&password = &characterEncoding = gb2312";
      try{
           con = DriverManager.getConnection(uri);
           Statement sql = con.createStatement(ResultSet.TYPE_SCROLL_SENSITIVE,
                                                ResultSet.CONCUR_READ_ONLY);
           ResultSet rs = sql.executeQuery(condition);
           rowSet = new CachedRowSetImpl();            //创建行集对象
           rowSet.populate(rs);
           dataBean.setRowSet(rowSet);                 //行集数据存储在 dataBean 中
           con.close();                                //关闭连接
      }
      catch(SQLException exp){}
      response.sendRedirect("byPageShow.jsp");        //重定向到 byPageShow.jsp
   }
   public void doGet(HttpServletRequest request,HttpServletResponse response)
                        throws ServletException,IOException{
      doPost(request,response);
   }
   public void fail(HttpServletRequest request,HttpServletResponse response,
                        String backNews) {
       response.setContentType("text/html;charset = GB2312");
       try {
        PrintWriter out = response.getWriter();
```

```
            out.println("<html><body>");
            out.println("<h2>" + backNews + "</h2>") ;
            out.println("返回：");
            out.println("<a href = searchMobile.jsp>查询手机</a>");
            out.println("</body></html>");
          }
          catch(IOException exp){}
      }
  }
```

10.9 查询订单

本模块用到了10.5节给出的模型loginBean。用户可以通过该模块视图部分lookOrderForm.jsp查看自己的订单。

10.9.1 视图(JSP页面)

视图部分由一个JSP页面lookOrderForm.jsp构成,负责显示用户的订单信息。效果如图10-17所示。

注册　登录　浏览手机　查询手机　查看购物车　查看订单　退出　主页

订单号	信息	价格
245	1:(DD111,2016苹果新款,苹果公司,9987.0)	9987
246	1:(B8978,三星A98手机,三星公司,8976)2:(AA98F,小米手机908,小米公司,6785.0)	15761

图10-17　显示订单

lookOrderForm.jsp(效果如图10-17所示)

```
<%@ page contentType = "text/html;charset = GB2312" %>
<jsp:useBean id = "loginBean" class = "mybean.data.Login" scope = "session"/>
<%@ page import = "java.sql.*" %>
<HTML><HEAD><%@ include file = "head.txt" %></HEAD>
<div align = "center">
<% if(loginBean == null){
        response.sendRedirect("login.jsp");           //重定向到登录页面
    }
    else {
       boolean b = loginBean.getLogname() == null||
                   loginBean.getLogname().length() == 0;
       if(b)
         response.sendRedirect("login.jsp");          //重定向到登录页面
    }
    Connection con;
    Statement sql;
    ResultSet rs;
    try{ Class.forName("com.mysql.jdbc.Driver");
    }
    catch(Exception e){}
```

```
    try { String uri = "jdbc:mysql://127.0.0.1/mobileshop";
          String user = "root";
          String password = "";
          con = DriverManager.getConnection(uri,user,password);
          sql = con.createStatement();
          String cdn =
         "SELECT id,mess,sum FROM orderform where logname = '" + loginBean.getLogname() + "'";
          rs = sql.executeQuery(cdn);
          out.print("<table border = 2>");
          out.print("<tr>");
            out.print("<th width = 100>" + "订单号");
            out.print("<th width = 100>" + "信息");
            out.print("<th width = 100>" + "价格");
          out.print("</TR>");
          while(rs.next()){
            out.print("<tr>");
              out.print("<td>" + rs.getString(1) + "</td>");
              out.print("<td>" + rs.getString(2) + "</td>");
              out.print("<td>" + rs.getString(3) + "</td>");
              out.print("</tr>") ;
          }
          out.print("</table>");
          con.close();
    }
    catch(SQLException e){
          out.print(e);
    }
 %>
</div">
</BODY></HTML>
```

10.9.2 模型(JavaBean)

本模块用的模型是 id 为 loginBean,生命周期是 session 的 JavaBean(见 10.5.2 节)。

10.9.3 控制器(servlet)

本模块无控制器。

10.10 退出登录

该模块只有一个名字为 exitServlet 的 servlet 控制器,exitServlet 负责销毁用户的 session 对象,导致登录失效。

HandleExit.java

```
package myservlet.control;
import javax.servlet.*;
import javax.servlet.http.*;
import java.io.*;
```

```
public class HandleExit extends HttpServlet {
    public void init(ServletConfig config) throws ServletException{
        super.init(config);
    }
    public void doPost(HttpServletRequest request,HttpServletResponse response)
                          throws ServletException,IOException {
        HttpSession session = request.getSession(true);
        session.invalidate();                          //销毁用户的 session 对象
        response.sendRedirect("index.jsp");            //返回主页
    }
    public void doGet(HttpServletRequest request,HttpServletResponse response)
                          throws ServletException,IOException {
        doPost(request,response);
    }
}
```

图书资源支持

感谢您一直以来对清华版图书的支持和爱护。为了配合本书的使用，本书提供配套的资源，有需求的读者请扫描下方的“书圈”微信公众号二维码，在图书专区下载，也可以拨打电话或发送电子邮件咨询。

如果您在使用本书的过程中遇到了什么问题，或者有相关图书出版计划，也请您发邮件告诉我们，以便我们更好地为您服务。

我们的联系方式：

地　　址：北京海淀区双清路学研大厦 A 座 707

邮　　编：100084

电　　话：010－62770175－4604

资源下载：http://www.tup.com.cn

电子邮件：weijj@tup.tsinghua.edu.cn

QQ：883604（请写明您的单位和姓名）

资源下载、样书申请

书圈

用微信扫一扫右边的二维码，即可关注清华大学出版社公众号“书圈”。